Feature Papers of Oceans 2024

Feature Papers of Oceans 2024

Guest Editors

Alexander Werth
João Silva

MDPI

Basel • Beijing • Wuhan • Barcelona • Belgrade • Novi Sad • Cluj • Manchester

Guest Editors
Alexander Werth
Department of Biology
Hampden-Sydney College
Hampden-Sydney
USA

João Silva
CCMAR—Centre of Marine Sciences
University of Algarve
Faro
Portugal

Editorial Office
MDPI AG
Grosspeteranlage 5
4052 Basel, Switzerland

This is a reprint of the Special Issue, published open access by the journal *Oceans* (ISSN 2673-1924), freely accessible at: https://www.mdpi.com/journal/oceans/special_issues/LQ5LPM1HK9.

For citation purposes, cite each article independently as indicated on the article page online and as indicated below:

Lastname, A.A.; Lastname, B.B. Article Title. *Journal Name* **Year**, *Volume Number*, Page Range.

ISBN 978-3-7258-4897-3 (Hbk)
ISBN 978-3-7258-4898-0 (PDF)
https://doi.org/10.3390/books978-3-7258-4898-0

Contents

oceans

Editorial

Feature Papers of *Oceans* 2024

Alexander J. Werth

Department of Biology, Hampden-Sydney College, Hampden-Sydney, VA 23943, USA; awerth@hsc.edu

1. Introduction

As the noted futurist and author Arthur C. Clarke famously remarked, our world of Planet Earth should be more appropriately named Planet Ocean [1]. The global ocean covers nearly three-quarters of the planet's surface and, considering its great depths, makes up roughly 99% of the planet's total habitable space [2]. Life is presumed to have originated in the sea, and the ocean remains the home of an estimated 94% of global wildlife, including 78% of the planet's animal biomass [2–4]. Although the ocean houses most of our planet's life, we know far more about terrestrial ecosystems. Fewer than a million marine species have been described, yet according to one estimate [5], the total number of undescribed marine species may be as high as ten million. Of the 33 known animal phyla, only 12 are found on land and only 1 is endemic to land, whereas 32 occur in the oceans and 14 are endemic there [3]. The ocean has an outsize impact on land, with at least half of Earth's atmospheric oxygen coming from the sea and roughly 30% of its carbon dioxide absorbed by the sea, being the planet's single greatest carbon sink [4]. Every bit of Earth's global climate, even on land far from the sea, is governed by the ocean's mass of water, its remarkable thermal inertia, and its powerful conveyor currents [5–8].

In terms of humans, the ocean's impact is no less profound, even if we often forget this impact on our daily lives. The ocean is a major source of food, especially protein in the form of finfish and shellfish. It provides biological materials such as agarose gel and carrageenan (both from seaweed), which are used as thickeners and stabilizers in shampoo, toothpaste, other toiletry products, and foods such as ice cream. Many medicines and biofuels come from the sea [4]. It is a prime source of energy, including extracted oil, gas, and other petroleum compounds, as well as tidal, wave, wind, and thermal gradient energy. The ocean is a major contributor of valuable resources, including sand and gravel, salt and freshwater (from desalination), and polymetallic nodules mined from the seafloor. The ocean is a key resource for transport/travel, trade, and commerce, not to mention recreation, tourism, and valuable real estate. Coastal habitats protect people's lives and structures from hurricanes and other storm winds and wave surges.

Despite the ocean's tremendous importance to humans and other organisms, it largely remains a mystery. We know distressingly little about basic marine processes and even less about the topography and geology of deep-sea basins [2]. Although we have grasped that our existence depends on marine processes, we still have much to learn about marine ecosystems, their operation, and how they relate to life on land [5,8]. Research into basic marine science and oceanography—including geology, chemistry, physics, meteorology, microbiology, botany, and zoology—has never been more important, yet it continues to lag behind studies of terrestrial regions. In a slow but profound shift, however, this is beginning to change.

 https://doi.org/10.3390/oceans6030040

Any deep or notable transformation is often described by the term "sea change," and this phrase perfectly captures what is happening at this moment regarding marine science: not just a monumental transformation in how we study the ocean but a profound and newfound awareness of why a clearer understanding of our ocean matters to everyone (and everything) on the planet.

As distressing as our longstanding ignorance of basic marine processes is, our dawning understanding of humanity's increasingly harmful impact on the ocean and its biosphere is even more distressing. With the sea once seen as a limitless expanse, a boundless source of food, and an infinite dumping ground for trash and other waste, we now recognize our heavy anthropogenic influence on the sea. We realize that the sea's pH has quickly become more acidic than in pre-industrial times [4], and that it has silently absorbed more than 90% of the climate's excess heat generated chiefly by carbon emissions [4]. We know that noise pollution dramatically affects marine life, while plastic pollution, accumulating from an influx of roughly ten million metric tons of new plastic entering the ocean each year (89% of which involves single-use plastics), increasingly clogs seaways, imperils marine trophic webs, and clearly threatens human health [4,9].

In the face of so much ignorance and need, what can be done? Clearly, the first step is to learn much more about the ocean and its role in regulating Earth's fundamental activities. From basic information to detailed knowledge, we need to know more. We need more research into marine science, at every level and from multiple funding sources. We need more education for and engagement by an informed global populace. We need to understand that we are all citizens of Planet Ocean and that the future of our world, and humanity's future, hangs in the balance.

2. Overview of Special Issue

Arriving at this crucial time, this Special Issue, the **Feature Papers of *Oceans* 2024**, helps in filling many of the gaps in our understanding of marine science. At the same time, this Special Issue showcases a variety of research methodologies with high-quality, quantitative datasets. This collection of open access publications from leading researchers in marine science demonstrates some of the interdisciplinary ways that mathematical modeling, biodiversity exploration, and advances in engineering and technology continue to unlock some of the ocean's mysteries.

At the same time, this collection features a wide array of topics that illustrate many of the key ways in which humans continue to influence the ocean for good and ill [10]. The ocean regions featured and discussed in these papers, including the Caribbean, Mediterranean, and Black Seas, are sites of tremendous human activity and hence anthropogenic impact. They offer test cases on how to study and alter ocean processes to aid all of Earth's species, including our own.

The nine publications assembled here offer a broad-ranging introduction to current issues affecting marine science, including a trio of offerings each on oceanography, marine biology, and marine policy—along the way, each publication reveals how all three of these areas are intertwined.

The three papers collected in this Special Issue that involve oceanography demonstrate how ongoing marine research continues to reveal fundamental aspects of Earth science and meteorology. The paper by Pelegri et al. (contribution 1) on ocean mixing shows how perturbations from shear and flow instability derive from gradients of temperature, density, and other basic aspects of the ocean and atmosphere. The paper by Rudan and Sviličić (contribution 2) shows how physical experiments and wave modeling from computational fluid dynamics (CFD) inform our understanding of how breakwaters or other man-made barriers impact wind-induced waves to provide physical protection from storms, thereby saving beaches and homes from coastal erosion. The paper from Bisinicu and Lazar (contribution 3) straddles the boundaries between basic oceanographic research on how seawater quality (including nutrients and pollution) affects plankton biogeography, biodiversity, and biomass, with an awareness of how human environmental policies and other choices interact with seasonal changes and invasive species. These factors determine plankton species richness and marine distribution, both vertically and horizontally, in both shallow coastal and deep pelagic international waters.

The three papers collected in this Special Issue that focus on marine biology also demonstrate a variety of methods and study organisms, but all show that life persists along with, and often despite, stark human impacts [11]. The paper by Lucey et al. (contribution 4) shows how climate change and ocean acidification lead to mass mortality events of Caribbean corals from hypoxia and bleaching, but it confirms that coral populations can adapt their metabolic rates and show stress tolerance to hypoxia, offering hope for the restoration of damaged reefs. The paper by Lombardo and Marletta (contribution 5) on heterobranch gastropod mollusks (sea slugs) near a petrochemical hub in the Mediterranean reveals how marine biodiversity adapts to temperature and other essential factors of habitat, as well as to industrial pollution from human sources, showing that we share the ocean with diverse life that is generally resilient even in fragile, delicate ecosystems. Likewise, the paper by Ayala and Méndez-Tejada (contribution 6) shows not only how blooms of *Sargassum* macroalgae (seaweed) respond to sea surface temperatures, cloud cover, and solar radiation but also that these ocean factors are easily influenced by human activity, such as erosion from agricultural tilling practices in distant arid lands. This generates particulate deposition that affects human health on land just as it affects seaweed health in the sea, and *Sargassum* blooms in turn affect humans in multiple ways, such as by posing navigation hazards.

Finally, the three remaining papers focus primarily on human technologies and policies that both help and hinder our ability to understand care for the sea. The paper by Bassolillo et al. (contribution 7) concerns unmanned underwater vehicles (UUVs, or marine drones) and remotely operated underwater vehicles (ROUVs) for autonomous navigation for shipping and exploration, with algorithms from mother ships helping in the detection and planning of optimal routes for adaptive weather routing. The paper from Prieto et al. (contribution 8) considers government controls and international agreements of risk indicators of port inspections, maritime transport, and accidents at sea and how these factors together determine hazards and safety controls of ocean shipping and commerce. The paper by Henderson et al. (contribution 9) looks at the Cultural Heritage Framework Programme (CHFP) of the UN Decade of Ocean Science (2021–2030) and how it relates to such programs as the UNESCO SeaVoice and Blue Economy initiatives, offering Western and traditional/indigenous scientific and cultural education on such matters as marine protected areas (MPAs), marine archaeology, sustainable development, and general ocean literacy.

3. Conclusions

Each feature paper in this Special Issue stands alone as a valuable contribution to our ever-increasing body of knowledge about how the ocean operates and how these functions are affected by human impact. Taken together, these diverse research reports provide a testament to the essential value of conducting, analyzing, and communicating findings about our global ocean. Despite a striking variety of topics and methodologies, these interdisciplinary reports join a chorus to tell the same story: We live on Planet Ocean, and despite its resilience in the face of the many risks we pose to it [11], there are limits to this resilience [10].

New opportunities create new threats. The shipping of oil in huge tankers, which began in the late 19th century, helped carry valuable new fuel sources around the world relatively easily and inexpensively. Unfortunately, this also created entirely new hazards, such as coastal oil spills caused by these tankers sinking or running aground. Many of the most severe threats to whales and other marine life, including ship strikes, entanglement in nets or longlines, and bycatch in giant trawls or purse seines, did not become problematic until the 20th century, such that marine life has had little time to adapt to new threats. Other dangers posed to marine life, such as intense noise pollution from high-intensity SONAR and seismic testing, arrived even more recently [4,12]. The human use of the ocean has benefited our species, but often at the expense of other species and also, as we are increasingly learning, by posing threats to human health as well [9,13,14].

However, threats also present new opportunities. The climatic and health threats from fossil fuels have spurred tremendous innovation in renewable energy sources. These threats have also created urgency to innovate newer, more energy-efficient vessels and marine propulsion technologies. Overfishing has spurred newer, more sustainable harvesting and mariculture practices, not to mention key international agreements. This has led to the better management of marine resources and has created protected areas to preserve fragile habitats. Plastic pollution is driving innovation in biodegradable materials as well as ushering in new practices for packing and shipping goods. Climate change is leading scientists and futurists to consider bold advances in geoengineering. Most of all, these threats are driving major changes in education and awareness, so that there is hope for future generations of our world's citizens and researchers.

We must continue to study the basic Earth and environmental science underlying the sea and its life forms and apply these lessons learned to protecting that life and changing how we interact with it. Our world is distinguished by a massive, ancient, and thriving ocean, but we can no longer deny our impacts on it, nor can we overlook its finite fragility. For the sake of our world and our human future, we must learn to live in harmony with, rather than merely near, the ocean. The fight continues, and the sea change accelerates. The editors of this journal, ***Oceans***, hope to report additional important findings as feature papers in years to come.

Funding: This research received no external funding.

Conflicts of Interest: The author declares no conflicts of interest.

List of Contributions:

1. Pelegrí, G.L.; Claret, C.; Sangrà, P. Vertical Shear, Diapycnal Shear and the Gradient Richardson Number. *Oceans* **2024**, *5*, 785–804.
2. Rudan, S.; Sviličić, S. Determination of Submerged Breakwater Efficiency Using Computational Fluid Dynamics. *Oceans* **2024**, *5*, 742–757.
3. Bisinicu, E.; Lazar, L. Exploring Mesozooplankton Insights by Assessing the Ecological Status of Black Sea Waters Under the Marine Strategy Framework Directive. *Oceans* **2024**, *5*, 923–950.
4. Lucey, N.; César-Ávila, C.; Eckert, A.; Veintimilla, P.; Collin, R. Locally Adapted Coral Species Withstand a 2-Week Hypoxic Event. *Oceans* **2025**, *6*, 5.
5. Lombardo, A.; Marletta. G. An Unexpected Small Biodiversity Oasis of Sea Slugs (*Mollusca, Gastropoda, Heterobranchia*) in the Largest Petrochemical Hub of Italy (Central Mediterranean). *Oceans* **2024**, *5*, 695–719.
6. Hernández Ayala, J.J.; Méndez-Tejeda. R. Analyzing Trends in Saharan Dust Concentration and Its Relation to Sargassum Blooms in the Eastern Caribbean. *Oceans* **2024**, *5*, 637–646.

7. Bassolillo, S.R.; D'Amato, E.; Iacono, S.; Pennino, S.; Scamardella, A. Trajectory Planning of a Mother Ship Considering Seakeeping Indices to Enhance Launch and Recovery Operations of Autonomous Drones. *Oceans* **2024**, *5*, 720–741.
8. Prieto, J.M.; Almorza, D.; Amor-Esteban, V.; Muñoz-Perez, J.J.; Jigena-Antelo, B. Identification of Risk Patterns by Type of Ship Through Correspondence Analysis of Port State Control: A Differentiated Approach to Inspection to Enhance Maritime Safety and Pollution Prevention. *Oceans* **2025**, *6*, 15.
9. Henderson, J.; Holly, G.; Rey Da Silva, A.; Trakadas, A. The Cultural Heritage Framework Programme: Highlighting the Contribution of Marine Cultural Heritage to the UN Decade of Ocean Science for Sustainable Development (2021–2030). *Oceans* **2025**, *6*, 1.

References

1. Lovelock, J.A. *Gaia: A New Look at Life on Earth*; Oxford University Press: Oxford, UK, 1979; p. 84.
2. Garrison, T. *Oceanography: An Introduction to Marine Science*, 10th ed.; Cengage: Independence, KY, USA, 2021.
3. Nybakken, J.W.; Bertness, M.D. *Marine Biology: An Ecological Approach*, 6th ed.; Pearson/Benjamin Cummings: San Francisco, CA, USA, 2005.
4. United Nations Sustainability Goal 14 (Life Below Water): Conserve and Sustainably Use the Oceans, Seas, and Marine Resources. Available online: https://www.un.org/sustainabledevelopment/oceans/ (accessed on 14 May 2025).
5. Grassle, J.F.; Maciolek, N. Deep sea species richness: Regional and local diversity estimates from quantitative bottom samples. *Am. Nat.* **1992**, *139*, 313–341. Available online: https://www.journals.uchicago.edu/doi/abs/10.1086/285329 (accessed on 15 May 2025). [CrossRef]
6. Stow, D. *Oceans: An Illustrated Reference*; University of Chicago Press: Chicago, IL, USA, 2005.
7. Kennish, M.J. *CRC Practical Handbook of Marine Science*; CRC Press: Boca Raton, FL, USA, 1989.
8. Czerski, H. *The Blue Machine: How the Ocean Works*; W.W. Norton: New York, NY, USA, 2024.
9. Sharma, S.; Chatterjee, S. Microplastic pollution, a threat to marine ecosystem and human health: A short review. *Environ. Sci. Pollut. Res.* **2017**, *24*, 21530–21547. [CrossRef] [PubMed]
10. Crain, C.M.; Halpern, D.S.; Beck, M.W.; Kappel, C.V. The year in ecology and conservation biology: Understanding and managing human threats to the coastal marine environment. *Ann. N. Y. Acad. Sci.* **2009**, *1162*, 39–62. [CrossRef] [PubMed]
11. Dube, K. A comprehensive review of climatic threats and adaptation of marine biodiversity. *J. Mar. Sci. Eng.* **2024**, *12*, 344. [CrossRef]
12. Simmonds, M.P.; Dolman, S.J.; Jasny, M.; Parsons, E.C.M.; Weilgart, L.; Wright, A.J.; Leaper, R. Marine noise pollution—Increasing recognition but need for more practical action. *J. Ocean Technol.* **2014**, *9*, 71–90.
13. Landrigan, P.J.; Stegeman, J.J.; Fleming, L.E.; Allemand, D.; Anderson, D.M.; Backer, L.C.; Brucker-Davis, F.; Chevalier, N.; Corra, L.; Czerucka, D.; et al. Human Health and Ocean Pollution. *Ann. Glob. Health* **2020**, *86*, 151. [CrossRef] [PubMed]
14. Fleming, L.E.; Broad, K.; Clement, A.; Dewailly, E.; Elmir, S.; Knap, A.; Pomponi, S.A.; Smith, S.; Solo Gabriele, H.; Walsh, P. Oceans and human health: Emerging public health risks in the marine environment. *Mar. Poll. Bull.* **2006**, *53*, 545–560. [CrossRef] [PubMed]

oceans

Article

Vertical Shear, Diapycnal Shear and the Gradient Richardson Number

Josep L. Pelegrí [1,*], Mariona Claret [1] and Pablo Sangrà [2,†]

1 Institut de Ciències del Mar, CSIC, Unidad Asociada ULPGC-CSIC, 08003 Barcelona, Spain; mclaret@icm.csic.es
2 Instituto Universitario de Oceanografía y Cambio Global, Universidad de Las Palmas de Gran Canaria, Unidad Asociada ULPGC-CSIC, 35017 Las Palmas de Gran Canaria, Spain
* Correspondence: pelegri@icm.csic.es
† Deceased author.

Abstract: In Cartesian coordinates (x, y, z), the gradient Richardson number Ri is the ratio between the square of the buoyancy frequency N and the square of the vertical shear S, $Ri = N^2/S^2$, where $N^2 = -(g/\rho)\,\partial\rho/\partial z$ and $S^2 = (\partial u/\partial z)^2 + (\partial v/\partial z)^2$, with ρ potential density, (u, v) the horizontal velocity components and g gravity acceleration. In isopycnic coordinates (x, y, ρ), Ri is expressed as the ratio between $M^2 \equiv N^{-2}$ and the squared diapycnal shear ${S_\rho}^2 = (\rho/g)^2\left[(\partial u/\partial\rho)^2 + (\partial v/\partial\rho)^2\right]$, $Ri = M^2/{S_\rho}^2$. This could suggest that a decrease (increase) in stratification brings a decrease (increase) in dynamic stability in Cartesian coordinates, but a stability increase (decrease) in isopycnic coordinates. The apparently different role of stratification arises because S and S_ρ are related through the stratification itself, $S_\rho = S/N^2$. In terms of characteristic times, this is equivalent to $\tau \equiv S_\rho = {t_o}^2/t_d$, which is interpreted as a critical dynamic time τ that equals the buoyancy period $t_o \equiv N^{-1}$ normalized by the ratio t_d/t_o, where $t_d = S^{-1}$ is the deformation time. Here we follow simple arguments and use field data from three different regions (island shelf break, Gulf Stream and Mediterranean outflow) to endorse the usefulness of the isopycnal approach. In particular, we define the reduced squared diapycnal shear ${\sigma_\rho}^2 = {S_\rho}^2 - M^2$ and compare it with the reduced squared vertical $\sigma^2 = S^2 - N^2$, both being positive (negative) for unstable (stable) conditions. While both Ri and σ^2 remain highly variable for all stratification conditions, the mean ${\sigma_\rho}^2$ values approach ${S_\rho}^2$ with increasing stratification. Further, the field data follow the relation ${\sigma_\rho}^2 = (1 - Ri)/(N^2 Ri)$, with a subcritical $Ri = 0.22$ for both the island shelf break and the Mediterranean outflow. We propose ${\sigma_\rho}^2$ and ${S_\rho}^2$ to be good indexes for the occurrence of effective mixing under highly stratified conditions.

Keywords: vertical mixing; diapycnal mixing; isopycnic coordinates; Richardson number; flow instability

1. Introduction

The analysis of small flow perturbations, for the case of uniformly stratified and sheared flow, goes back to Geoffrey Taylor's 1915 essay for the Adams Prize [1], later followed by many other researchers [2–5]. Other authors employed energetic arguments to assess under which conditions the energy of fluctuations will increase [6–8], for a brief historical review see [9]. The conclusion from these works is that the growth of flow perturbations may be expressed in terms of the (gradient) Richardson number, Ri, with perturbations remaining stable when the Richardson number is above some critical value Ri_c (supercritical, $Ri > Ri_c$) but becoming unstable when it goes below (subcritical, $Ri < Ri_c$). Further, the nonlinear stability analysis of three-dimensional stratified shear flow leads to a critical value of one [5], in agreement with energy arguments [8]. Despite this, there are still some theoretical and observational uncertainties on the actual Ri threshold value for the development of turbulence [10,11].

The most common representation for the Richardson number comes in the Cartesian (x, y, z) coordinate system,

$$Ri = \frac{-\left(\frac{g}{\rho}\right)\frac{\partial \rho}{\partial z}}{\left(\frac{\partial u}{\partial z}\right)^2 + \left(\frac{\partial v}{\partial z}\right)^2} = \frac{N^2}{S^2} \tag{1}$$

with ρ the potential density and (u, v) the horizontal velocity components in the horizontal (x, y) directions. Both the vertical-oscillation or buoyancy frequency $N = [-(g/\rho)\,\partial\rho/\partial z]^{1/2}$ and vertical shear $S = \left[(\partial u/\partial z)^2 + (\partial v/\partial z)^2\right]^{1/2}$ have frequency units.

An alternative, possibly more natural representation of the Richardson number is in isopycnic coordinates. We apply the chain rule $\partial u/\partial z = (\partial u/\partial\rho)(\partial\rho/\partial z)$ as well as the inverse function rule $\partial\rho/\partial z = (\partial z/\partial\rho)^{-1}$ and introduce the (ρ/g) factor to obtain

$$Ri = \frac{-\left(\frac{\rho}{g}\right)\frac{\partial z}{\partial \rho}}{\left(\frac{\rho}{g}\frac{\partial u}{\partial \rho}\right)^2 + \left(\frac{\rho}{g}\frac{\partial v}{\partial \rho}\right)^2} = \frac{M^2}{{S_\rho}^2} \tag{2}$$

where $M \equiv N^{-1} = [-(\rho/g)\,\partial z/\partial\rho]^{1/2}$ is the vertical-oscillation or buoyancy period and $S_\rho = (\rho/g)\left[(\partial u/\partial\rho)^2 + (\partial v/\partial\rho)^2\right]^{1/2}$ is the diapycnal shear, both with units of time. We introduce the "diapycnal shear" terminology to emphasize that the derivatives are with respect to density; when multiplied by (ρ/g), the result has time units.

As a result of the above change of coordinates, a paradox arises. From the Cartesian perspective, *Ri* is directly proportional to stratification, while from the isopycnic viewpoint, *Ri* is inversely proportional to stratification [12–14]. The answer lies in the fact that the vertical and diapycnal shear are related through stratification, $S = N^2 S_\rho$. For constant vertical shear, an increase in stratification enhances dynamic stability. In contrast, for constant diapycnal shear, the same increase in N actually reduces dynamic stability as it implies a quadratic increase in vertical shear. We are undoubtedly most used to the vertical perspective, so perhaps we are missing some lessons that may be learned from the isopycnic view. In this work, we will pursue such an isopycnic inspection, by means of both simple conceptual arguments and through the analysis of real data for different dynamic situations.

The appearance and extensive use of instruments that sense small-scale turbulence, hence providing data that can be translated into turbulent dissipation rates and vertical diffusivities, has progressively reduced the early emphasis on the Richardson number as an index for vertical instability. However, there are still many observational and numerical studies that, in the absence of microstructure measurements, resort to hydrographic-velocity data and the standard parameterizations of vertical diffusivity in terms of the Richardson number [15–19]. Hence, it seems timely to follow other authors [18,20–24] and further explore *Ri*-related alternative indexes that could possibly be used to assess the presence of vertical instability.

In the next section, we explore some relevant differences between the vertical and isopycnic perspectives, and introduce the concepts of natural oscillation time, reduced squared vertical shear and reduced squared diapycnal shear. We present the non-dimensional variables in Section 3, examining the interdependence between background stratification, vertical shear and diapycnal shear. The three datasets used in our study are briefly described in Section 4, and in Section 5 we look at these data in two different ways: first as cloud points plotted in terms of two variables (among background stratification, vertical shear and diapycnal shear) and next examining the dependence of both the reduced squared vertical shear and the reduced squared diapycnal shear in terms of the background stratification. Finally, in Section 6, we summarize the main conclusions.

2. Isopycnic Versus Vertical Perspectives

2.1. Basic Considerations

In vertical coordinates, the critical Richardson number criterion is commonly interpreted as if sufficiently high stratification (high N) leads to stable conditions ($Ri > 1$), provided the vertical shear remains moderate, while low stratification is prone to unstable conditions ($Ri < 1$) (Equation (1)). In a layered ocean, with near-constant density layers bounded by relatively thin steps of rapid density changes, this vertical-coordinate perspective would usually associate the steps with stable regions and the layers with unstable zones (Figure 1). This is true for a static ocean, but dynamic stability depends on the vertical shear. A layer will have most properties well mixed so that both N and S tend to zero; weak vertical shear will lead to instabilities, but these will only redistribute near-homogeneous water. In a step, on the other hand, N is large and instabilities can effectively redistribute density (and other water properties), but these will only develop for large enough vertical shear S.

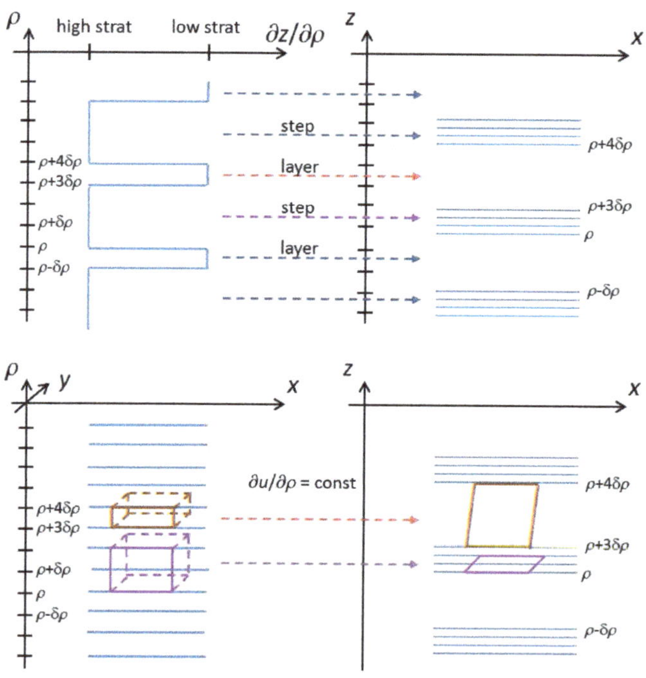

Figure 1. Schematics of flow deformation in Cartesian and isopycnic coordinates. (**Top left**) Imagine the water column is split in high- and low-stratified regions, characterized by small and large values of $\partial z/\partial\rho$ in the isopycnic representation, (**Top right**) which corresponds to the steps and layers in Cartesian coordinates. Let the entire water column experience a constant diapycnal shear $\partial u/\partial\rho$, and (**Bottom left**) explore what will happen to material water volumes defined in the isopycnic representation. (**Bottom right**) When viewed in the vertical domain relatively large material volumes located in layers will experience low deformation while much smaller volumes within steps will experience high deformation.

Equation (2) in isopycnic coordinates, on the other hand, suggests that a highly stratified region (a step, $M = N^{-1} \to 0$) may be prone to mixing as long as the diapycnal shear S_ρ, which depends on the velocity of water parcels in adjacent layers, remains finite. Consider for example a 2D symmetrically stable front with the density field $\rho(x, z)$, and further consider that the along-front flow is in geostrophic balance. The thermal wind equation in Cartesian and isopycnic coordinates are $\rho f\, \partial v/\partial z = g\, \partial\rho/\partial x$ and $\rho f\, \partial v/\partial\rho = g\, \partial z/\partial x$, respectively, and the corresponding Richardson numbers are $Ri = C\,(\partial\rho/\partial z)/(\partial\rho/\partial x)^2$ and $Ri = C\,(\partial z/\partial\rho)/(\partial z/\partial x)^2$, where $C = -(f^2\rho/g)$. In a frontogenetic situation where the isopycnals get steeper with time (both $\partial\rho/\partial x$ and $\partial z/\partial x$ increase) and vertical stratification does not change ($\partial\rho/\partial z$ remains constant), Ri will decrease in both reference frames. However, in those locations where frontogenesis also leads to increased vertical stratification, the Cartesian expression suggests that stability may increase while the isopycnic perspective points to further instability.

The above simple arguments suggest that diapycnal shear may be a convenient variable to analyse whether sheared oceanic flows are prone to mixing. As an illustrative example, consider the two-layer flow between parallel planes such as that set up in the laboratory to simulate interfacial Kelvin–Helmholtz instabilities, e.g., [25] and references therein. The upper and lower layers have densities $\rho = \rho_0(1-\Delta\rho)$ and $\rho = \rho_0(1+\Delta\rho)$, respectively, and the corresponding velocities are $u = \Delta\rho\; g\; t\; sin\theta$ and $-\Delta\rho\; g\; t\; sin\theta$, where θ is the tilting of the table in the experimental setup. The vertical shear between both layers is not uniquely defined but, in contrast, the isopycnic derivative of the horizontal velocity is easily calculated from $\rho = \rho_0(1 - u/(g\; t\; sin\theta))$, as $\partial u/\partial\rho = -(g\; t\; sin\theta)/\rho_0$; notice this derivative is equal to the finite-difference fraction, $\delta u/\delta\rho = (2\; g\; t\; \Delta\rho\; sin\theta)/(2\;\rho_0\;\Delta\rho) = (g\; t\; sin\theta)/\rho_0$. The resulting expression for the diapycnal shear is simply $S_\rho = t\; sin\theta$, with time units, implying that instabilities will develop independently of stratification if sufficient time is allowed.

2.2. Characteristic Times

As mentioned in the Introduction, the Richardson number criterion arises from either flow-stability or energy arguments. A complementary interpretation comes in terms of characteristic times associated with the inverse of the buoyancy frequency and the vertical shear, as well as with the diapycnal shear itself. The inverse of the buoyancy frequency is the time characteristic for the vertical oscillation of a water parcel in a stratified medium, hereafter the vertical-oscillation or buoyancy period, $t_o \equiv N^{-1}$. The inverse of the vertical shear gives a characteristic time for the horizontal deformation of a water parcel (or temporal rate of strain), henceforth the vertical-deformation time, $t_d \equiv S^{-1}$.

The diapycnal shear is proportional to the derivative of the horizontal velocity with respect to density, e.g., $\partial u/\partial\rho$; when we multiply this derivative by ρ/g, we get the diapycnal shear with units of time, so we set $\tau \equiv S_\rho$ to simply remind us of its time units. Because the vertical and diapycnal shears are related through the stratification, $S_\rho = S/N^2$, the diapycnal shear is related to the buoyancy and deformation times as $\tau = {t_o}^2/t_d$. We envision the diapycnal shear as reflecting adjacent layers that maintain their velocity, so it naturally incorporates both stratification (through t_o) and shear deformation (through t_d). We may hence interpret τ as a critical dynamic time for the system, equal to the vertical-oscillation period t_0 normalized by the ratio (t_d/t_o).

Equations (1) and (2) may be written in terms of these characteristic times as

$$Ri = \frac{N^2}{S^2} = \left(\frac{t_d}{t_o}\right)^2 \tag{3}$$

$$Ri = \frac{1}{N^2 {S_\rho}^2} = \left(\frac{t_0}{\tau}\right)^2 \tag{4}$$

or alternatively as

$$Ri = \frac{1}{S\,S_\rho} = \frac{t_d}{\tau} \tag{5}$$

We may now see that Equations (3) and (5) are not really that different. According to Equation (3), the Richardson number is subcritical (less than one) when $t_d < t_o$, i.e., when the vertical excursion of the water parcel takes long enough for it to become greatly distorted. Similarly, Equation (5) tells us that subcritical conditions require a short deformation time t_d as compared with the critical dynamic time τ, $t_d < \tau$. The main difference lies on the linear and quadratic dependences: in vertical coordinates the dependence is quadratic $(t_d/t_0)^2$ while in isopycnal coordinates it is linear (t_d/τ). Equations (3) and (4) further tell us that the instability condition implies $t_d < t_0 < \tau$.

2.3. Reduced Squared Shears

The reduced vertical shear was introduced as a complementary way to understand the relevance of subcritical conditions for mixing [20,21]. By considering a critical value $Ri = 1/4$, the subcritical condition would correspond to $S^2 - 4N^2 = (S+2N)(S-2N) > 0$ or, alternatively, to a reduced shear greater than zero, $S - 2N > 0$. A formal advantage of the reduced shear is that it is approximately proportional to the growth rate of Kelvin–Helmholtz instabilities [20,26]. A practical advantage is that reduced shear helps assess when effective mixing occurs. As mentioned above, growing instabilities are effective blenders in well-stratified conditions; in poorly stratified conditions, however, they can only redistribute water which is already well mixed. The Richardson number Ri is not capable of distinguishing cases of low stratification and low shear from those of high stratification and high shear, but the reduced shear does, i.e., for two different situations (high and low stratification) with equal Ri subcritical values, it turns out that $S - 2N = N\left(Ri^{-1/2} - 2\right)$ is larger for the well-stratified condition.

Let us pursue these ideas and define a reduced squared vertical shear σ^2 and a reduced squared diapycnal shear ${\sigma_\rho}^2$, but using the critical Richardson value of one [5]:

$$\sigma^2 \equiv S^2 - N^2 = S^2(1 - Ri) = N^2(1 - Ri)/Ri \quad (6)$$

$${\sigma_\rho}^2 \equiv {S_\rho}^2 - M^2 = {S_\rho}^2(1 - Ri) = (1 - Ri)/\left(N^2 Ri\right) \quad (7)$$

The definitions in (6) and (7) show that both squared shears are positive for unstable conditions ($Ri < 1$) and negative for stable conditions ($Ri > 1$). For constant shears, σ^2 decreases with stratification from S^2 to increasingly negative values, and ${\sigma_\rho}^2$ increases with stratification up to a maximum value of ${S_\rho}^2$. Equations (6) and (7) illustrate how, for constant Ri values, the reduced squared shears change with either stratification or vertical/diapycnal shear. Both reduced squared shears are equal to the corresponding squared shear but with a $(1 - Ri)$ reduction factor, i.e., the reduced squared shears increase with decreasing Ri up to a maximum value equal to the squared shear, with σ^2 approaching S^2 and ${\sigma_\rho}^2$ approaching ${S_\rho}^2$. When expressed as a function of stratification N^2, both variables increase as Ri decreases but with σ^2 and ${\sigma_\rho}^2$ directly and inversely proportional to stratification, respectively.

3. Data Analysis

To see how the above ideas fit in the real world, we consider several different cases of stratification and vertical/diapycnal shear. In order to compare these cases, it is convenient to consider the nondimensional form of the dependent variables in Equations (3)–(7). For this purpose we use a background stratification, N_b, to set the following relations: $N = N_b(N/N_b) \equiv N_b N'$, $S = N_b(S/N_b) \equiv N_b S'$, $M = (1/N_b)(N_b M) \equiv (1/N_b)M'$ and $S_\rho = (1/N_b)(N_b S_\rho) \equiv (1/N_b){S_\rho}'$; notice there are only two independent variables as $M' = 1/N'$ and ${S_\rho}' = S'/N'^2$.

Hereafter, we drop primes for all dependent variables and will always refer (except where indicated) to their nondimensional forms. In particular, notice that N^{-1}, S^{-1} and S_ρ, respectively, are the nondimensional forms of the vertical-oscillation period, the deformation time and the critical dynamic time. The Richardson number may hence be expressed in terms of the nondimensional variables exactly as in Equations (3)–(5). Analogously, the nondimensional reduced squared shears have the same form as in Equations (6) and (7).

The subcritical condition ($Ri < 1$) translates into the following instability conditions for the reduced squared shears:

$$\sigma^2 = \left(S^2 - N^2\right) > 0 \tag{8}$$

$$\sigma_\rho{}^2 = \left(S_\rho{}^2 - \frac{1}{N^2}\right) > 0 \tag{9}$$

To illustrate the relations among the nondimensional dependent variables, we have produced several sets of plots, where we explore the relations between two variables as a function of the third variable and the Richardson number (Figure 2). Specifically, we look at the following dependences:

- in the (S, N) domain, $S = S_\rho N^2 = \left(1/Ri^{1/2}\right)N$,
- in the (S_ρ, N) domain, $S_\rho = S/N^2 = \left(1/Ri^{1/2}\right)(1/N)$, and
- in the (S_ρ, S) domain, $S_\rho = S/N^2 = (1/Ri)(1/S)$.

In all these plots, we show an overall perspective that goes all the way to stratification values as large as 10 times the background stratification (Figure 2, upper panels), and display a close-up on that region with stratification values between well mixed and twice the background stratification (Figure 2, lower panels). As justified above, we are most interested in what leads to mixing in relatively well-stratified conditions, e.g., for $N \in [1, 2]$.

In the left panels of Figure 2, the domain is halved in regions of supercritical and subcritical conditions, illustrating that subcritical conditions may only be reached if the vertical shear is sufficiently large, $S > N$. For low stratification values, subcritical conditions ($Ri \leq 1$) can only be attained for relatively large diapycnal shears (Equation (1)), e.g., $S_\rho \geq 2$ when $N \leq 0.5$; however, as stratification increases, we find that subcritical conditions are achieved with a progressively smaller diapycnal shear, e.g., when $N \geq 2$ it is sufficient to have $S_\rho \leq 0.5$.

The middle panels of Figure 2 show a quite different partition between subcritical and supercritical conditions, which is the result of the quadratic inverse dependence of Ri on both stratification and diapycnal shear (Equation (2)). Only a small portion of the domain corresponds to supercritical conditions, tending to zero as either N or S_ρ become large. In particular, for well-stratified conditions, the domain of subcritical conditions reaches moderate and even low diapycnal shears, necessarily coincident with high values of vertical shear.

Finally, the right panels in Figure 2 show the inverse linear dependence of Ri on both vertical and diapycnal shear (Equation (3)), with subcritical conditions requiring sufficiently large values of either one or both variables, depending on the size of vertical stratification. In particular, subcritical flow requires only moderate diapycnal shear during high stratification conditions.

The principal idea arising from Figure 2 is that moderate diapycnal shear in well-stratified conditions is a guarantee for actual mixing of distinct waters. We may further explore this idea by examining the dependence of the nondimensional reduced squared shears as a function of the nondimensional stratification N for different S, S_ρ and Ri values:

- in the (σ^2, N) domain, $\sigma^2 = S^2 - N^2 = \left(S_\rho{}^2 N^2 - 1\right)N^2 = N^2(1 - Ri)/Ri$, and
- in the $(\sigma_\rho{}^2, N)$ domain, $\sigma_\rho{}^2 = S_\rho{}^2 - 1/N^2 = (S^2 - N^2)N^4 = (1 - Ri)/\left(N^2 Ri\right)$.

The corresponding plots are shown in Figure 3, again up to $N = 10$ (upper panels) and for $N \in [0, 2]$ (lower panels). Recall that subcritical conditions correspond to positive values of either reduced squared shear.

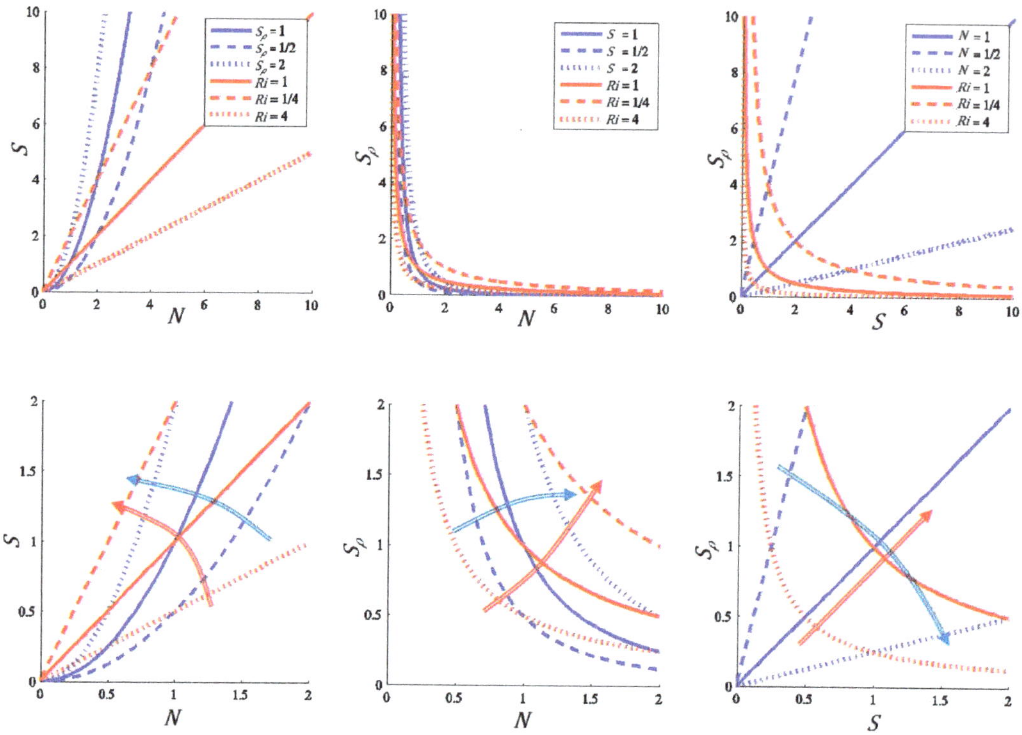

Figure 2. Relations among nondimensional dependent variables. (**Left panels**) Vertical shear as a function of stratification, (**Middle panels**) diapycnal shear as a function of stratification, (**Right panels**) diapycnal shear as a function of vertical shear, for different Ri values (red curves) and different values of the third variable (blue curves) as shown; the dashed, solid and dotted lines, respectively, correspond to values less than, equal to and greater than one. In the top panels, we present the results in the [0:10, 0:10] domain, while in the bottom panels we zoom into the [0:2, 0:2] domain; the colored arrows illustrate the direction for decreasing Ri (red curves in all panels), increasing S_ρ (blue in left panels), increasing S (blue in middle panels) and increasing N (blue in right panel).

The plots for the reduced squared vertical shear (Figure 3, left panels) illustrate that the same Richardson number may correspond to many different σ^2 values, each of them with a different combination of N, S and S_ρ values (except $Ri = 1$, which corresponds to $\sigma^2 = 0$). In particular, positive values of σ^2 are obtained under highly stratified conditions for moderate and even low diapycnal shear but require high values of vertical shear; for example, a value as low as $S_\rho = 0.5$ guarantees the subcritical conditions for $N \geq 2$, implying $S \geq 2$.

The plots for the reduced squared diapycnal shear (Figure 3, right panels) serve to emphasize the idea that, under well-stratified conditions, vertical instability and mixing will only occur if the vertical shear is very high, but even a small diapycnal shear can cause such stirring. In particular, for relatively large stratifications, the approximation ${\sigma_\rho}^2 = {S_\rho}^2 - 1/N^2 \cong {S_\rho}^2 > 0$ holds always, which suggests that the diapycnal shear is a good stability index. In these conditions, the critical dynamic time is large enough to guarantee that the flow deformation will create instability.

In the following sections, we will use three very different datasets to explore the relevance of the isopycnic perspective. Our approach will be empirical: we will display scatter plots of the sort shown in Figures 2 and 3 and will explore the different Ri-regimes in each dynamical system.

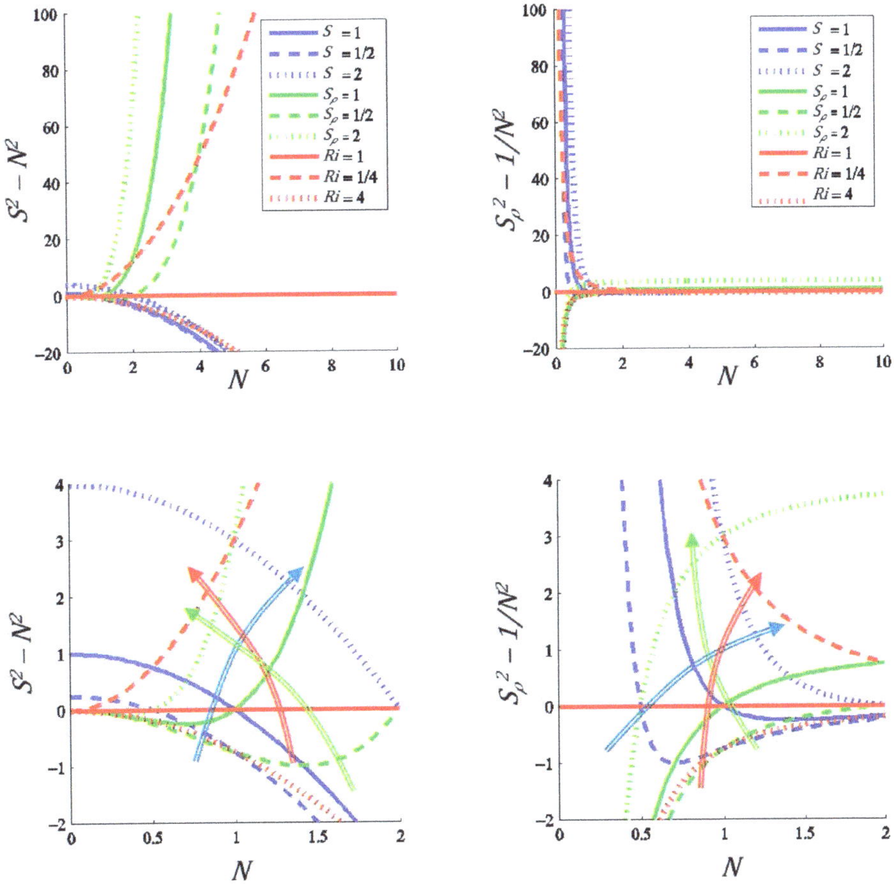

Figure 3. (**Left panels**) Nondimensional reduced squared vertical shear $S^2 - N^2$ and (**Right panels**) nondimensional reduced squared diapycnal shear $S_\rho{}^2 - 1/N^2$, in both cases plotted as a function of stratification N. Red curves stand for different Ri values, blue curves for different S values and green curves for different S_ρ values; the dashed, solid and dotted lines correspond, respectively, to values less than, equal to and greater than one. In the top panels, we present the results in the [0:10, 0:100] domain, while in the bottom panels we zoom into the [0:2, 0:4] domain; the colored arrows illustrate the direction where the flow becomes subcritical (Ri decreasing) and the shear increases.

4. Datasets

The three different datasets represent very diverse oceanographic settings and dynamic conditions: (1) the shelf break of Gran Canaria Island (GCI), a deep-ocean island where internal waves are commonly present over the slope and shelf break (Figure 4); (2) the Gulf Stream (GS), an intense western boundary baroclinic current (Figure 5); and (3) the Mediterranean outflow (MO), a density-driven intense jet (Figure 6). All three datasets share the characteristic that a significant fraction of the data corresponds to well-stratified waters (Figure 7) which are experiencing an intense flow. Further, in all cases the velocity field is actually sampled, therefore allowing a good assessment of vertical and diapycnal shears. Therefore, these datasets may be quite useful for examining the relationships between stratification, vertical shear and diapycnal shear and to explore how the Richardson number and the reduced squared (vertical and diapycnal) shears change with stratification.

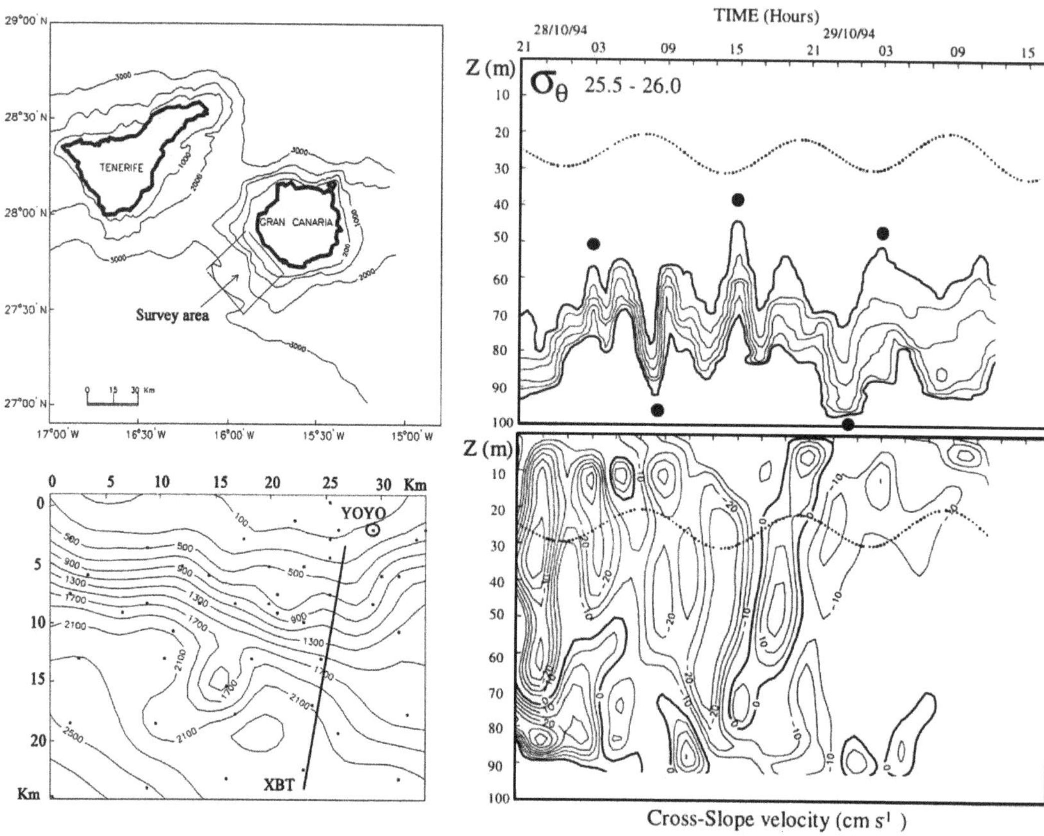

Figure 4. (**Left panels**) A map of the central Canary Islands, (**Top left**) showing the location of the survey area in the shelf break southwest of Gran Canaria Island and (**Bottom left**) a detail of the shelf-break bathymetry with the location of the yo-yo repeated vertical sampling (YOYO station). (**Right panels**) Time–depth plots over two days of (**Top panel**) potential density (isopycnals at 0.1 σ_θ-intervals with 25.5 and 26.0 shown as bold contours) and (**Bottom panel**) cross-slope speed (in cm s^{-1}, positive values onshore). In the right panels, the barotropic tide is denoted with a dotted line amplified by a factor of 10. Reproduced with permission from Sangrà et al., *Sci. Mar.*, published by CSIC Press, 2001 [27].

The first set of data corresponds to repeated vertical sampling at a fixed location on the 100 m isobath (yo-yo type) in the shelf break south of GCI (Figure 4, left panels) in fall 1994 [27]. Sampling consisted of a conductivity–temperature–depth (CTD) probe which included an acoustic Doppler point current meter. Density and velocity profiles were obtained at 1 m vertical resolution every hour during nearly four days, two days of neap tides and two days of spring tides, for a total of about 9000 samples.

Gran Canaria Island has a steep slope (about 0.15) that goes down to depths over 2500 m (Figure 4, bottom left). The dominant semidiurnal tide propagates northwards into the slope and, as a result of the interaction with the topography, packets of internal waves with semidiurnal and lower periodicities are generated. The structure of the water column consisted of a 50–70 m surface mixed layer, a 20–30 m thick seasonal thermocline where the potential density increased from 1025.5 to 1026.0 kg m^{-3}, and a relatively thin bottom mixed layer (Figure 4). The buoyancy frequency reflects the fairly well-mixed region in the upper half of the water column, on top of varyingly stratified waters in the bottom half, with a mean buoyancy frequency equal to 0.010 s^{-1} (Figure 7, left panel). The cross-slope speed often exceeds 0.2 m s^{-1} over the entire water column (Figure 4, bottom right), leading to subcritical conditions not only in the mixed layers but also in the thermocline [24].

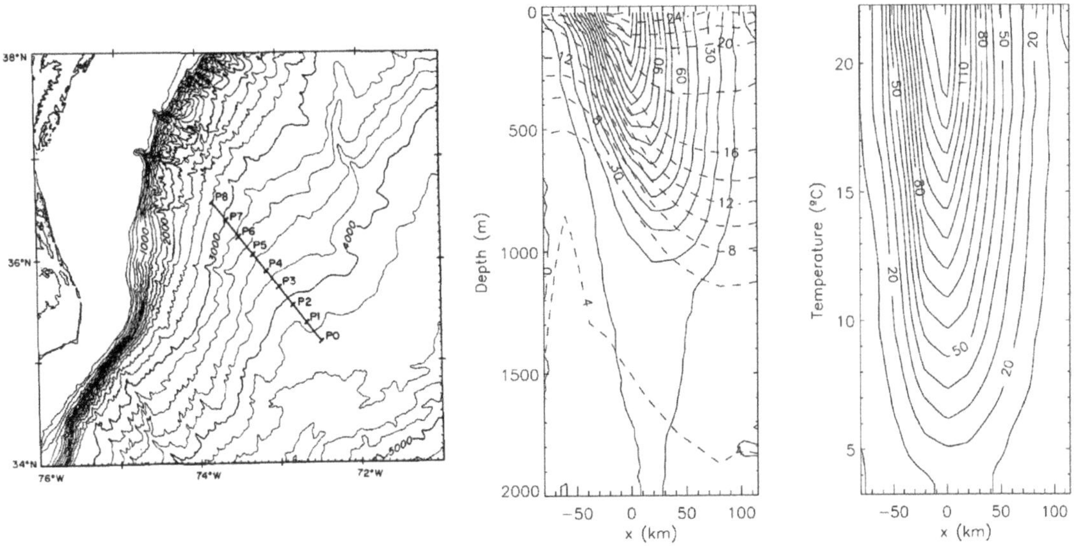

Figure 5. (**Left panel**) Location of the velocity and temperature stations (P0 through P8) across the Gulf Stream, off the east coast of North America, with the bottom isobaths. Reproduced with permission from Halkin and Rossby, *J. Phys. Oceanogr*; published by American Meteorological Society, 1985 [28]. (**Middle panel**) Vertical section of the along-stream mean velocity (solid contours in cm s^{-1}) and the temperature in (dashed contours in °C). (**Right panel**) Along-stream velocity as a function of temperature. Reproduced from [29].

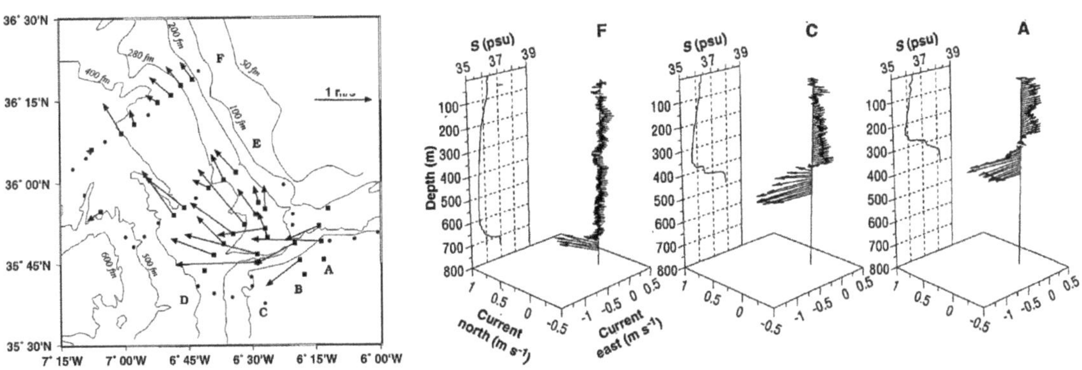

Figure 6. (**Left panel**) Location of most stations with conductivity and velocity data used in our analysis, with vectors illustrating the peak Mediterranean outflow velocity; some additional stations inside the Strait of Gibraltar are not shown. (**Right panels**) Examples of velocity and salinity vertical profiles along the outflow axis of sections A, C and F (see left panel for the location of each section). Reproduced with permission from Price et al., *Sci.*; published by AAAS, 1993 [30].

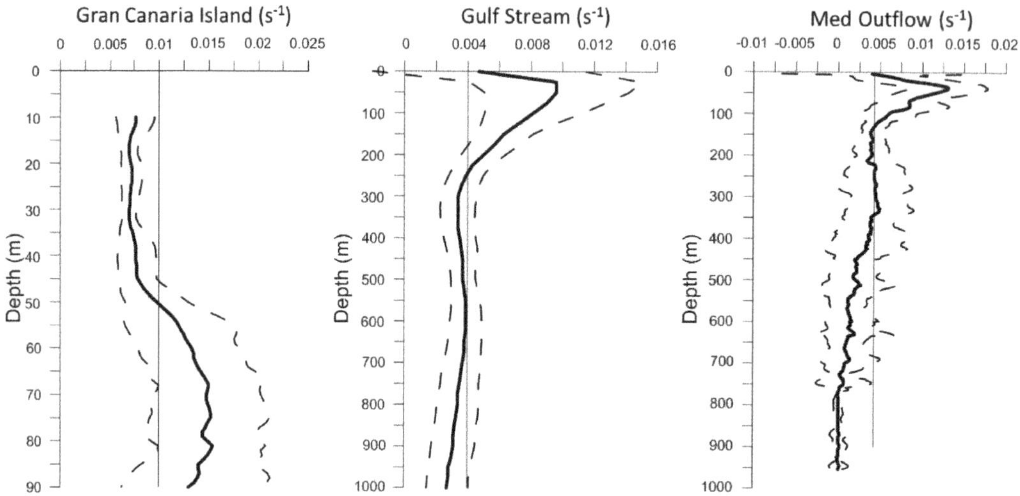

Figure 7. Buoyancy frequency as a function of depth for (**Left panel**) the shelf break of Gran Canaria Island, (**Middle panel**) the Gulf Stream permanent thermocline and (**Right panel**) the region west of Gibraltar which includes the Mediterranean outflow. Note the changing scales between figures and the lack of data from the top and bottom 10 m in Gran Canaria Island because of instrumental limitations. Thick solid lines represent mean profiles while dashed lines indicate the one-standard deviation limit from mean profiles. Thin solid lines indicate the mean buoyancy frequency for each dataset, used to obtain the nondimensional variables.

The second dataset consists of 20 repeated hydrographic sections across the GS (Figure 5, left panel), carried out between September 1980 and May 1983. These sections include temperature data obtained with expendable bathythermographs (XBT) and velocity data obtained from free-falling Pegasus instruments [31], originally reported and analyzed by [28], see also [29] and references therein. Temperature values were converted into potential density by means of a recursive utilization of algorithms [32,33]. Each section contained between four and ten stations of velocity and potential density from the sea surface down to 2000 m at 25 m intervals, for a total of some 10,000 samples.

The permanent thermocline of the GS, with temperatures between about 6 and 16 °C, slopes sharply towards the continental slope, rising some 600–700 m in 100 km. As a result, the GS core velocities increase from about 0.25 to more than 1.5 m s^{-1} over this same temperature range. On both sides of the GS core, the mean vertical stratification remains similar but the velocities are substantially smaller (Figure 5, middle panel). The relatively low vertical data resolution limits the sampling of high-shear events associated to internal waves, which would be responsible for a significant increase in the gradient Richardson values, and hence restricts our analysis to near-geostrophic flow [9]. Despite this, the background geostrophic velocity field is intense enough for the core of the GS to experience several instances of subcritical conditions ($Ri < 1$) which are responsible for vertical mixing [12,34]. The buoyancy frequency decreases largely in the top 300 m and remains approximately constant down to 1000 m, with low variability characteristics for the subtropical permanent thermocline. The mean buoyancy frequency for this second dataset is 0.004 s^{-1} (Figure 7, middle panel).

The third set of data consists of 56 stations with full-depth CTD profiles and simultaneous velocity measurements with expendable current profilers (XCP) that sampled the eastern Gulf of Cádiz during fall 1988 [30,35,36], with about half of the stations over the MO (Figure 6, left panel). Velocity and density data were recorded every 2 m down to an average depth of about 500 m so that some 14,000 measurements were collected.

As the MO exits the Strait of Gibraltar, it experiences three different dynamic regions over a distance of less than 50 km. In the first phase, located between Spartel Sill (360 m) and Western Spartel Sill (420 m, located at about 6°20′W, 35°47′N), the MO follows west along a relatively narrow (about 5 km wide) and gently sloping channel that deepens from 360 to 420 m in about 20 km. In the second phase, the MO continues west along a channel of variable amplitude (typically 5 to 10 km) and undergoes three abrupt topography-driven accelerations (the bathymetry deepens from 420 to 700 m over a distance of another 20 km) with core velocities in excess of 1 m s^{-1}, resulting in substantial dilution of the Mediterranean water properties. During these two phases, the MO approximately occupies the bottom third of the water column (Figure 6, right panels). In the third phase, the MO follows northwest, its diluted core slowing down and leaning against the Iberian continental slope [30,37–39]. Therefore, the local profile of the buoyancy frequency depends on whether the station is close to Gibraltar and whether it is found on the path or away from the path of the MO; for this reason, the variability is largest at depths between 200 and 750 m, and decreases in the bottom 100 m simply because of the few stations we have for deep waters. The mean buoyancy frequency for this third dataset is 0.004 s^{-1} (Figure 7, right panel).

5. Results

We nondimensionalize the variables with a different constant background frequency for each dataset, as explained in Section 3. This background frequency, which is defined as the mean buoyancy frequency over each region (Table 1), allows consideration of the dynamic changes within each single region. Additionally, the utilization of different background stratifications (one per region) facilitates a unified interpretation of all three, potentially quite different, cases.

Table 1. Dimensional mean buoyancy frequency N (s^{-1}) and best-fit parameters to the nondimensional mean ${\sigma_\rho}^2$ values for two different functions, with the R^2 correlation coefficients.

		Gran Canaria Island	Gulf Stream	Mediterranean Outflow
Number of samples		9000	10,000	14,000
Mean N (s^{-1})		0.010	0.004	0.004
best exponential adjustment ${\sigma_\rho}^2 = {\sigma_0}^2 exp(-rN)$	${\sigma_0}^2$	174.2	−1.9	250.3
	r	2.90	1.05	4.03
	R^2	0.66	0.83	0.97
best adjustment as ${\sigma_\rho}^2 = (1 - Ri)/(N^2 Ri)$	Ri	0.217	2.067	0.222
	R^2	0.70	0.86	0.81

Consider first the cloud points for the nondimensional vertical stratification, vertical shear and diapycnal shear (Figure 8), in particular the zoom for the high stratification and high shear domain (Figure 9). The overall shape of S against N changes greatly in all three cases, with no noticeable dependence of vertical shear on stratification. The common denominator in all three cases is that S takes a relatively high range of values for any N (Figures 8 and 9, top panels). In contrast, all S_ρ against N cloud points have similar shapes, with relatively high values and high variability in low stratification conditions and the opposite for well-stratified waters ($N > 3$) (Figures 8 and 9, middle panels); these cloud points actually suggest a decay of the maximum diapycnal shears with increasing vertical stratification. Finally, the S_ρ against S scatter plots also display very large differences between all three cases, with the GCI and GS cases displaying completely different distributions and the MO possibly as an intermediate situation (Figures 8 and 9, bottom panels).

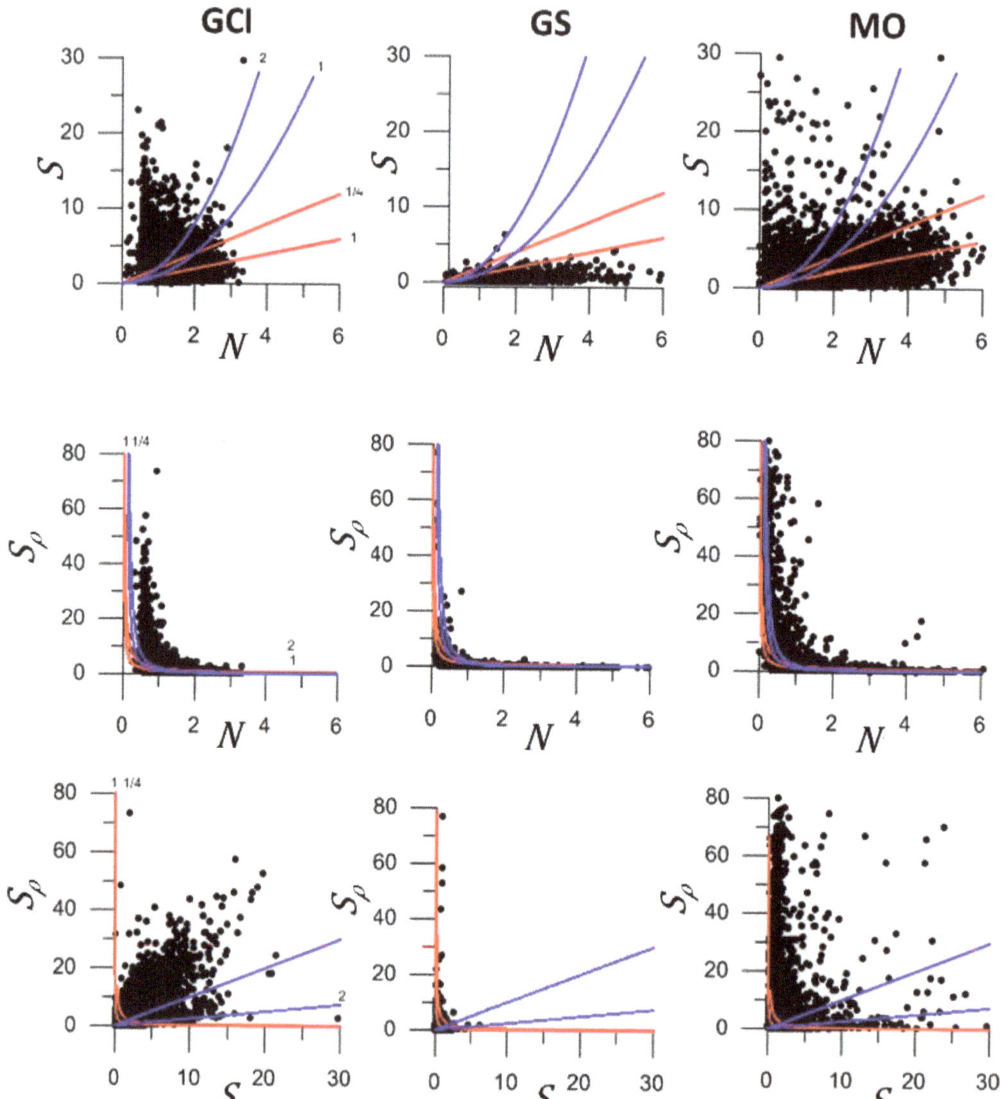

Figure 8. Scatter plots of nondimensional variables: (**Top panels**) S as a function of N, (**Center panels**) S_ρ as a function of N and (**Bottom panels**) S_ρ as a function of S. The three columns correspond to the three datasets: (left panels, GCI) Gran Canaria Island shelf break, (middle panels, GS) Gulf Stream and (right panels, MO) Mediterranean outflow. The red lines correspond to contours of $Ri = 0.25$ and 1. The blue lines correspond to contour values of 1 and 2 for (**Top panels**) S_ρ, (**Center panels**) S and (**Bottom panels**) N.

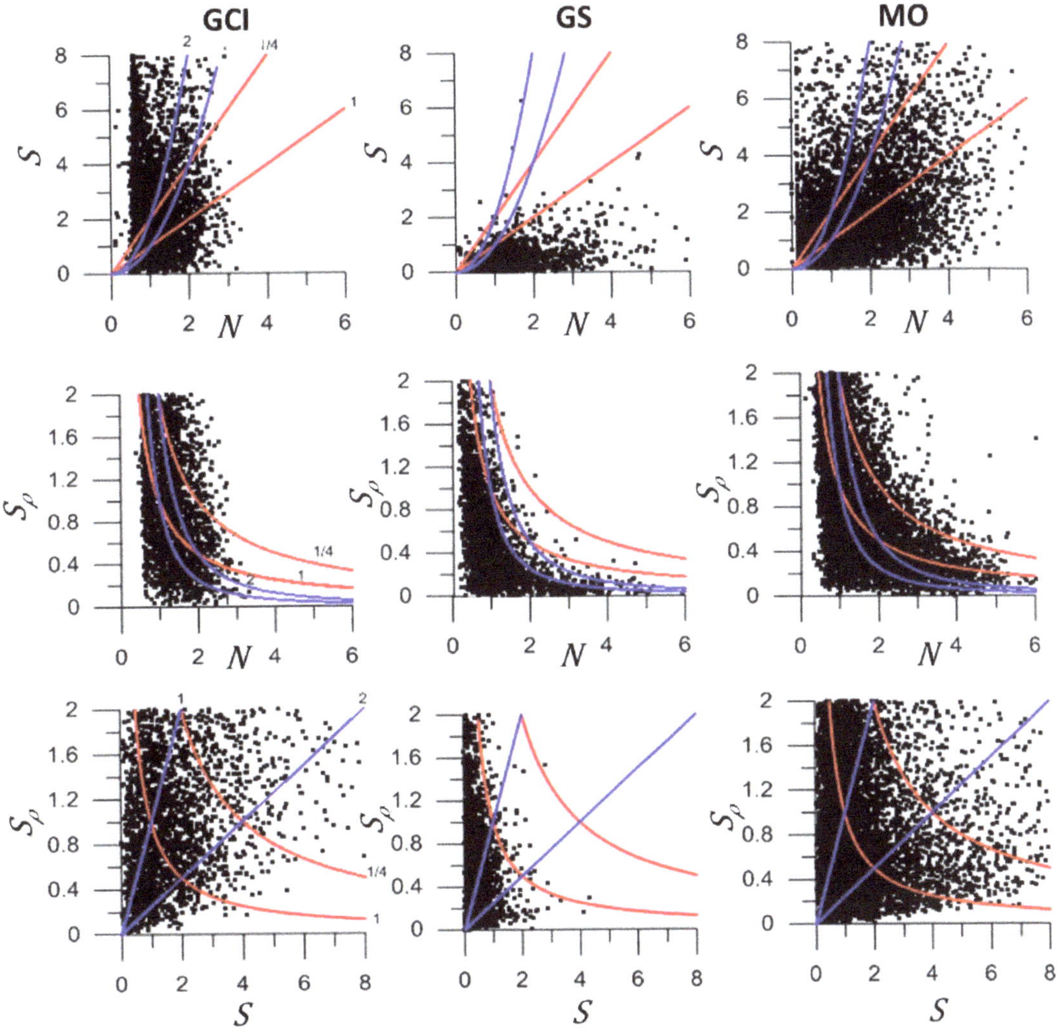

Figure 9. As in Figure 8 but zooming at $N \in [0, 6]$, $S \in [0, 8]$ and $S_\rho \in [0, 2]$.

These results may be interpreted as if each flow has its own dynamics, setting a range of characteristic diapycnal shears. We may step forward and explore if each dynamical system is characterized by a controlling parameter. In Section 2, we argued that the reduced squared diapycnal shear ${\sigma_\rho}^2$ may be such parameter. This variable is an index for the stability of the system, in the same way as the Richardson number, but with the characteristic that its maximum value is the squared diapycnal shear ${S_\rho}^2$, to be attained during high stratification conditions. These ideas are sustained by the cloud points of σ^2 and ${\sigma_\rho}^2$ as a function of stratification (Figures 10 and 11). The distribution of σ^2 is quite different in each case, with large scattering for all N values. In contrast, the distribution of ${\sigma_\rho}^2$ does hint at a dependence on N, with its maximum (positive) values decreasing with increasing vertical stratification, with a threshold value ${S_\rho}^2$ that is itself conditioned by N, i.e., ${\sigma_\rho}^2 = {S_\rho}^2 - N^{-2} = S^2 N^{-4} - N^{-2}$.

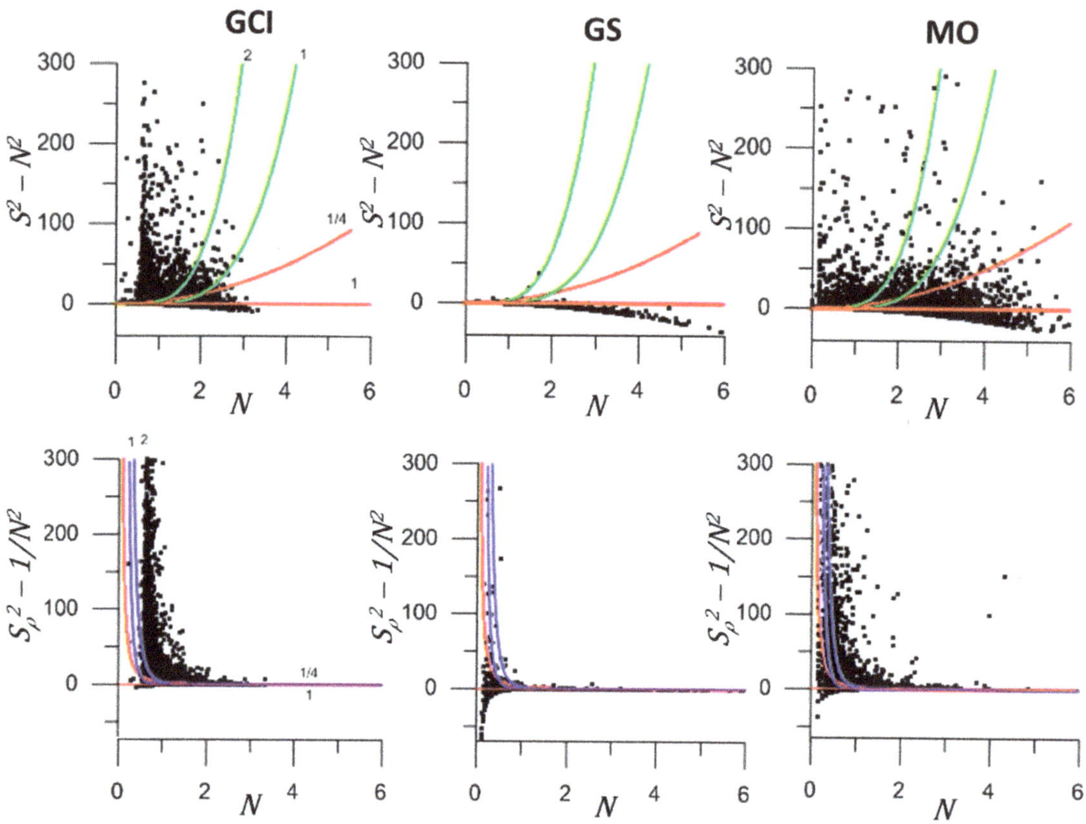

Figure 10. Scatter plots of nondimensional variables: (**Top panels**) σ^2 and (**Bottom panels**) $\sigma_\rho{}^2$, both as a function of N. The three columns correspond to the three datasets: (left panels, GCI) Gran Canaria Island shelf break, (middle panels, GS) Gulf Stream and (right panels, MO) Mediterranean outflow. The red lines correspond to contours of $Ri = 0.25$ and 1. The green lines in the top panels correspond to contours of $S_\rho = 1$ and 2, while the blue lines in the bottom panels correspond to contours of $S = 1$ and 2.

In order to further examine these ideas, we look at the way the mean values and standard deviations of all three stability indices (Ri, σ^2 and $\sigma_\rho{}^2$) change with stratification (Figure 12; for this calculation the mean values and standard deviations are calculated over 0.2 intervals of N). The results show no clear relation of σ^2 and Ri with stratification, but suggest the existence of a univocal dependence of $\sigma_\rho{}^2$ with stratification for each particular flow, with $\sigma_\rho{}^2$ proportional to N^{-1} in GCI and the MO, and $\sigma_\rho{}^2$ proportional to $-N^{-1}$ in the GS (Figure 13).

Finally, we explore if there are simple functions that can provide a good fit to the observed mean $\sigma_\rho{}^2$ dependence on stratification N (Figure 13; mean values are here calculated over $0.1N$ intervals). Our first attempt involves simple exponential decay with stratification, given by $\sigma_\rho{}^2 = \sigma_0{}^2 \, exp(-rN)$. It turns out that the adjustment is fairly good in all cases, with the best-fit parameters (as calculated separately for each dataset) presented in Table 1; because of the excellent fit for low and high N values, the highest correlation with the data corresponds to the MO case. For GCI and the MO, the exponential coefficient is relatively large (2.90 and 4.03) as compared with the one for the GS (1.05); the main difference, however, is that for GCI and the MO the mean $\sigma_\rho{}^2$ values remain always positive, indicative of unstable conditions, while for the GS they are negative, indicative of stable conditions for shear instabilities. The negative values in the GS are likely the consequence of the low vertical resolution of these data, which cannot resolve the relevant turbulent scales.

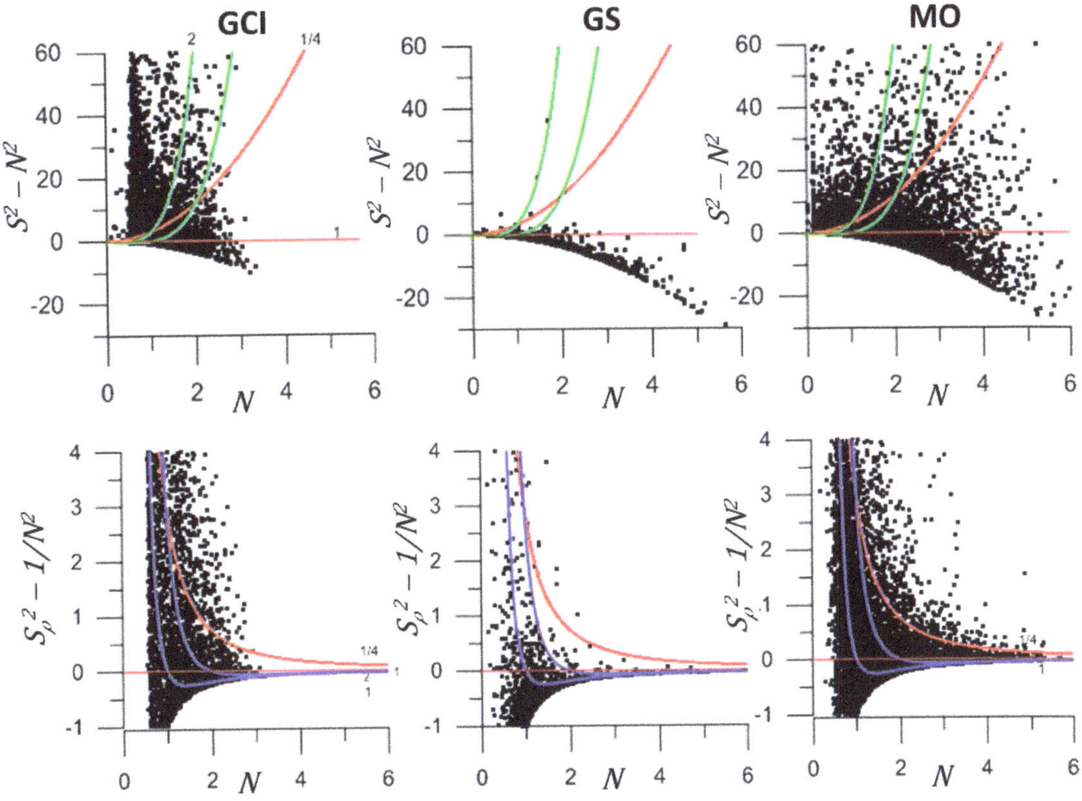

Figure 11. As in Figure 10 but zooming at $N \in [0, 6]$, $\sigma^2 \in [-30, 60]$ and $\sigma_\rho{}^2 \in [-1, 4]$.

An equally good $\sigma_\rho{}^2$ fit corresponds to the *Ri*-dependence given by Equation (7). We may again appreciate a similar behavior for GCI and the MO, quite different from what happens for the GS. Both former cases are best adjusted with a subcritical $Ri = 0.22$, while the GS reflects a supercritical flow $Ri = 2.1$ (Table 1). It is remarkable that this happens despite the mean Richardson values being almost always significantly greater than one (top panels of Figure 12). Our interpretation is that, despite all the variability reflected by the scattered points in Figure 11, each case responds to a different flow dynamic. This dynamic setting is characterized by a $\sigma_\rho{}^2(N)$ relation with a characteristic flow-dependent *Ri* value, in agreement with the idea that strongly stratified flows self-organize around some local critical state [40].

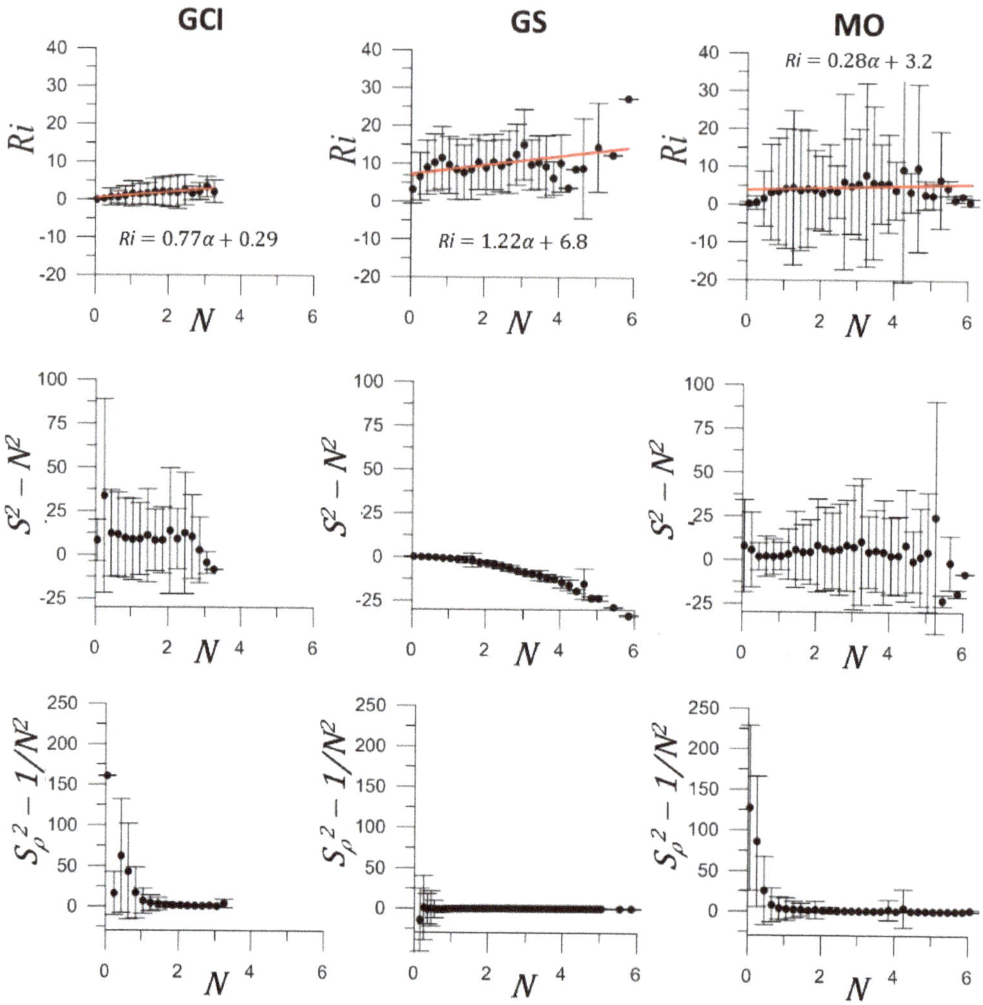

Figure 12. Nondimensional mean values, together with standard deviations, of (**Top panels**) *Ri*, (**Center panels**) σ^2 and (**Bottom panels**) $\sigma_\rho{}^2$, plotted as a function of *N*. The three columns correspond to the three datasets: (left panels, GCI) Gran Canaria Island shelf break, (middle panels, GS) Gulf Stream and (right panels, MO) Mediterranean outflow.

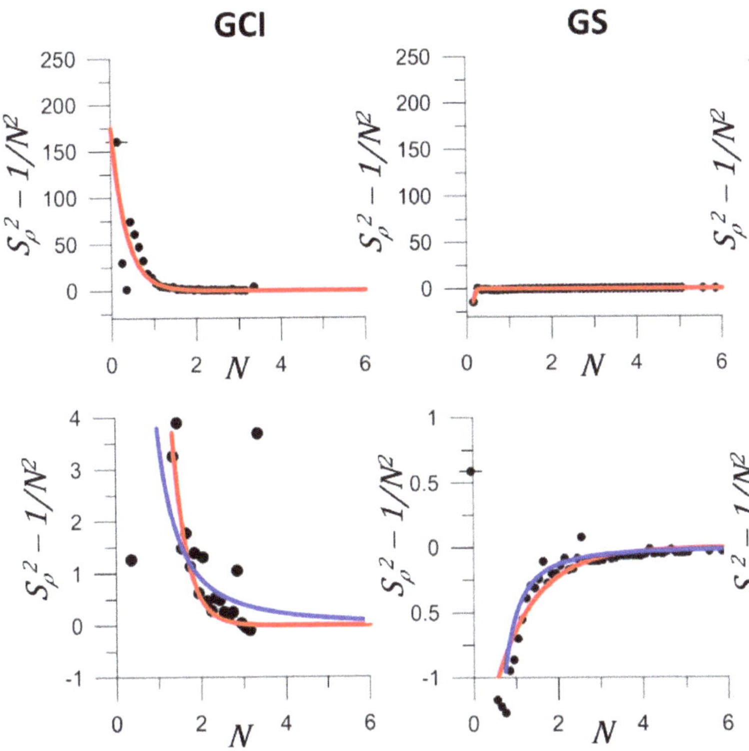

Figure 13. Scatter plot of nondimensional mean $\sigma_\rho{}^2$ plotted as a function of N; the lower panels reproduce the top panels but zooming into the (**Left and Right panels**) $\sigma_\rho{}^2 \in [-1, 4]$ and (**Middle panel**) $\sigma_\rho{}^2 \in [-1, 1]$ intervals. The red and blue lines respectively correspond to the exponential and Ri-based fits (see text for explanation). The three columns correspond to the three datasets: (left panels, GCI) Gran Canaria Island shelf break, (middle panels, GS) Gulf Stream and (right panels, MO) Mediterranean outflow.

6. Conclusions

In this work, we have endorsed the advantages of isopycnic thinking to better understand the dynamic conditions leading to effective mixing in stratified sheared flows. A nondimensional analysis, in terms of the background stratification, proves to be useful to explore the interdependences between stratification (expressed in terms of the buoyancy frequency N), vertical shear S, diapycnal shear S_ρ and the (gradient) Richardson number Ri. In particular, we have discussed vertical instability in terms of characteristic times, showing that instability occurs when the critical dynamic time τ—which is equal to the diapycnal shear and hence includes both stratification and vertical shear, $\tau \equiv S_\rho = t_o{}^2/t_d$—is longer than the vertical-oscillation period $t_0 = N^{-1}$, which itself is longer than the vertical-deformation time $t_d = S^{-1}$.

The Richardson number Ri is the classical index to assess the occurrence of instability, but it cannot differentiate between instances when mixing effectively redistributes density (and other properties, which occurs when stratification is high) from cases when instabilities simply displace water that is already homogeneous. Here, we have extended previous work [20,21] by defining the reduced squared vertical shear σ^2 in vertical coordinates and the reduced squared diapycnal shear $\sigma_\rho{}^2$ in isopycnic coordinates. The three indices for instability (Ri, σ^2 and $\sigma_\rho{}^2$) are sensitive to stratification, but $\sigma_\rho{}^2$ has the special property of decreasing with increasing stratification towards a limiting value equal to the squared diapycnal shear ${S_\rho}^2$.

Finally, we have examined the distribution of the nondimensional dependent variables (N, S, S_ρ) and indices (Ri, σ^2, ${\sigma_\rho}^2$) for three quite different dynamic regimes: the Gran Canaria Island shelf break characterized by intense internal waves, the strongly baroclinic Gulf Stream and the gravity-current Mediterranean outflow. The most remarkable feature is the decay of both the mean and maximum ${S_\rho}^2$ and ${\sigma_\rho}^2$ values with increasing stratification, which changes for each different flow regime, endorsing the idea that ${\sigma_\rho}^2 > 0$ is a good index for assessing flow stability: ${\sigma_\rho}^2$ has to be large for low stratification values, but small positive values are sufficient for high stratification. The non-dimensional expression ${\sigma_\rho}^2 = (1 - Ri)/(N^2 Ri)$ fits well to the mean ${\sigma_\rho}^2$ experimental values, leading to a family of instability functions with a characteristic Ri value for each flow regime.

Author Contributions: Conceptualization, J.L.P.; methodology, J.L.P., M.C. and P.S.; software, M.C. and P.S.; validation, J.L.P. and P.S.; formal analysis, J.L.P. and M.C.; Investigation, J.L.P.; writing—original draft preparation, J.L.P.; writing—review and editing, J.L.P. and M.C.; supervision, J.L.P. All authors have read and agreed to the published version of the manuscript.

Funding: This work has been funded by the Spanish Government through Project SACO (Ministerio de Ciencia e Innovación, reference PID2022-139403NB-C22). MC is grateful to the MSCA programme, funded by the European Union (HORIZON-MSCA-2021-PF MOORING, grant agreement no. 101064423). This article is a publication of the Unidad Océano y Clima of the Universidad de Las Palmas de Gran Canaria, an R+D+I CSIC-associate unit. The authors also recognize the institutional support of the Spanish Government through the Severo Ochoa Center of Excellence accreditation (CEX2019-000928-S).

Institutional Review Board Statement: Not applicable.

Informed Consent Statement: Not applicable.

Data Availability Statement: The data presented in this study are available on request from the corresponding author.

Acknowledgments: The authors wish to sincerely thank Tom Rossby and Jim Price for sharing, respectively, the Gulf Stream and Mediterranean outflow datasets.

Conflicts of Interest: The authors declare no conflicts of interest.

References

1. Taylor, G.I. Effect of variation in density on the stability of superposed streams of fluid. *Proc. R. Soc. Lond. A* **1931**, *132*, 499–523.
2. Miles, J.W. On the stability of heterogenous shear flows. *J. Fluid Mech.* **1961**, *10*, 496–508. [CrossRef]
3. Howard, L.N. Note on a paper of John W. Miles. *J. Fluid Mech.* **1961**, *10*, 509–512. [CrossRef]
4. Abarbanel, H.D.I.; Holm, D.D.; Mardsen, J.E.; Ratiu, R. Richardson number criterion for the nonlinear stability of three-dimensional stratified flow. *Phys. Rev. Lett.* **1984**, *52*, 2352–2355. [CrossRef]
5. Abarbanel, H.D.I.; Holm, D.D.; Mardsen, J.E.; Ratiu, R. Nonlinear stability analysis of stratified fluid equilibria. *Philos. Trans. R. Soc. Lond.* **1986**, *318*, 349–409.
6. Richardson, L.F. The supply of energy from and to atmospheric eddies. *Proc. R. Soc. Lond. A* **1920**, *97*, 354–373.
7. Taylor, G.I. Internal waves and turbulence in a fluid of variable density. *Rapp. P-v Réun.-Cons. Int. Explor. Mer.* **1931**, *76*, 35–43.
8. Miles, J.W. Richardson's criterion for the stability of stratified shear flow. *Phys. Fluids* **1986**, *29*, 3470–3471. [CrossRef]
9. Van Gastel, P.; Pelegrí, J.L. Estimates of gradient Richardson numbers from vertically smoothed data. *Sci. Mar.* **2004**, *68*, 459–482. [CrossRef]
10. Grachev, A.A.; Andreas, E.L.; Fairall, C.W.; Guest, P.S.; Persson, P.O.G. The critical Richardson number and limits of applicability of local similarity theory in the stable boundary layer. *Boundary Layer Meteorol.* **2013**, *147*, 51–82. [CrossRef]
11. Roget, E.; Pelegrí, J.L.; Planella-Morató, J.; Puigdefàbregas, J.; Emelianov, M.; Vallès-Casanova, I.; Orúe-Echevarria, D. Diapycnal mixing in the Brazil-Malvinas confluence front. *Progr. Oceanogr.* **2023**, *211*, 102968. [CrossRef]
12. Pelegrí, J.L.; Csanady, G.T. Diapycnal mixing in western boundary currents. *J. Geophys. Res.* **1994**, *99*, 18275–18304. [CrossRef]
13. Pelegrí, J.L.; Sangrá, P. A mechanism for layer formation in stratified geophysical flows. *J. Geophys. Res.* **1998**, *103*, 30679–30693. [CrossRef]
14. Pelegrí, J.L.; Rodríguez-Santana, A.; Sangrá, P.; Marrero-Díaz, A. Modeling of shear-induced diapycnal mixing in frontal systems. *Appl. Sci. Res.* **1998**, *59*, 159–175. [CrossRef]

15. Alford, M.H.; Pinkel, R. Observations of overturning in the thermocline: The context of ocean mixing. *J. Phys. Oceanogr.* **2000**, *30*, 805–832. [CrossRef]
16. Hallberg, R. Time integration of diapycnal diffusion and Richardson number–dependent mixing in isopycnal coordinate ocean models. *Mon. Weather Rev.* **2000**, *128*, 1402–1419. [CrossRef]
17. Burchard, H.; Craig, P.D.; Gemmrich, J.R.; van Haren, H.; Mathieu, P.P.; Meier, H.M.; Smith, W.A.; Prandke, H.; Rippeth, T.P.; Skyllingstad, E.D.; et al. Observational and numerical modeling methods for quantifying coastal ocean turbulence and mixing. *Prog. Oceanogr.* **2008**, *76*, 399–422. [CrossRef]
18. Jampana, V.; Ravichandran, M.; Sengupta, D.; D'Asaro, E.A.; Rahaman, H.J.; Sreelekha, J.S.; Chaudhuri, D. Shear flow instabilities and unstable events over the North Bay of Bengal. *J. Geophys. Res. Oceans* **2018**, *123*, 8958–8969. [CrossRef]
19. Lewin, S.F.; Caulfield, C.P. Stratified turbulent mixing in oscillating shear flows. *J. Fluid Mech.* **2022**, *944*, 1469–7645. [CrossRef]
20. Kunze, E.; Williams, A.J.; Briscoe, M.G. Observations of shear and vertical stability from a neutrally buoyant float. *J. Geophys. Res.* **1990**, *95*, 18127–18142. [CrossRef]
21. Sun, C.; Smyth, W.D.; Moum, J.N. Dynamic instability of stratified shear flow in the upper equatorial Pacific. *J. Geophys. Res.* **1998**, *103*, 10323–10337. [CrossRef]
22. Fox-Kemper, B.; Ferrari, R.; Hallberg, R. Parameterization of mixed layer eddies. Part I: Theory and diagnosis. *J. Phys. Oceanogr.* **2008**, *38*, 1145–1165. [CrossRef]
23. Liu, E.; Thorpe, S.A.; Smyth, W.D. Instability and hydraulics of turbulent stratified shear flows. *J. Fluid Mech.* **2012**, *695*, 235–256. [CrossRef]
24. Smith, K.M.; Caulfield, C.P.; Taylor, J.R. Turbulence in forced stratified shear flows. *J. Fluid Mech.* **2021**, *910*, A42. [CrossRef]
25. Cisneros-Aguirre, J.; Pelegrí, J.L.; Sangrà, P. Experiments on layer formation in stratified shear flow. *Sci. Mar.* **2001**, *65* (Suppl. 1), 117–126. [CrossRef]
26. Hazel, P.G. Numerical studies of the stability of inviscid parallel shear flows. *J. Fluid Mech.* **1972**, *51*, 39–62. [CrossRef]
27. Sangrà, P.; Basterretxea, G.; Pelegrí, J.L.; Arístegui, J. Chlorophyll increase by internal waves in the shelf-break of Gran Canaria (Canary Islands). *Sci. Mar.* **2001**, *65* (Suppl. 1), 89–97. [CrossRef]
28. Halkin, D.; Rossby, T. The structure and transport of the Gulf Stream at 73°W. *J. Phys. Oceanogr.* **1985**, *15*, 1439–1452. [CrossRef]
29. Ratsimandresy, A.W.; Pelegrí, J.L. Vertical alignment of the Gulf Stream. *Tellus A* **2005**, *57*, 691–700. [CrossRef]
30. Price, J.F.; Baringer, M.O.; Lueck, R.G.; Johnson, G.C.; Ambar, I.; Parrilla, G.; Cantos, A.; Kennelly, M.A.; Sanford, T.B. Mediterranean outflow mixing and dynamics. *Science* **1993**, *259*, 1277–1282. [CrossRef]
31. Spain, P.F.; Dorson, D.L.; Rossby, H.T. PEGASUS, a simple, acoustically tracked, velocity profiler. *Deep-Sea Res.* **1981**, *28*, 1553–1567. [CrossRef]
32. Armi, L.; Bray, N.A. A standard analytic curve of potential temperature versus salinity for the western North Atlantic. *J. Phys. Oceanogr.* **1982**, *12*, 384–387. [CrossRef]
33. Rodríguez-Santana, A.; Pelegrí, J.L.; Sangrà, P.; Marrero-Díaz, A. Diapycnal mixing in Gulf Stream meanders. *J. Geophys. Res.* **1999**, *104*, 25891–25912. [CrossRef]
34. Pelegrí, J.L.; Csanady, G.T. Nutrient transport and mixing in the Guf Stream. *J. Geophys. Res. Oceans.* **1991**, *96*, 2577–2583. [CrossRef]
35. Baringer, M.O.; Price, J.F. Mixing and spreading of the Mediterranean outflow. *J. Phys. Oceanogr.* **1997**, *27*, 1654–1677. [CrossRef]
36. Baringer, M.O.; Price, J.F. Momentum and energy balance of Mediterranean outflow. *J. Phys. Oceanogr.* **1997**, *27*, 1678–1692. [CrossRef]
37. Gasser, M.; Pelegrí, J.L.; Nash, J.; Peters, H.; García-Lafuente, J. Topographic steering of the early plunging Mediterranean outflow. *Geo-Mar. Lett.* **2011**, *31*, 301–314. [CrossRef]
38. Nash, J.D.; Peters, H.; Kelly, S.M.; Pelegrí, J.L.; Emelianov, M.; Gasser, M. Turbulence and High-Frequency Variability in a Deep Gravity Current Outflow. *Geophys. Res. Lett.* **2012**, *39*, L18611. [CrossRef]
39. Gasser, M.; Pelegrí, J.L.; Emelianov, M.; Bruno, M.; Gracia, E.; Pastor, M.; Peters, H.; Rodríguez-Santana, A.; Salvador, J. Tracking the Mediterranean outflow in the Gulf of Cádiz. *Progr. Oceanogr.* **2017**, *157*, 47–71. [CrossRef]
40. Zhou, Q. Threshold behavior of local gradient Richardson number in strongly stratified nonequilibrium turbulence. *Phys. Rev. Fluids* **2022**, *7*, 104802. [CrossRef]

oceans

MDPI

Article

Determination of Submerged Breakwater Efficiency Using Computational Fluid Dynamics

Smiljko Rudan and Šimun Sviličić *

Faculty of Mechanical Engineering and Naval Architecture, University of Zagreb, Ivana Lučića 5, 1000 Zagreb, Croatia; smiljko.rudan@fsb.unizg.hr
* Correspondence: ssvilicic@fsb.hr; Tel.: +385-99-504-5317

Abstract: Wind-induced waves can lead to the partial or complete wash-over of beaches, causing erosion that impacts both the landscape and tourist infrastructure. In some regions of the world, e.g., Croatia, this process, which usually occurs during a harsh winter, has a major impact on the environment and the economy, and preventing or reducing this process is highly desirable. One of the simplest methods to reduce or prevent beach erosion is the use of innovative underwater structures designed to decrease wave energy by reducing wave height. In this study, submerged breakwaters are numerically investigated using various topologies, positions, and angles relative to the free surface. Not only is the optimal topology determined, but the most efficient arrangement of multiple breakwaters is also determined. The advantage of newly developed submerged breakwaters over traditional ones (rock-fixed piers) is that they do not require complex construction, massive foundations, or high investment costs. Instead, they comprise simple floating bodies connected to the seabed by mooring lines. This design makes them not only cheap, adaptable, and easy to install but also environmentally friendly, as they have little impact on the seabed and the environment. To evaluate wave damping effectiveness, the incompressible computational fluid dynamics (ICFD) method is used, which enables the use of a turbulence model and the possibility of accurate wave modelling.

Keywords: beach erosion; wave modelling; incompressible flow; submerged breakwaters

1. Introduction

Beach erosion is a natural process that can lead to significant land loss, often exacerbated by high tides and strong wave action. To mitigate the effects of wave-induced erosion, structures known as breakwaters are employed in both deep and shallow waters. The function of these structures is to dissipate the kinetic energy of waves and reduce their amplitude, thereby preserving the natural beach slope [1]. Traditional breakwaters, often referred to as wave breakers, face numerous challenges, including complex construction, high costs, and significant environmental impact. They are also rigid in design, making them unsuitable for varying environmental conditions and seabed types. In contrast, non-traditional floating breakwaters consist of buoyant bodies [2] tethered by mooring systems. These floating structures offer numerous benefits: they are adaptable to both deep and shallow water environments, exert minimal impact on sediment and local ecosystems, and their costs are less dependent on water depth and foundational structures. The physical foundation of breakwater is based on wave-breaking theory [3] which states that as a wave propagates through a barrier, it is dissipated, transmitted, and reflected.

Experimental research conducted by Matsunaga et al. [4] has shown that floating breakwaters effectively manage larger waves exceeding 1.2 m in height. Further studies, such as that by Pena et al. [5], have included mooring systems in their analyses, further confirming the effectiveness of these systems. Another experimental study of double-row floating breakwaters under regular wave action conducted by Ji et al. [6] showed that including an additional second row of breakwaters significantly reduced the transmission coefficient, which is attributed to dissipation caused by eddies and the moon pool effect. An analytical approach by Kim et al. [7], which incorporated small wave amplitude theory to solve wave damping in the porous seabed, concluded that the seabed should be included as a boundary condition to obtain accurate results. A more recent study conducted by Chalmoukis et al. [8] applied numerical modelling for porous seabeds to analyse the influence of vegetation on coastal waves. A more detailed analysis of the wave breaking of a rubble-mound model was presented by Setiyawan [9] where analysis showed that the slope of the breakwater is the most important parameter. Numerical modelling has become a preferred method for studying breakwaters due to its cost-effectiveness and ability to optimise design parameters. In early studies, two-dimensional (2D) linear analysis using techniques such as the high-order boundary method [10] and the Green method [11] was often used to solve these problems. A similar study by Cheng et al. [12] compared experimental tests with 2D numerical simulations using the Smoothed Particle Hydrodynamics (SPH) method. The study found that the absence of turbulence modelling and limited degrees of freedom in mooring force calculations led to discrepancies between experimental and numerical results, though the numerical model still provided reliable outcomes. However, recent research by Fitriadhy et al. [13] and He et al. [14] suggested that three-dimensional (3D) analysis is essential for accurately modelling the complex interactions between submerged breakwaters and the wave dynamics of complex design parameters, indicating that 3D analysis should be used.

For modelling floating breakwater systems and their interaction with waves, the SPH method is becoming increasingly popular. For instance, Guo et al. [15] conducted a study focusing on tautly moored floating breakwaters to give insight into the engineering design of the mooring systems of floating breakwaters. They concluded that taut mooring systems exhibit superior performance compared to those with ropes and slack chains, in both low- and high-tide conditions. Another study by He et al. [16] introduced a novel multi-float breakwater structure. They employed the SPH method to examine wave interactions, solving the structure's motion using Newton's second law for rigid bodies. Their results were validated through laboratory experiments. As accurate wave modelling is a crucial parameter in assessing accurate breakwater efficiency, an accurate wave generation method should be applied. Regular waves are the most straightforward approach to applying wave conditions, and Stokes' second-order theory is capable of accurately predicting their behaviour and energy, as shown in Lloyd et al. [17]. An extensive study on adequate mesh for wave modelling conducted by Huang et al. [18] showed that the results converge if at least 15 cells per wave height and 100 cells per wavelength are applied. As waves propagate through breakwaters, turbulence is expected at the submerged breakwater boundary layer due to its geometry. Ge et al. [19] concluded that RANS turbulence models produce accurate results in wave problems.

This study focuses on the numerical modelling of submerged breakwaters using ICFD via LS-DYNA R11.1. solver [20]. The ICFD solver, based on finite element formulation developed by LS-DYNA, handles wave–structure interactions. It incorporates Stokes' wave theory and turbulence models, unlike the Arbitrary Lagrangian–Eulerian (ALE) method, which lacks these features. The objective of this research is to identify the optimal breakwater system that maximises wave height reduction while minimising the number of breakwater elements required. A two-stage analysis is carried out for this purpose: first assessing the most efficient topology out of three possible solutions and later finding the best arrangement of the chosen solution. The structure of this article is as follows:

- Section 2.1 presents a detailed explanation of the ICFD method used for breakwater analysis, including a description of the theoretical background and its advantages over another method for solving fluid–structure interaction (FSI) problems, namely the ALE method.
- Three possible topologies of breakwater systems with geometrical particulars are presented in Section 2.2.

- Section 3 details the numerical setup for assessing the most efficient topology, including comprehensive information about the mesh and boundary conditions of the breakwater system.
- Section 4 presents the results of both analyses, emphasising wave height reduction and drag estimation.
- The conclusion in Section 5 focuses on the findings obtained and points out the potential issues and/or limitations of the applied methods.

This study marks the first phase in the development of a system to protect beaches from erosion. The aim is to determine the optimal topology and angle from all the solutions proposed by third parties, thus reducing costly and time-consuming experimental tests. Given the high computational cost of each breakwater analysis, the design matrix was kept concise. Initially, the solutions were evaluated taking into account numerous constraints such as the manufacturing capabilities and dimensions of the towing tank, and finally, three optimal solutions were selected.

Future testing will be conducted in the towing tank shown in Figure 1, which imposes additional constraints on the system layout. The primary constraint is the distance variation between rows; in the current setup, this would place submerged breakwaters too close to the tank sides, causing wave reflection. A larger scale than the chosen 1:20 would reduce breakwater dimensions and allow for varying row distances (Figure 1b, X dimension). However, this would result in wave amplitudes smaller than 7.5 cm, making it impractical for both numerical modelling and real wave generation in the towing tank.

(a)

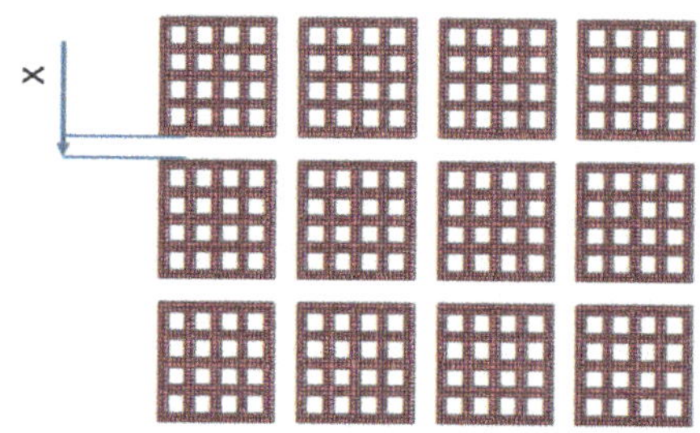

(b)

Figure 1. (**a**) Towing tank at the University of Zagreb, Faculty of Mechanical Engineering; (**b**) width between breakwater rows.

2. Material and Methods

2.1. ICFD

ICFD uses the Navier–Stokes equations and the continuity equation for numerical analysis. For incompressible flow, the Mach number should be less than 0.3, i.e.,

$$M = \frac{V}{a} \leq 0.3, \tag{1}$$

where V is the flow velocity relative to a fixed object, and a is the speed of sound in the medium. The conservation of mass and momentum for incompressible fluids in the Eulerian conventional form is represented by the Navier–Stokes equations combined with the continuity equations:

$$\rho\left(\frac{\partial u_i}{\partial t} + u_j \frac{\partial u_i}{\partial x_j}\right) = \frac{\partial \sigma_{i,j}}{\partial x_j} + \rho f_i \tag{2}$$

where ρ represents density, $\frac{\partial u_i}{\partial t}$ represents local acceleration, $u_j \frac{\partial u_i}{\partial x_j}$ is convective acceleration, $\frac{\partial \sigma_{i,j}}{\partial x_j}$ is the divergence of the stress tensor, and ρf_i represents external forces. While the more common Finite Volume Method is used to approximate discrete fields for fluid dynamics, spatial discretisation is conducted here by the Finite Element Method. The final set of equations that are solved by the ICFD solver is written in compact matrix form:

$$\rho M + \Delta t K(\mu) u_i^* + S\left(\rho u_i^{n+1}\right) u_i^* = \rho M u_i^n - \Upsilon \Delta t G p^n + \Delta t F \tag{3}$$

$$\Delta t L\left(\frac{1}{\rho}\right) p^{n+1} = D u_i^* + \Upsilon \Delta t L\left(\frac{1}{\rho}\right) p^n - \dot{\tilde{U}} \tag{4}$$

$$\rho M u_i^{n+1} = \rho M u_i^* - \Delta t G\left(p^{n+1} - \Upsilon p^n\right) \tag{5}$$

where ρM represents the mass term, Δt is the time step, $K(\mu)$ is the stiffness matrix, $S(\rho {u_i}^{n+1})$ is a convective term for non-linear advection, Y is the time integration parameter, G is the gradient operator matrix, F is the external force vector, D is the divergence operator matrix, $\dot{\tilde{U}}$ is the source term vector, and u_i and p_n represent velocity and pressure.

To solve Equations (3)–(5), LS-DYNA employs a predictor–corrector scheme. In this process, intermediate variables ${u_i}^*$ and $p^{(n+1)}$ are computed from Equations (4) and (5), respectively, and then reintroduced into the iterative loop to determine the next step velocity ${u_i}^{(n+1)}$. This approach enhances the accuracy of the final pressure and velocity values, although it is more time-consuming. LS-DYNA employs the Finite Element (FE) Method for spatial discretisation, adopting strategies similar to those in the ALE method but with a few notable differences. In particular, the element type used in the ICFD cases is tetrahedral, while the ALE method typically uses hexahedral elements. While the solver of the ALE method is designed for compressible fluids and uses an explicit time integration method, the ICFD solver is an implicit solver optimised for incompressible fluids and is able to perform double-precision calculations required for turbulence modelling and boundary layer definition. In addition to the previously mentioned advantages of ICFD over ALE for wave damping problems, the ICFD solver incorporates wave theory [21], offering options for Stokes' 1st-, 2nd-, and 3rd-order theories. The choice of Stokes' theory over linear theory is based on the ratio of wave height to wavelength. For 2nd-order Stokes' theory, this ratio lies in the range $0.04 < H/L < 0.141$ [22], which corresponds with the measured average wave height and length [23,24], making it the appropriate choice.

Finally, the ICFD approach benefits from advanced turbulence modelling capabilities, whereas ALE does not have implemented turbulence models. A comparative study conducted by Perin et al. [25] demonstrated that the ICFD solver is particularly effective in scenarios where turbulence is a significant factor, providing results that align closely with experimental data.

2.2. Breakwater Geometry

The breakwater system is costly due to several complex aspects. These include the anchoring system, proper placement on the seabed without interfering with sea lanes, and challenges such as the corrosion and fouling of the system. These factors make the system expensive to manufacture, install, and test. In the choice of geometry, there exists an infinite number of topologies, while in manufacturing, several other factors such as manufacturing capacities as well as towing tank dimensions and the numerical cost of the analysis, mentioned in the introductory section are limitations. Therefore, three different topologies were selected for the development of the beach erosion system based on their porosity percentage, as shown in Figure 2, and the most efficient solution was determined from these options. All three geometries have identical external dimensions, $0.66 \times 0.66 \times 0.06$ m, with only the internal hole size and numbers varying. For the geometry with cylindrical holes, six holes with a diameter of 0.12 m each and a porosity of 15% were chosen (Figure 2a). In contrast, the other geometries (Figure 2b,c) have square openings of 0.1×0.1 m with porosities of 36 and 70%.

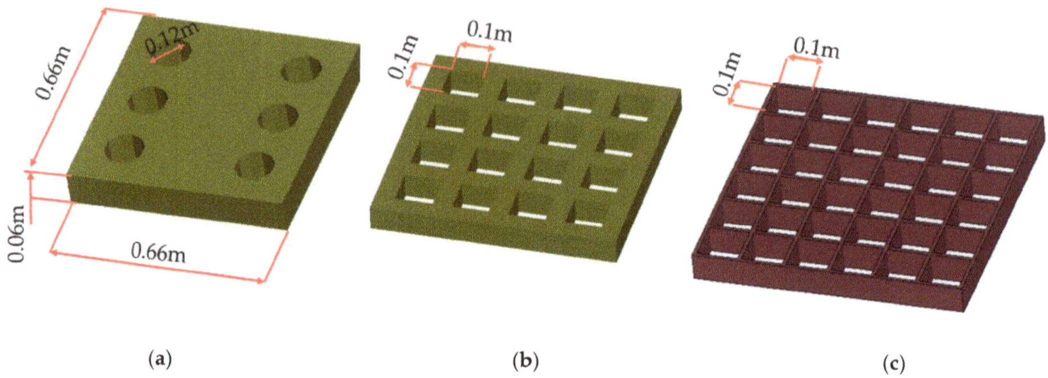

Figure 2. Breakwater topology: (**a**) cylindrical 6-hole, (**b**) Quadratic 16-hole, and (**c**) Quadratic 36-hole topologies.

The breakwaters are manufactured at a 1:20 scale of their actual dimensions in preparation for future testing in the towing tank at the University of Zagreb, Faculty of Mechanical Engineering and Naval Architecture. Given that the towing tank has a length of 32.8 m, a width of 3.6 m, and a depth of 1.8 m, this scale ratio allows for multiple breakwater elements to be positioned across the tank's width while facilitating easy installation and fixation.

3. Numerical Setup

3.1. Mesh Particulars

This beach anti-erosion system was developed for water deeper than 10 m, further away from beaches, and thus, only deep-water conditions were examined. While the literature, according to Conell and Cashman [26], suggests a minimum of 20 elements per wave height, the LS-DYNA solver did not converge with this number. Therefore, a minimum of 40 elements per height, or 20 elements per amplitude, was applied. For the wavelength, 200 elements were used. Because ICFD uses an implicit solver, a sufficiently small time step must be provided to ensure that the Courant–Friedrichs–Lewy condition is met, with the chosen time step of 0.1 s resulting in a condition number around 0.1. Turbulence was modelled while using the Large Eddy Simulation (LES) sub-grid model where the sub-grid scale represents the effect of small-scale motion. The Smagorinsky constant was set to 0.13 [27], and the boundary layer thickness was determined at 0.01 m with the first cell height being equal to 0.007 m. The LES model showed a good correlation with the experimental test conducted by Calero et al. [28] in a wave flume under different wave generation scenarios. Because the numerical domain was shortened in the X direction compared to the actual length of the tank, the beach structure did not exist in the model as it did in the towing tank, and waves in this case would not be damped but reflected off the wall. To prevent this, the damping layer was introduced at the last metre of the domain to stop any wave reflections from the wall. For the meshing procedure, surface elements were created which were later automatically extruded by LS-DYNA into 3D tetrahedral elements.

3.2. Numerical Setup for Assessing Most Efficient Topology

The initial studied case consists of a smaller domain that efficiently reduces the simulation time. A fluid domain, as shown in Figure 3a, consists of 207 k surface elements, with 15 elements per wave amplitude, used for an appropriate wave definition (Figure 3b). The waves are defined using Stokes' 2nd-order wave definition with an amplitude of 0.075 m and a length of 1.5 m for each of the three different topologies. The wave boundary condition is applied at the inlet, and a prescribed pressure boundary condition is used at the outlet. This condition is essentially the same as a pressure outlet, which imposes a hydrostatic fluid pressure that ensures ambient environment conditions and results in a better convergence rate as it prevents backflow occurrence. The submerged breakwaters are modelled both as a surface model (ICFD boundary part) and as a solid (structural part), and their anchoring system consists of a wired chain system that is self-stiffening and allows for just enough movement to reduce the hydromechanical forces while keeping the submerged breakwater system practically in the same position. For this reason, anchoring is neglected, and all six degrees of freedom of the submerged breakwater movement are fixed in space. The influence of the submerged breakwaters is evaluated at four different positions, ranging from 20 to 5 cm from the free surface, Figure 3b, as below 20 cm, it is assumed that the submerged breakwater system has no significant influence.

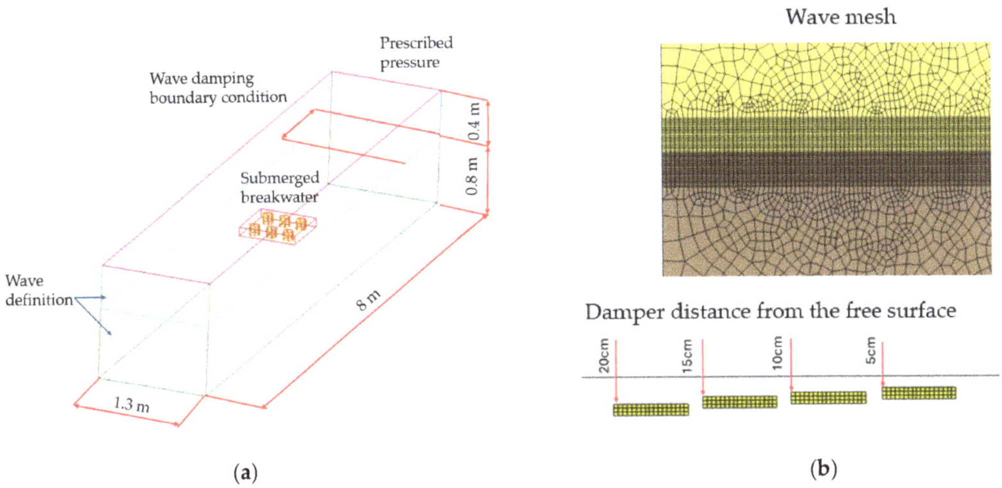

Figure 3. Numerical setup: (**a**) domain and (**b**) wave mesh with four different positions.

3.3. Convergence Analysis

A convergence study was conducted to assess the accuracy of the ICFD method. In previous research [29], a similar wave damping study was performed by comparing the influence of cylindrical topology on the wave profile, Figure 4a, using the LS-DYNA ICFD method with the Ansys Fluent results which are presented in Figure 4b.

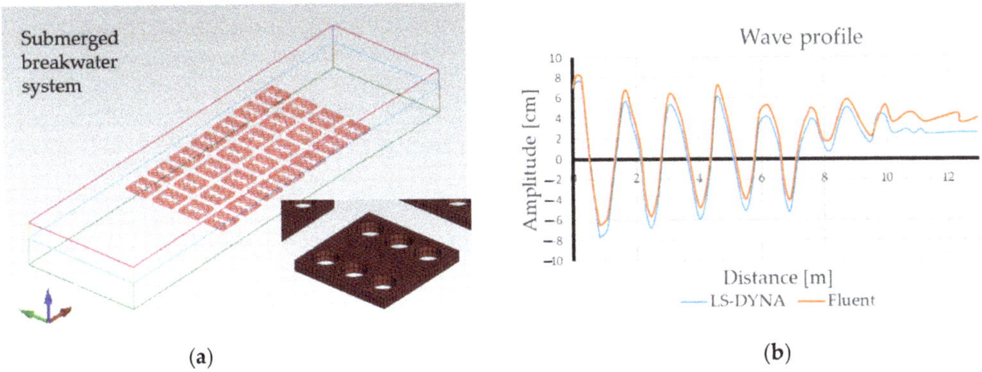

Figure 4. Comparison of LS-DYNA and Fluent: (**a**) numerical setup, (**b**) wave profile [29].

For convergence, a Quadratic 16-hole geometry was tested on four different meshes, 10, 13, 15, and 18 elements per wave amplitude, Figure 5a, and comparing the calculated total drag force between them, Figure 5b, while using the same boundary conditions as in Figure 4. The drag force was calculated using the database output option *ICFD_DATABASE_DRAG, and LS-DYNA calculates the drag using the following two equations:

$$F_P = \int P dA \tag{6}$$

where F_P is the drag force due to pressure, with P being pressure and dA being the differential area:

$$F_v = \int \mu \frac{\partial u}{\partial y} dA \tag{7}$$

where F_v is the viscous component of drag force, μ is viscosity, and $\frac{\partial u}{\partial y}$ is the shear velocity at the wall. The convergence diagram showed that results with less than 15 elements per amplitude diverge significantly from the results obtained with 15 or more elements as wave generation is irregular, Figure 5b.

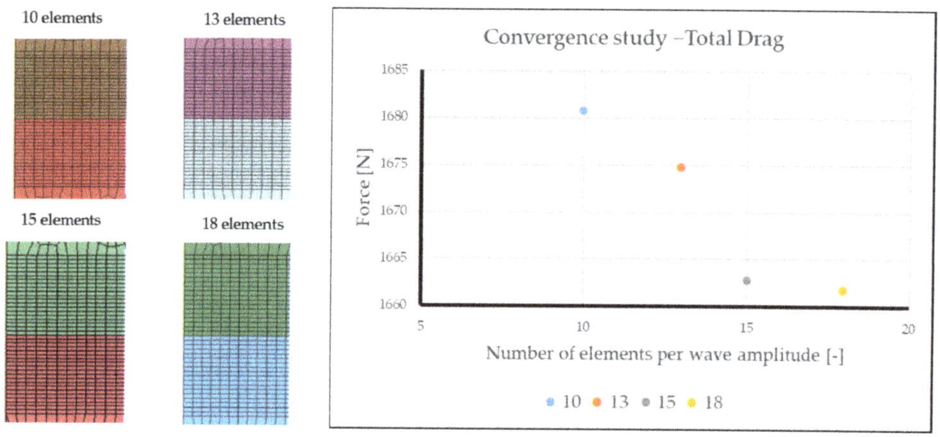

(a) (b)

Figure 5. Numerical setup: (**a**) four different wave meshes, (**b**) convergence diagram.

3.4. Numerical Setup for Submerged Breakwater System

Once the most efficient topology of the submerged breakwaters is determined, a submerged breakwater system is created by varying the angle and position of each breakwater element relative to its free surface. The objective of this system is to attenuate different wave heights and lengths, maximising efficiency under expected environmental conditions. Figure 6a shows the numerical setup for the system analysis. The boundary conditions are the same as for the evaluation of the topology. At the inlet, the wave is defined as a 2nd-order Stokes wave, while the pressure is defined at the outlet. Figure 6b illustrates the submerged breakwater system, consisting of several breakwater elements arranged in three rows, with six elements in a line (as shown in Figure 6b). The angle ° between individual elements is varied. Three rows are chosen due to the towing tank's limited width, consistent with future experiments.

To determine the most efficient and cost-effective submerged breakwater system, four different angle variations are tested. These variations provide insight into the optimal number of breakwater elements and the threshold depth below which the breakwater becomes ineffective. The angle is defined by the height difference between the elements in a row, resulting in a different number of elements in the system, as shown in Figure 7. Angle variations could reduce costs and save space once determined, as steeper angles require fewer elements in a submerged breakwater system to achieve the same wave damping.

Consequently, the system becomes more efficient and economical, requiring fewer materials and less installation effort.

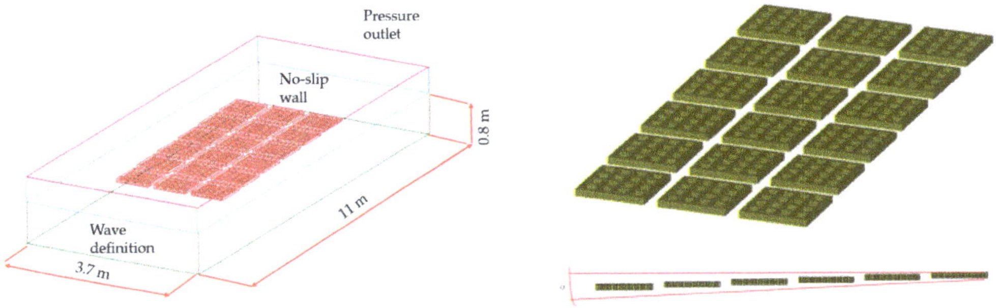

(a) (b)

Figure 6. The numerical setup for a submerged breakwater system: (**a**) the global domain, (**b**) a breakwater system with 6 elements in a line (angle 3 case).

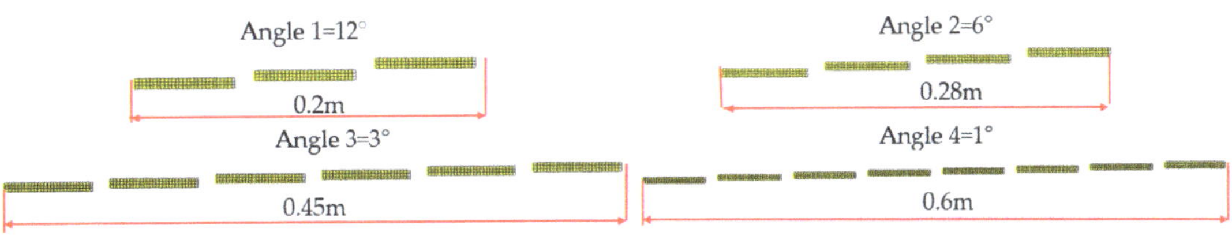

Figure 7. Variation in submerged breakwater system arrangement angle.

4. Results

As outlined in the article structure, two analyses are shown separately. First, an assessment of the most efficient topology consisting of a single submerged breakwater element is shown, and later, a complete submerged breakwater system analysis is presented. Since these simulations are intended to serve as the basis for future towing tank experiments, the wave amplitudes are shown on a centimetre scale. To determine the most efficient breakwater topology, four different wave steepness ratios (wave height/wavelength) and δ are chosen, 1/10, 1/15, 1/20, and /1/33, while simulating 12 s of real time for each analysis. Analysis particulars for each topology are shown in Table 1 for several wave steepness ratios δ and wave periods T. Each topology is analysed on the aforementioned four different wave periods and additionally four different positions for the common wave in the Adriatic Sea [24], $\delta = 1/20$, $T = 5$ s, meaning that there are 24 simulation runs in total.

Table 1. Assessing the efficiency of breakwater topology analysis particulars.

	Model Name	Number of ICFD Elements	Number of Solid Elements
Cylindrical holes	$\delta = 1/10$, $T = 1.75$ s	206 k	500
	$\delta = 1/15$, $T = 3.75$ s	206 k	500
	$\delta = 1/20$, $T = 5$ s	206 k	500
	$\delta = 1/33$, $T = 7.5$ s	206 k	500
Quadratic 16 holes	$\delta = 1/10$, $T = 1.75$ s	206 k	1284
	$\delta = 1/15$, $T = 3.75$ s	206 k	1284

Table 1. *Cont.*

	Model Name	Number of ICFD Elements	Number of Solid Elements
Quadratic 16 holes	$\delta = 1/20, T = 5$ s	206 k	1284
	$\delta = 1/33, T = 7.5$ s	206 k	1284
Quadratic 36 holes	$\delta = 1/10, T = 1.75$ s	206 k	2148
	$\delta = 1/15, T = 3.75$ s	206 k	2148
	$\delta = 1/20, T = 5$ s	206 k	2148
	$\delta = 1/33, T = 7.5$ s	206 k	2148

At the submerged breakwater system level, three different wave amplitudes with an additional three wave steepness ratios are presented, which were generated using three different meshes. A summary of the breakwater system with information about the mesh for both the ICFD fluid domain and the number of solid elements for the breakwater system topology and simulation time can be found in Table 2. Each simulation was run on 32 cores of an Intel Xeon Gold 6248 CPU using the LS-DYNA mpp_d_13_0 solver, with 15 s of wave dampening simulated. The table reveals a significant discrepancy in simulation time between the 7.5 cm amplitude wave and the other two amplitudes, 10 cm and 12.5 cm. The reason why these amplitudes are chosen lies not only in the measured average wave height in the Mediterranean basin [23] but also in wave modelling and the aspect ratio, which is especially challenging for the smallest 7.5 cm amplitude. The ICFD solver is sensitive to poor surface element quality, where elements with an aspect ratio greater than 3 can cause numerical divergence. Thus, a fine wave mesh and a small time step were required for accurate wave modelling. In this study, 15 elements per wave amplitude converged to an accurate shape, but this meant that small elements were used in the other two directions to respect the aspect ratio which resulted in almost double the mesh size compared to the other two amplitudes which finally resulted in an $8\times$ bigger 3D mesh size once the solver extruded initial surface elements to tetrahedral elements. To reduce computational time, three additional wave steepness ratios were tested with the 12.5 cm amplitude.

Table 2. Analysis information for submerged breakwater system.

	Model Name	Number of ICFD Elements	Number of Solid Elements	Real-World Time
7.5 cm amplitude, 1.5 m wavelength, $\delta = 1/20$, $T = 3.75$ s	Angle 1	435 k	22 k	12 h, 29 min
	Angle 2	435 k	33 k	12 h, 43 min
	Angle 3	435 k	45 k	13 h, 13 min
	Angle 4	435 k	60 k	13 h, 50 min
10 cm amplitude, 2 m wavelength, $\delta = 1/20$, $T = 5$ s	Angle 1	284 k	22 k	3 h, 50 min
	Angle 2	284 k	33 k	3 h, 55 min
	Angle 3	284 k	45 k	4 h, 6 min
	Angle 4	284 k	60 k	4 h, 15 min
12.5 cm amplitude, 2.5 m wavelength, $\delta = 1/20$, $T = 6.25$ s	Angle 1	197 k	22 k	2 h, 58 min
	Angle 2	197 k	33 k	3 h
	Angle 3	197 k	45 k	3 h, 2 min
	Angle 4	197 k	60 k	3 h, 12 min
12.5 cm amplitude, 1.25 m wavelength, $\delta = 1/10$, $T = 3.1$ s	Angle 1	197 k	22 k	4 h, 3 min
	Angle 2	197 k	33 k	4 h, 15 min
	Angle 3	197 k	45 k	4 h, 54 min
	Angle 4	197 k	60 k	8 h, 21 min

Table 2. *Cont.*

	Model Name	Number of ICFD Elements	Number of Solid Elements	Real-World Time
12.5 cm amplitude, 1.875 m wavelength, $\delta = 1/15$, $T = 6.25$ s	Angle 1	197 k	22 k	4 h, 45 min
	Angle 2	197 k	33 k	6 h, 40 min
	Angle 3	197 k	45 k	7 h, 3 min
	Angle 4	197 k	60 k	7 h, 14 min
12.5 cm amplitude, 3.6 m wavelength, $\delta = 1/33$, $T = 9$ s	Angle 1	197 k	22 k	5 h, 46 min
	Angle 2	197 k	33 k	6 h, 36 min
	Angle 3	197 k	45 k	6 h, 42 min
	Angle 4	197 k	60 k	7 h, 24 min

4.1. Breakwater Efficiency

The generated waves and the streamlines around the breakwater are shown in Figure 8. The effect of wave damping can be seen even when a single breakwater is used (Figure 8a). Regular waves are generated before they reach the breakwater, and after passing through it, both the amplitude and wavelength of the wave are altered. This alteration occurs due to the deformation of the wave segment that passes directly over the system, resulting from vorticity and friction between the wave and each element of the system. The physical effect of the submerged breakwater is depicted in Figure 8b, where a section of the domain is examined. Vorticity is generated around each hole, and the fluid velocity streamlines indicate that the vertical component of the velocity is significant, meaning that the breakwater redirects water in the vertical direction (*Z*-axis) rather than in the wave direction (*X*-axis).

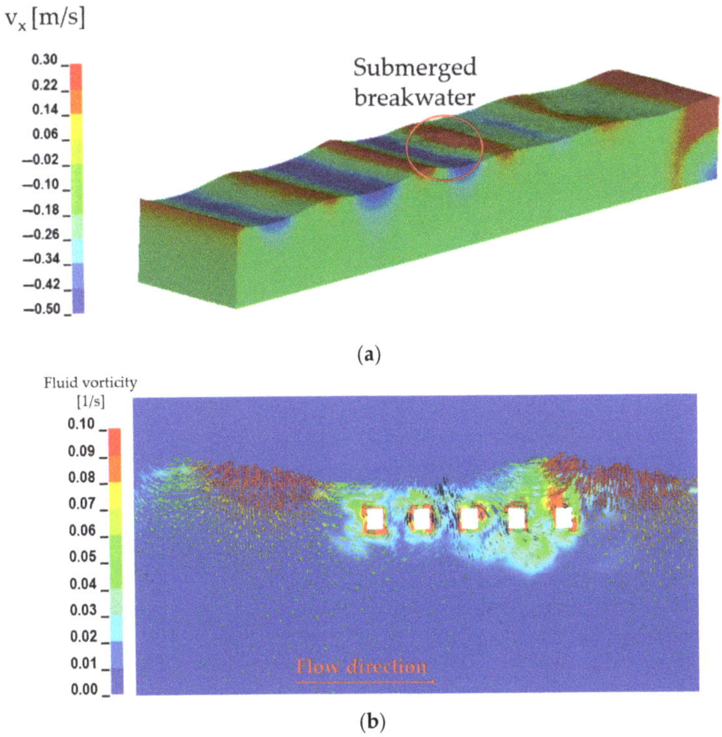

Figure 8. Results: (**a**) 15 cm wave amplitude generation, (**b**) streamline around breakwater.

When evaluating the performance of breakwaters, two crucial factors—the drag force and the curve of the wave profile—are of paramount importance. The drag force measures the resistance that a breakwater offers to incoming waves, which has a direct impact on its stability and effectiveness. The wave profile curve provides information on how the breakwater changes the energy and shape of the waves travelling through or around it. By comparing these parameters of different designs, researchers can determine which configurations reduce wave energy the most efficiently. This comparative analysis is essential for the development of submerged breakwater systems that optimally protect coastal areas from erosion, Figure 9. The highest drag force, Figure 9a, and the lowest wave transmission coefficients, Figure 9b, were obtained with the Quadratic 16-hole topology. The cylindrical topology also showed a flattening of the waves, but it was less efficient, while the Quadratic 36-hole topology proved to be too porous. A comparison of drag force for each topology over the different steepness ratios is presented in Figure 9c. It can be seen that the Quadratic 16-hole topology results in the biggest drag for different wave steepness ratios with especially superior performance for steep waves, $\delta = 1/33$.

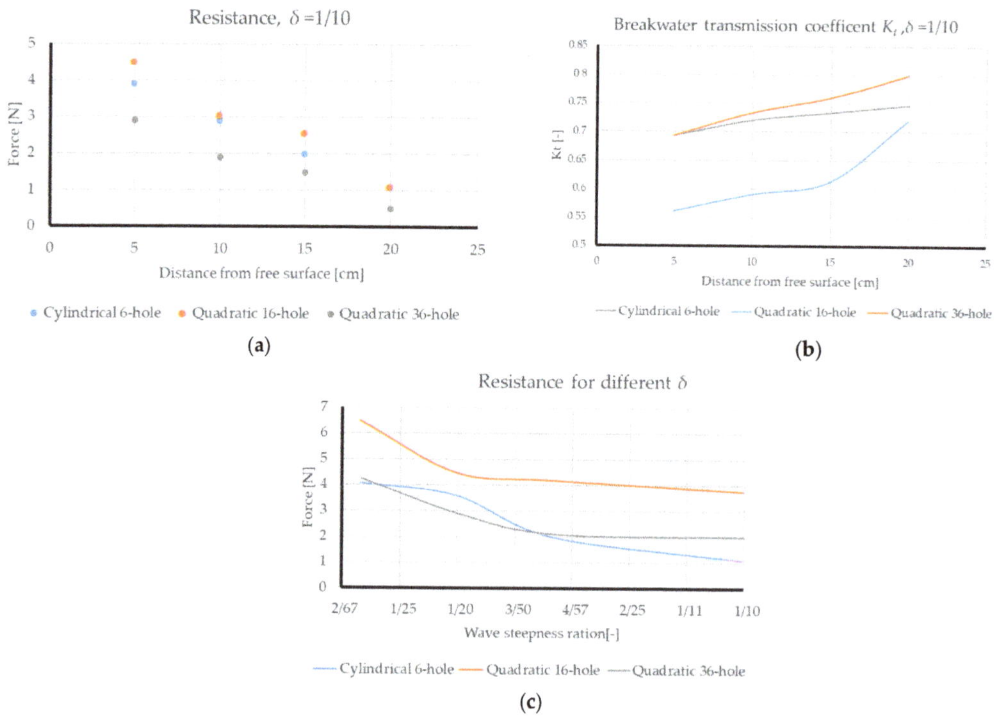

Figure 9. Comparison of different breakwater topologies based on (**a**) drag, (**b**) wave profile, and (**c**) drag for different wave steepness ratios.

A visualisation of wave flattening for each breakwater topology can be seen in Figure 10. The wave velocity, shown in Figure 10a, does not significantly decrease in the cylindrical six-hole topology, and the amplitude remains almost unchanged as the waves pass through the breakwater system. Similarly, the amplitude of the waves in the Quadratic 36-hole topology also stays nearly the same. The effect of wave flattening is the most evident in Figure 10b, which shows a part of the domain just before the submerged breakwater, with a regular wave approaching. In the case of the Quadratic 36-hole topology, the wave remains almost identical after passing through the breakwater due to its high porosity. In contrast, the other two topologies exhibit noticeable deformation in the wave shape.

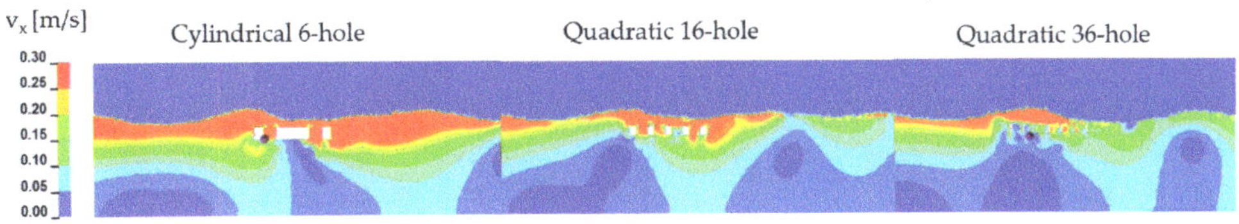

(a)

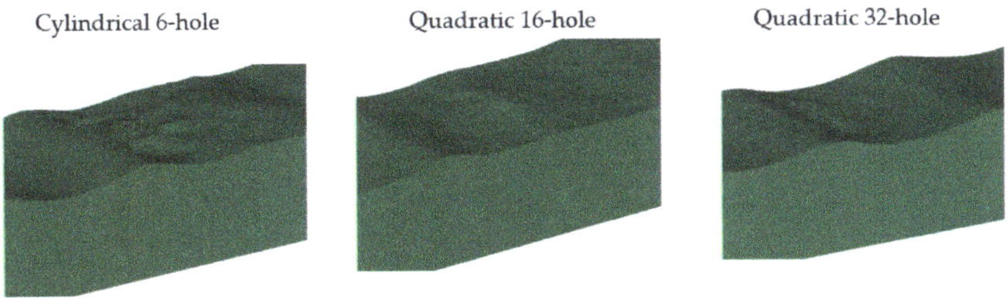

(b)

Figure 10. Comparison of different breakwater topologies based on (**a**) wave velocity and (**b**) wave profile.

4.2. Submerged Breakwater System Analysis

The Quadratic 16-hole topology proved to be the most efficient in terms of drag resistance, so it was chosen for the submerged breakwater system. As a wave passes the submerged breakwater system, it is slowed down and deformed not only in amplitude and length, but it is also bent, as shown in Figure 11, marked red.

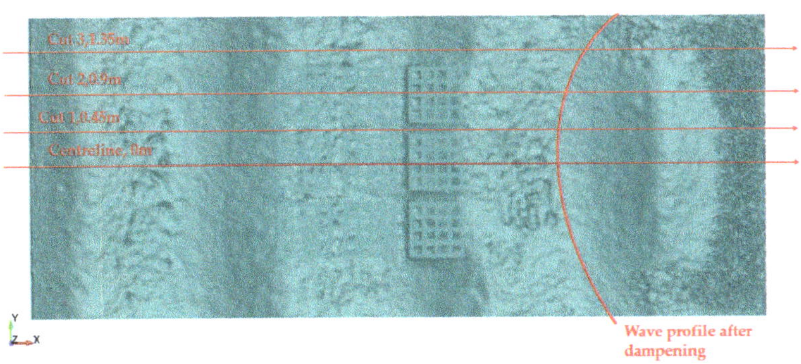

Figure 11. Wave pattern, 12.5 cm amplitude, angle 1.

Due to the irregular shape of the wave in the XY plane, wave amplitude will be measured at four different points: along the centreline, where the wave directly interacts with the submerged breakwater system, and at three additional positions spaced 0.45 m apart. This setup will help determine the system's influence on both the centre and the periphery. The sideline location will indicate the breakwater's influence on velocity in the Y direction and provide the wave position, which will be used for direct comparison once experimental tests are conducted. A comparison of wave transmission coefficient K_t for different positions along the Y-axis is presented in Table 3. While the centreline passes above the wall of the breakwater, Cut 1 passes between two neighbouring breakwater elements, and Cut 2 directly above holes in the breakwater, Cut 3 passes further away from the submerged breakwater system. The transmission coefficient shows small variations between the centreline and Cuts 1 and 2, with Cut 2 having the lowest K_t, meaning that breakwater holes reduce the wave the most but only for a small margin.

Table 3. Wave transmission coefficients for different wave amplitudes.

δ = 1/20	Model Name	Centreline K_t [-]	Cut 1 K_t [-]	Cut 2 K_t [-]	Cut 3 K_t [-]
7.5 cm amplitude	Angle 1	0.34	0.36	0.33	0.4
	Angle 2	0.3	0.32	0.29	0.33
	Angle 3	0.07	0.09	0.067	0.16
	Angle 4	0.04	0.05	0.032	0.12
10 cm amplitude	Angle 1	0.46	0.48	0.45	0.53
	Angle 2	0.33	0.35	0.32	0.42
	Angle 3	0.22	0.24	0.21	0.33
	Angle 4	0.15	0.17	0.14	0.28
12.5 cm amplitude	Angle 1	0.53	0.55	0.53	0.6
	Angle 2	0.28	0.29	0.27	0.46
	Angle 3	0.2	0.22	0.19	0.37
	Angle 4	0.13	0.15	0.12	0.18

Next, Figure 12 illustrates the effect of velocity changes. A significant reduction in wave crest height is observed only after the wave has passed through the last sixth row of breakwater elements, which are positioned 5 cm from the free surface.

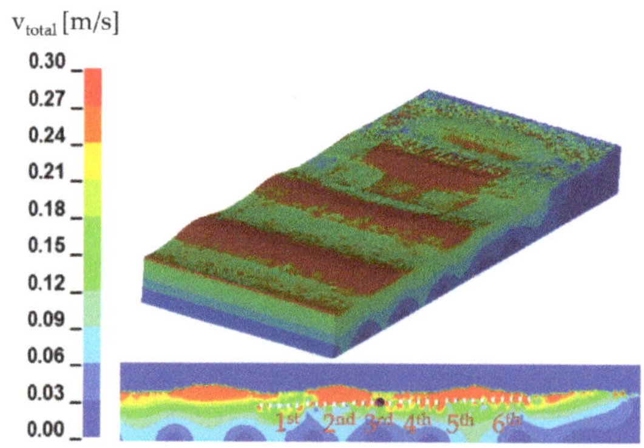

Figure 12. Wave velocity profile for 10 cm amplitude, angle 3.

This becomes clear by analysing the turbulent kinetic energy, Figure 13. The results show that the kinetic energy of turbulence is the highest at an amplitude of 12.5 cm both at the wave crest and in the wave trough. Most of the turbulence is generated at the edges of the breakwater with a significant difference between the first row of breakwaters, Figure 13a, and the last row of breakwaters, Figure 13b. This means that the first row is far below the zone of influence and does not contribute significantly to the reduction in wave amplitude.

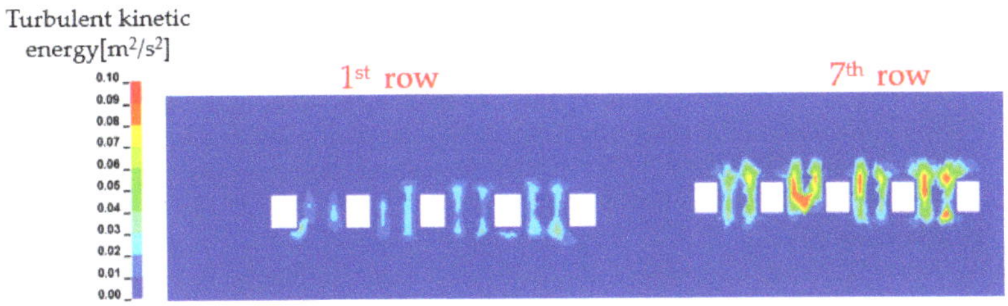

(**a**) (**b**)

Figure 13. Maximum turbulence creation at both wave crest and trough at (**a**) first row of breakwater and (**b**) last row of breakwater.

Next, the reduction in wave amplitude is compared for four different wave steepness ratios at the same amplitude of 12.5 cm and the same system arrangement of angle 4, Figure 14. It is visible that submerged breakwater efficiency decreases with the decrease in steepness, with the transmission coefficient being equal to 0.16 for the steepest wave and up to 0.64 for the most gradual wave.

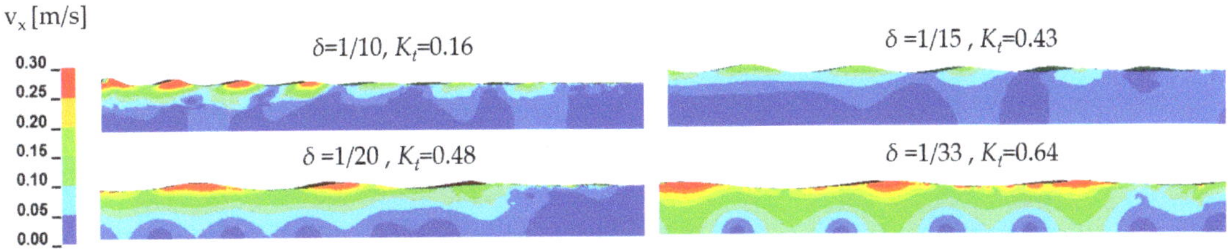

Figure 14. Comparison of wave amplitude reduction for different δ ratios—angle 4.

Finally, Figure 15a compares the drag force in the wave (x-axis direction) for different amplitudes, Figure 15b, for different wave steepness ratios. The efficiency of the system increases as the angle increases, with drag values being consistent across the three tested wave amplitudes. This drag predominantly consists of the viscous component, as shown in Figure 15a. Regarding the relationship between the drag force and steepness ratio, the angle 4 arrangement is the most effective at damping steeper waves. For the average wave steepness of 1/20, efficiency slightly increases with the system arrangement angle, as illustrated in Figure 15b.

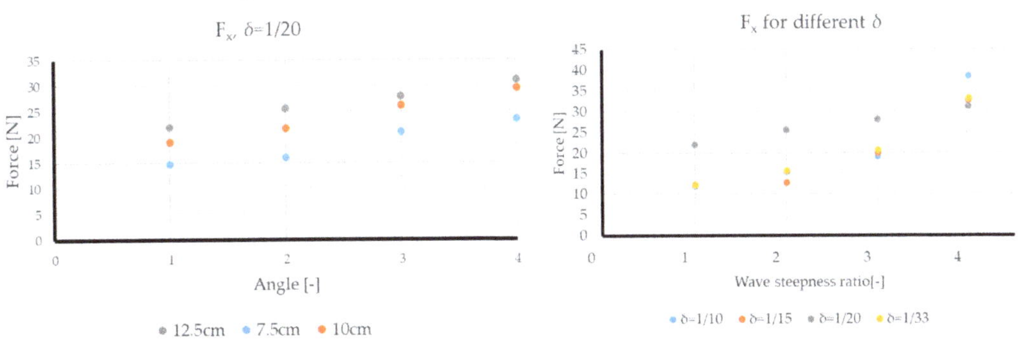

(**a**) (**b**)

Figure 15. Force comparison for different wave amplitudes in (**a**) drag force in X direction for $\delta = 1/20$; (**b**) relation between drag force in X direction and steepness ratio.

5. Conclusions

The ICFD analysis demonstrated its capability to model and solve wave interactions effectively. The initial setup using 20 elements per wave height did not produce an adequate wave at the inlet and resulted in divergence further along the domain. Consequently, 30 elements per wave height and 15 elements per wave amplitude were selected to improve accuracy.

Research consisting of a two-stage analysis of submerged breakwater is presented. Testing the topology of the submerged breakwater itself, the Quadratic 16-hole topology proved to be the most efficient in terms of wave damping. The more porous option with 36 holes is less effective due to the smaller surface area, which generates both a lower drag, Figure 9a, and lower wave profile, Figure 9b, as well as an overall lower efficiency over different wave steepness ratios compared to the Quadratic 16-hole topology. This is also demonstrated when the wave profile is visualised, as shown in Figure 10a,b. In comparison, the cylindrical 6-hole topology also dampens the wave but to a lesser extent than the Quadratic 16-hole topology, while the Quadratic 36-hole topology has almost no effect on wave amplitude.

The analysis of a submerged breakwater system has shown that the damping effect becomes more pronounced by reducing the angle and increasing the length of the overall system and by connecting more elements in series. Significant damping only occurred after the fourth row of a breakwater was reached, namely at the position 10 cm from the free surface at an amplitude height of 7.5 cm. Here, amplitude shift happened, and the amplitude decreased to only 7% of the initial value. When observing system angles, it is evident that angle 1 and angle 2 are too steep, as the first row is positioned too low even for the largest wave amplitudes. A significant increase in wave damping efficiency occurs at angle 3, while angle 4 shows only a marginal improvement. Turbulence creation, Figure 13, is visible close to the surface where fluid passes through breakwater holes, and partial backflow is created. When examining single system angle across different wave steepness ratios, submerged breakwater demonstrates a more pronounced advantage of wave dampening at higher wave steepness ratios, as depicted in Figure 14.

The analysis indicates that to significantly dampen the wave amplitude, the lowest position of the breakwater should ideally be no lower than 40% of the maximum expected wave amplitude. Therefore, the first row of breakwaters can be omitted in future towing tank tests.

Author Contributions: Š.S.: investigation, methodology, validation, formal analysis, visualisation, and writing—original draft preparation and editing. S.R.: investigation, methodology, supervision, and writing—review and editing. All authors have read and agreed to the published version of the manuscript.

Funding: This research received no external funding.

Institutional Review Board Statement: Not applicable.

Informed Consent Statement: Not applicable.

Data Availability Statement: Data are contained within this article.

Conflicts of Interest: The authors declare that they have no known competing financial interests or personal relationships that could have appeared to influence the work reported in this paper.

References

1. Zheng, Y.; Liu, M.; Chen, C.P.; Jiang, Y.P. Experimental study on the wave dissipation performance and mooring force of porous floating breakwater. *IOP Conf. Ser. Earth Environ. Sci.* **2018**, *189*, 022058. [CrossRef]
2. Chybowski, L.; Grządziel, Z.; Gawdzińska, K. Simulation and Experimental Studies of a Multi-Tubular Floating Sea Wave Damper. *Energies* **2018**, *11*, 1012. [CrossRef]
3. Coastal Engineering Manual-CEM. *Coastal Engineering Manual Part II*; US Army Coastal Engineering Research Centre: Washington, DC, USA, 2002.

4. Matsunaga, N.; Hashida, N.; Uzaki, K.; Kanzaki, T.; Uragami, Y. Performance of Wave Absorption by a Steel Floating Breakwater with Truss Structure. In Proceedings of the Twelfth International Offshore and Polar Engineering Conference, Kitakyushu, Japan, 26–31 May 2002; pp. 768–772.
5. Pena, E.; Ferreras, J.; Sanchez, F. Experimental study on wave transmission coefficient, mooring lines and module connector forces with different designs of floating breakwaters. *Ocean Eng.* **2011**, *38*, 1150–1160. [CrossRef]
6. Ji, C.; Deng, X.; Cheng, Y. An experimental study of double-row floating breakwaters. *J. Mar. Sci. Technol.* **2018**, *54*, 359–371. [CrossRef]
7. Kim, N.H.; Young, Y.L.; Yang, S.B.; Park, K.I. Wave damping analysis in a porous seabed. *KSCE J. Civ. Eng.* **2006**, *10*, 305–310. [CrossRef]
8. Chalmoukis, I.A.; Leftheriotis, G.A.; Dimas, A.A. Large-Eddy Simulation of Wave Attenuation and Breaking on a Beach with Coastal Vegetation Modelled as Porous Medium. *J. Mar. Sci. Eng.* **2023**, *11*, 519. [CrossRef]
9. Setiyawan. Transmission Coefficient (kt) and Reflection Coefficient (kr) on Breakwater Type Cube. *Int. J. Innov. Res. Adv. Eng.* **2018**, *5*, 158–166. [CrossRef]
10. Ji, C.; Cheng, Y.; Yang, K.; Oleg, G. Numerical and experimental investigation of hydrodynamic performance of a cylindrical dual pontoon-net floating breakwater. *Coast. Eng.* **2017**, *129*, 1–16. [CrossRef]
11. Syed, S.A.; Mani, J.S. Performance of rigidly interconnected multiple floating pontoons. *J. Nav. Archit. Mar. Eng.* **2004**, *1*, 3–17. [CrossRef]
12. Cheng, X.; Liu, C.; Zhang, Q.; He, M.; Gao, X. Numerical Study on the Hydrodynamic Characteristics of a Double-Row Floating Breakwater Composed of a Pontoon and an Airbag. *J. Mar. Sci. Eng.* **2021**, *9*, 983. [CrossRef]
13. Fitriadhy, A.; Faiz, M.A.; Abdullah, S.F. Computational fluid dynamics analysis of cylindrical floating breakwater towards reduction of sediment transport. *J. Mech. Eng. Sci.* **2017**, *11*, 3072–3085. [CrossRef]
14. He, M.; Gao, X.; Xu, W.; Ren, B.; Wang, H. Potential application of submerged horizontal plate as a wave energy breakwater: A 2D study using the WCSPH method. *Ocean Eng.* **2019**, *185*, 27–46. [CrossRef]
15. Guo, W.; Zou, J.; He, M.; Mao, H.; Liu, Y. Comparison of hydrodynamic performance of floating breakwater with taut, slack, and hybrid mooring systems: An SPH-based preliminary investigation. *Ocean Eng.* **2022**, *258*, 111818. [CrossRef]
16. He, M.; Liang, D.; Ren, B.; Li, J.; Shao, S. Wave interactions with multi-float structures: SPH model, experimental validation and parametric study. *Coast. Eng.* **2023**, *184*, 104333. [CrossRef]
17. Lloyd, C.; O'Doherty, T.; Mason-Jones, A. Development of a wave-current numerical model using Stokes 2nd Order Theory. *Int. Mar. Energy J.* **2019**, *2*, 1–14. [CrossRef]
18. Huang, L.; Ren, K.; Li, M.; Tukovic, Z.; Cardiff, P.; Thomas, G. Fluid-structure interaction of a large ice sheet in waves. *Ocean Eng.* **2019**, *182*, 102–111. [CrossRef]
19. Ge, G.; Zhang, W.; Xie, B.; Li, J. Turbulence model optimization of ship wake field based on data assimilation. *Ocean Eng.* **2024**, *295*, 116929. [CrossRef]
20. Wiesent, L.; Geith, M.; Wagner, M. Simulation of Fluid-Structure Interaction between injection medium and balloon catheter using ICFD. In Proceedings of the 11th European LS-DYNA Conference, Salzburg, Austria, 9–11 May 2017.
21. *ICFD Theory Manual, Incompressible Fluid Solver in LS-DYNA*; Livermore Software Technology Corporation (LSTC): Livermore, CA, USA, 2014.
22. Maâtoug, M.A.; Ayadi, M. Numerical simulation of the second-order Stokes theory using finite difference method. *Alex. Eng. J.* **2016**, *55*, 3005–3013. [CrossRef]
23. Farkas, A.; Parunov, J.; Katalinić, M. Wave Statistics for the Middle Adriatic Sea. *J. Marit. Transp. Sci.* **2016**, *52*, 33–47. [CrossRef]
24. Zorović, D.; Mohović, R.; Mohović, Đ. Towards Determining the Length of the Wind Waves of the Adriatic Sea. *Naše More* **2003**, *50*, 145–150.
25. Perin, B.; Verdrel, P.; Bordenave, P.; Grippon, E.; Belloc, H.; Caldichoury, I. Computational Fluid Dynamic of NACA0012 with LS-DYNA® (ALE & ICFD) and Wind Tunnel Tests. In Proceedings of the 14th International LS-DYNA Users Conference, Detroit, MI, USA, 12–14 June 2016.
26. Connell, K.; Cashman, A. Development of a numerical wave tank with reduced discretization error. In Proceedings of the International Conference on Electrical, Electronics, and Optimisation Techniques (ICEEOT), Chennai, India, 3–5 March 2016. [CrossRef]
27. da Silva, C.B. The behavior of subgrid-scale models near the turbulent/nonturbulent interface in jets. *Phys. Fluids* **2009**, *21*, 081702. [CrossRef]
28. Calero, L.; Blanco, J.M.; Izquierdo, U.; Esteban, G. Performance Assessment of Three Turbulence Models Validated through an Experimental Wave Flume under Different Scenarios of Wave Generation. *J. Mar. Sci. Eng.* **2020**, *8*, 881. [CrossRef]
29. Šimun, S.; Smiljko, R. Development of wave dampers using computational fluid dynamics. In Proceedings of the 24th Numerical Towing Tank Symposium, Zagreb, Croatia, 15–18 October 2022; pp. 168–174.

oceans

MDPI

Article

Exploring Mesozooplankton Insights by Assessing the Ecological Status of Black Sea Waters Under the Marine Strategy Framework Directive

Elena Bisinicu [1] and Luminita Lazar [2,*]

[1] Ecology and Marine Biology Department, National Institute for Marine Research and Development "Grigore Antipa", 300 Mamaia Blvd., 900581 Constanta, Romania; ebisinicu@alpha.rmri.ro
[2] Chemical Oceanography and Marine Pollution Department, National Institute for Marine Research and Development "Grigore Antipa", 300 Mamaia Blvd., 900581 Constanta, Romania
* Correspondence: llazar@alpha.rmri.ro

Abstract: Mesozooplankton are highly responsive to environmental changes, making their population dynamics and species composition important indicators of large-scale oceanic conditions. This study investigates the mesozooplankton composition, emphasizing biomass values across all three marine reporting units in the Romanian Black Sea from 2013 to 2020, covering 45 stations that represent transitional, coastal, and marine waters during the warm season. It also examines environmental parameters, such as temperature, salinity, and nutrient levels associated with eutrophication, that impact the mesozooplankton, as well as the environmental status of the pelagic habitat in this region, influenced by various hydrological and anthropogenic factors. Statistical analyses, including multivariate methods, were employed to investigate correlations between mesozooplankton biomass and environmental parameters, facilitating ecological assessments in accordance with the Marine Strategy Framework Directive (MSFD) Descriptors 1 (Biodiversity) and 5 (Eutrophication) using indicators such as Mesozooplankton bBiomass, Copepoda Biomass, and *Noctiluca scintillans* iomass. The results indicated that Mesozooplankton and Copepoda Biomass did not reach good ecological status (GES). However, the *Noctiluca scintillans* Biomass indicator demonstrated that all marine reporting units achieved GES during the warm seasons. These findings underscore the dynamic nature of pelagic habitats and highlight the importance of ongoing monitoring to inform policy and conservation efforts.

Keywords: dynamics; biomass; pelagic habitat; good environmental status; policy

1. Introduction

Marine waters encompass a variety of habitats and organisms that deliver essential ecosystem services [1,2]. In the pelagic zone, many of these habitats and species have been impacted by human activities, both directly (e.g., fishing, deep-sea mining, mariculture, the ballast water spread of invasive species) [2,3] and indirectly through atmosphere–ocean interactions (e.g., climate change, ocean acidification) and land–ocean interactions (e.g., urban, industrial, and agricultural effluents, changes in runoff regimes due to dams) [4–6]. These pressures act across different temporal and spatial scales, leading to cumulative impacts and complicating the identification of causality [1,2,7]. The pelagic habitat, the largest biome on Earth, plays a crucial role in temperature regulation, oxygen production, and food supply [8]. Its physical and biological components and processes exhibit spatial and temporal variability influenced by numerous drivers [8]. Understanding this variability, along with the associated processes and interactions, is essential for identifying the drivers of change and accurately assessing pelagic habitats under the Marine Strategy Framework Directive (MSFD) [8,9].

The MSFD [9] is the principal legislation of the European Union aiming at safeguarding marine environments. The MSFD initially set a target for EU Member States to achieve a Good Environmental Status (GES) by 2020. However, the 2020 review highlighted that many marine regions had not fully met the required standards; consequently, the timeline was extended, now aiming for 2024. This extension allows for the implementation of additional measures and strategies, reflecting a continuous effort to address the ecological challenges identified in the earlier assessment cycles [10]. Adopted in June 2008, the directive encompasses various aspects of marine health using 11 descriptors, including biodiversity, invasive species, fisheries, food webs, eutrophication, hydrological conditions, contaminants, noise, and marine litter [9]. The MSFD mandates that each EU member state define what constitutes a good environmental status for their marine waters [9]. Pelagic habitats are inherently dynamic systems, influenced by various hydrological and anthropogenic drivers. Assessing these multi-scale systems is a policy priority under Descriptor 1 (Biodiversity, criterion D1C6) of the Marine Strategy Framework Directive (MSFD) and is also connected to other pressure descriptors such as Descriptor 2 (Non-Indigenous Species), Descriptor 5 (Eutrophication), and Descriptor 8 (Contaminants) [8]. For EU Member States, the MSFD aims to ensure that the biotic and abiotic structures and functions of pelagic habitats are not adversely affected by anthropogenic pressures, as defined by the Commission Decision (EU) 2017/848) [9]. GES is described by Article 1(3) of the MSFD [9] as the sustainable use of marine resources to ensure their availability for future generations. Its determination requires considering the spatial heterogeneity of regional sea characteristics and depends on the spatial and temporal attributes of the monitored marine area [1,8,11]. To achieve consistent and comparable results across European Seas, the assessment of pelagic habitats under the MSFD should involve a common determination, standardized evaluation criteria, and uniform methods, including data type and frequency, indicators, and analysis [8]. Marine Reporting Units (MRUs) are utilized within the reporting obligations of the Marine Strategy Framework Directive (MSFD) to connect the implementation of various articles to specific marine areas. The size of the MRUs can vary depending on the appropriate scale needed for different reports (e.g., region, sub-region, regional or sub-regional subdivision, Member State marine waters, WFD coastal waters, etc.), as specified in the decision [9].

Mesozooplankton, a fraction of the zooplankton community, consist of a diverse array of organisms that vary functionally, phylogenetically, and morphologically, ranging in size from 0.2 to 20 mm [12,13]. This group includes some of the most abundant animals on Earth, such as Copepoda and Euphausiacea [14]. Zooplankton organisms are sensitive to anthropogenic impacts and environmental fluctuations, making their study useful for predicting long-term changes [15–18]. Species diversity and community composition can indicate environmental changes or disturbances [19–22]. Furthermore, long-term changes in zooplankton biomass and species composition can significantly impact ecosystem function, as zooplankton are crucial for biogeochemical cycling [23]. Several researchers have reported that zooplankton can serve as indicators in studies of trophic dynamics and the ecological status of aquatic ecosystems [15,24–27]. This is particularly relevant for the Romanian Black Sea, which is affected by human activities such as river inputs (from the Danube, Dniester, and Dnieper Rivers), ports, and industrial operations [28–30].

Significant long-term changes in the Black Sea zooplankton community occurred between 1950 and 1970 and in 1995 due to increased anthropogenic impacts, affecting species composition [31,32]. Post-1970, notable shifts included the near disappearance of some species and the proliferation of others, such as *Noctiluca scintillans*, particularly in the nutrient-rich northwestern region influenced by the Danube, Dniester, and Dnieper Rivers' runoffs. Signs of ecosystem recovery appeared after 1994 due to reduced pollution and eutrophication, alongside changes in the hydro-climatic regime [32–35]. In the 1970s–1980s, the abundance of common species significantly declined, exacerbated by the invasion of the ctenophore *Mnemiopsis leidyi* after 1988 [36,37] which altered the biological properties of the ecosystem, promoting phytoplankton growth and decreasing the amounts of food for planktivorous and predatory fishes [38–40]. The introduction of the ctenophore predator *Beroe ovata* in 1997, from the Mediterranean or North Atlantic Seas via ballast waters, aided ecosystem recovery [39,41,42].

This study has three primary objectives: to analyze the mesozooplankton's composition, focusing on biomass values across all three marine reporting units of the Romanian Black Sea area during the warm season from 2013 to 2020; to assess the environmental parameters influencing mesozooplankton; and to evaluate the ecological status of the pelagic habitat in this region, using mesozooplankton as indicators of environmental changes.

2. Materials and Methods

2.1. Description of the Study Area

The Black Sea is a semi-enclosed sea, with its sole connection to the world's oceans being the narrow Bosphorus Strait. It spans latitudes 40°55′ to 46°32′ N and longitudes 27°27′ to 41°32′ E. To the south, it connects to the Mediterranean Sea via the Bosphorus, one of the world's narrowest straits, with an average width of 1.6 km, a depth of 36 m, and a length of 31 km [43]. To the north, it links to the Sea of Azov through the shallow Kerch Strait, which has a depth of less than 20 m. The Black Sea is bordered by six countries: Bulgaria, Georgia, Romania, Russia, Turkey, and Ukraine. Its dynamics are influenced by 17 countries within its basin, 13 capital cities, and approximately 160 million people [43]. Additionally, the Danube, Dnieper, and Don Rivers—Europe's second, third, and fourth longest rivers—flow into the Black Sea, affecting nearly one-third of continental Europe's land area [43].

Based on the statistical analysis of key physical parameters, the degree of anthropogenic impact, chlorophyll-a concentration, bathymetry, and scientific studies covering the entire basin, four categories of marine reporting units (MRUs) were identified for the Romanian Black Sea coast [44]:

- RO-TT03 (transitional waters)—the northern part, consisting of marine waters directly influenced by the Danube River, at depths of at least 30 m;
- RO-CT01 (coastal waters)—marine waters from the central to the southern area (Portita down to Vama Veche), from the baseline to 30 m isobaths;
- RO-MT01 (marine waters)—corresponding to the inner and outer shelf marine area from 30 m to 100 m isobaths;
- RO-MT02—the offshore seawater located at least 100 m deep.

Mesozooplankton samples were collected from stations across the Romanian Black Sea monitoring network (Figure 1) during the warm season (May–October) from 2013 to 2020. This monitoring network covers the entire continental shelf along the Romanian Black Sea coast, across three marine reporting units. In total, 293 zooplankton and water samples were collected from these 45 monitoring stations.

2.2. Mesozooplankton Sampling and Analysis

Zooplankton samples were collected according to the methodology [45] using a Juday net (0.1 m^2 mouth opening area, 150 μm mesh size) equipped with a Hydro-Bios flow meter (438,110) for estimating the volume of water filtered, by vertically towing the net in the entire water column (bottom–surface).

The quantitative and qualitative processing of the zooplankton samples was performed under OlympusSZX10 stereomicroscope. When specifying the taxonomic affiliation of zooplankton and for species determination, we used proper manuals and guides [46,47], and classification for identified taxa was made according to the World Register of Marine Species (WoRMS) [48].

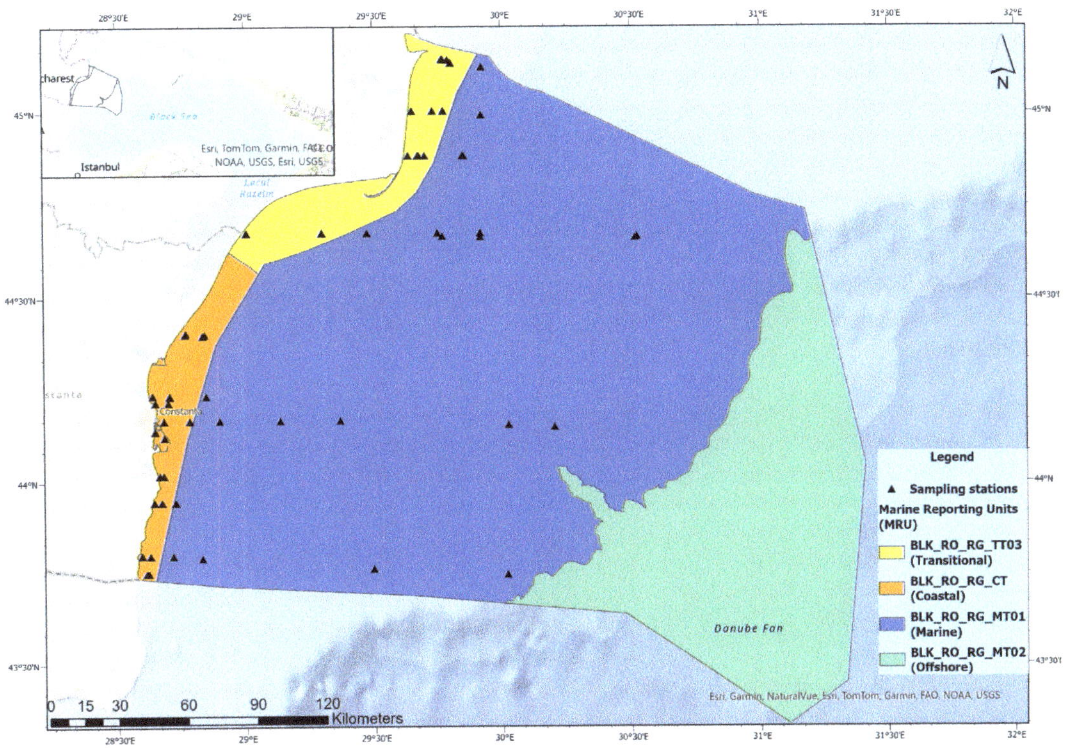

Figure 1. Map of the study area (left corner is the Romanian EEZ (white polygon) in the Black Sea region).

Not all organisms could be determined at the species level, with the meroplanktonic elements being identified only at the group level. In the subsample(s) examined in the Bogorov chamber, all organisms were counted until at least 100 individuals for each of the three dominant taxonomic groups were recorded. For the estimation of large animal numbers, the whole sample was examined in a Petri dish. The number of individuals were used for estimating the abundance as ind/m^3 and the biomass as mg/m^3 wet weight based on the tables of constant weight of the Black Sea zooplankters [45,49].

2.3. Environmental Parameters

Temperature and salinity were assessed using various methods including a reversible thermometer, titration, and the CastAway CTD multiparameter probe (SonTek CastAway-CTD, USA). Dissolved nutrient concentrations were determined according to established protocols for seawater analysis [50]. Nutrients were quantified using spectrophotometric analytical methods validated in the laboratory, following the guidelines outlined in the "Methods of Seawater Analysis" manual [50]. Specifically, nitrate concentrations were determined using the method described by Mullin and Riley (1955) [51], where nitrate was reduced to nitrite using hydrazine sulfate. The resulting nitrite reacted with sulfanilamide in an acidic solution, forming a diazonium compound that coupled with N-(1-Naphthyl)-ethylenediamine dihydrochloride to produce a colored azo dye, which was quantifiable using spectrophotometry.

Ammonia levels were determined using the indophenol blue method [50], wherein ammonia reacted in a moderately alkaline solution with hypochlorite to form monochloramine. In the presence of phenol, catalytic amounts of nitroprusside ions, and excess hypochlorite, indophenol blue was produced.

Inorganic phosphate ions were quantified by their reaction with ammonium molybdate in an acidic medium, forming a yellowish phosphomolybdenum complex. This complex was immediately reduced to a blue-colored compound, the intensity of which correlated with the concentration of phosphate ions. The intensity was measured at a wavelength of 885 nm using a spectrophotometer [50].

Similarly, silicon ions reacted with ammonium molybdate in an acidic medium to form a yellowish silicomolybdenum complex. In the presence of ascorbic acid as a reducing agent, this complex was reduced to a intensely blue-colored compound, the intensity of which was proportional to the silicon concentration. The measurement was performed at a wavelength of 810 nm using spectrophotometry [50].

2.4. Mesozooplankton Indicators

As an EU Member State, Romania has adopted a series of indicators to monitor the pelagic habitat of the Black Sea, based on the mesozooplankton component. These indicators encompass general metrics (biomass), which are derived from sampling strategies and methodologies to track changes in pelagic habitat status.

Under the Marine Strategy Framework Directive (MSFD), specifically Descriptor 1 (Biodiversity, criterion D1C6) and Descriptor 5 (Eutrophication, criterion D5C3), the assessment utilizes indicators such as Copepoda biomass, Mesozooplankton biomass, and *N. scintillans* biomass. These indicators provide crucial information on the health and diversity of pelagic habitats. Their biomass reflects the availability of food for higher trophic levels, such as fish and marine mammals, and can indicate changes in environmental conditions and ecosystem productivity [24,52].

N. scintillans biomass, on the other hand, serves as an indicator for Descriptor 5 (Eutrophication, criterion D5C3) (Table 1). *N. scintillans* is a dinoflagellate that can form large blooms, which are often associated with high nutrient loads in marine waters [53,54]. The biomass of this organism is a direct indicator of eutrophication, a condition caused by excessive nutrient inputs leading to the overgrowth of algae and subsequent oxygen depletion [53].

Table 1. Black Sea mesozooplankton indicators.

Descriptor	Criterion	Indicator
Descriptor 1: Biodiversity	D1C6 -primaryThe condition of the habitat type, including its biotic and abiotic structure and its functions (e.g., its typical species composition and their relative abundance, absence of particularly sensitive or fragile species or species providing a key function, size structure of species), is not adversely affected due to anthropogenic pressures.	Copepoda biomass (mg/m^3) Mesozooplankton biomass (mg/m^3)
Descriptor 5: Eutrophication	D5C3- secondary—Harmful algal blooms in the water column. The number, spatial extent, and duration of harmful algal bloom events are not at levels that indicate adverse effects of nutrient enrichment.	*Noctiluca scintillans* biomass (mg/m^3)

To evaluate the Black Sea environmental status by using mesozooplankton indicators and to determine the GES, reference periods were used to establish thresholds through descriptive statistics (Tables S1–S3). In Romania, these reference conditions are based on long-term data from 1960 to 2002, with two distinct reference periods: 1960–1969 (indicative of good conditions) and 1977–2002 (indicative of less-favorable conditions). The assessment of the reference conditions and limits for defining GES was conducted using statistical analysis of data from 1960 to 2002, complemented by expert judgment.

GES was determined by calculating the 90th percentile of values from the warm season and each marine reporting unit. The values obtained were then compared to the averages from the intervals of 1960–1969 (representing a Very Good Condition) and 1977–2002 (representing a Poor Condition) (Tables S1–S3). This comparative approach allowed for the establishment of threshold values for each marine reporting unit and indicator in the warm season (Tables S4–S6).

2.5. Data Analysis

Multivariate and univariate analyses were conducted using PRIMER v. 7.0, UK [55], and statistical analysis was performed via Statistica® 14.0.1.25 (TIBCO Software Inc., Palo Alto, CA, USA) [56].

Data analyzed with PRIMER v. 7.0 were square root transformed. Multivariate analyses compare two or more samples to determine the extent to which they share certain species at comparable levels of abundance, facilitating sample classification into mutually similar groups or creating an ordination diagram to reflect their relative differences in specific composition [57]. To characterize the differences in community quantitative structure across the three marine reporting units, we utilized non-metric multidimensional scaling (n-MDS). n-MDS is a non-linear method that arranges samples based on their rank-order dissimilarities, aiming to preserve the relative distances between samples rather than their absolute distances [55,58,59]. This approach provides insights into mesozooplankton's spatial distribution and community dynamics during the warm season, identifying regions with particularly high or low biomass. The resemblance metric used is the Bray–Curtis similarity index, which is commonly employed in ecological studies to measure the compositional dissimilarity between different sites based on quantitative data [60]. The Bray–Curtis metric is well-suited for ecological data because it considers both the presence/absence and abundance of species, making it ideal for analyzing community composition. This measure is particularly effective for detecting ecological gradients and is robust to variations in sample size, which are common in environmental studies. By using this metric, we ensured that the analysis accurately reflected the ecological dissimilarities between marine reporting units, providing a reliable basis for assessing spatial variations in mesozooplankton.

Diversity indices, shade plots, boxplots, and analysis were utilized for univariate analyses. Different diversity indices emphasize either species richness or equitability components, with the most common being the Shannon–Wiener diversity index (H′), species richness, and Pielou's evenness index (J′) [55].

$$H' = -Pi(\ln Pi)$$

where Pi is the proportion of each species in the sample, and å is the abundance of the ith species.

$$J' = \frac{H'}{H'max} = \frac{H'}{\log S}$$

where H′max is the maximum possible value of the Shannon diversity, i.e., that which would be achieved if all the species were equally abundant (namely, log S).

The Shannon–Wiener diversity index (H′) is an effective measure of water quality, where high values signify greater species diversity and even distribution, while lower values reflect poor species composition [61]. Additionally, this index can assess the pollution status of a water body. The Shannon–Wiener classification for the aquatic environment categorizes water quality as follows: >4 indicates very good quality, 3–4 denotes good quality, 2–3 signifies moderate quality, 1–2 represents poor quality, and less than 1 reflects very poor quality [61]. Pollution levels based on species diversity are classified as less than 1 for highly polluted, 1 to 3 for moderately polluted, and greater than 4 for unpolluted water bodies [61].

Shade plots, used for the identified mesozooplankton taxa, are visual displays of the data matrix itself. Shade plots are an effective visual tool used to represent the data matrix of identified mesozooplankton taxa. These plots provide a detailed graphical representation of the abundance or presence–absence data across various samples or time points. By using varying shades of color to denote different values, shade plots facilitate the quick identification of patterns, trends, within the dataset [62]. This method allows researchers to visually compare the distribution and abundance of different taxa, helping to reveal ecological dynamics and interactions [62].

Statistica® [56] was used to analyze the descriptive statistics of the environmental variables, including correlations, highlighting all significant correlation coefficients ($p < 0.05$, two-tailed) in red. T-tests were employed to compare environmental data by evaluating the differences in means between marine reporting units. The test returned *p*-values to check for statistical significance.

Spatial distribution maps were created using ArcGIS Desktop 10.7 USA [63]. GIS tools were employed for spatial interpolation, a technique that predicts values at unsampled points or fills data gaps within a study area. Spatial interpolation predicts the value of a variable at unmeasured locations based on data from known locations [64,65]. In our study, we used local interpolators, specifically inverse distance weighting (IDW).

3. Results

3.1. Environmental Parameters During the Warm Seasons of 2013–2020

Transitional waters were characterized by significant variability in all measured parameters, which reflects their dynamic nature as interfaces between freshwater and marine environments (Table S7). The temperature in transitional waters showed a considerable range, with a mean value of 20.89 °C. Salinity was also highly variable, with values ranging from 0.19‰ to 18.50‰, which is indicative of the mixing between the Danube's freshwater input and marine influences. The dissolved oxygen (O_2) concentration and saturation levels (mean values of 332.54 μM and 125.67%, respectively) suggested that these waters can support aerobic life but also experience fluctuations that might stress certain species [66]. Nutrient concentrations, particularly silicate (SiO_4) and nitrate (NO_3), showed high variability, indicating active biogeochemical processes likely driven by both natural and anthropogenic inputs. The high standard deviations in nutrient levels (e.g., NO_2 with a standard deviation of 7.89 μM) further emphasize the fluctuating conditions in these waters [66]. Coastal waters exhibited intermediate characteristics between transitional and marine waters (Table S7). The mean temperature of 21.70 °C and a relatively narrow range of salinity (mean 14.52‰) reflected a more stable environment compared to transitional waters. Dissolved oxygen levels were slightly lower on average than in transitional waters, with a mean concentration of 312.02 μM and a saturation level of 122.47%. Nutrient concentrations in coastal waters, such as those of phosphate (PO_4) and nitrate (NO_3), were lower than those in transitional waters but still showed significant variability, which could be attributed to inputs from both land and marine sources. The lower quartile values for nutrients suggest that at times, these waters can experience nutrient depletion, which may impact primary production and higher trophic levels.

Marine waters were the most stable in terms of the environmental variables measured (Table S7). The temperature was slightly higher on average (mean 22.52 °C) compared to coastal waters, and salinity was also more consistent (mean 15.36‰). Dissolved oxygen concentrations (mean 310.79 μM) were similar to those in coastal waters, but the lower minimum value (119.24 μM) indicated that some areas might experience hypoxic conditions. Nutrient concentrations were generally lower in marine waters, with phosphate and nitrate levels being the lowest among the three water types, reflecting the more oligotrophic nature of marine environments. The lower variability in nutrient levels (e.g., NO_2 with a standard deviation of 3.99 μM) suggests a more balanced and stable ecosystem, which is less subject to the fluctuations seen in transitional and coastal waters [66].

Data highlight the gradient of environmental conditions from transitional to marine waters, with transitional waters showing the greatest variability and extremes in environmental parameters. Coastal waters represented an intermediate state, while marine waters exhibited more stability and lower nutrient levels.

The temperature displayed a gradient of sea surface temperatures along the Romanian Black Sea coast (Figure 2). Cooler temperatures were observed near the coast, particularly around the Danube Delta, likely due to freshwater inflows from the Danube River, which mix with the seawater, creating cooler conditions. As we move offshore, temperatures increase to the maximum in the southern parts, indicating a warming trend that is likely influenced by solar radiation and water stratification.

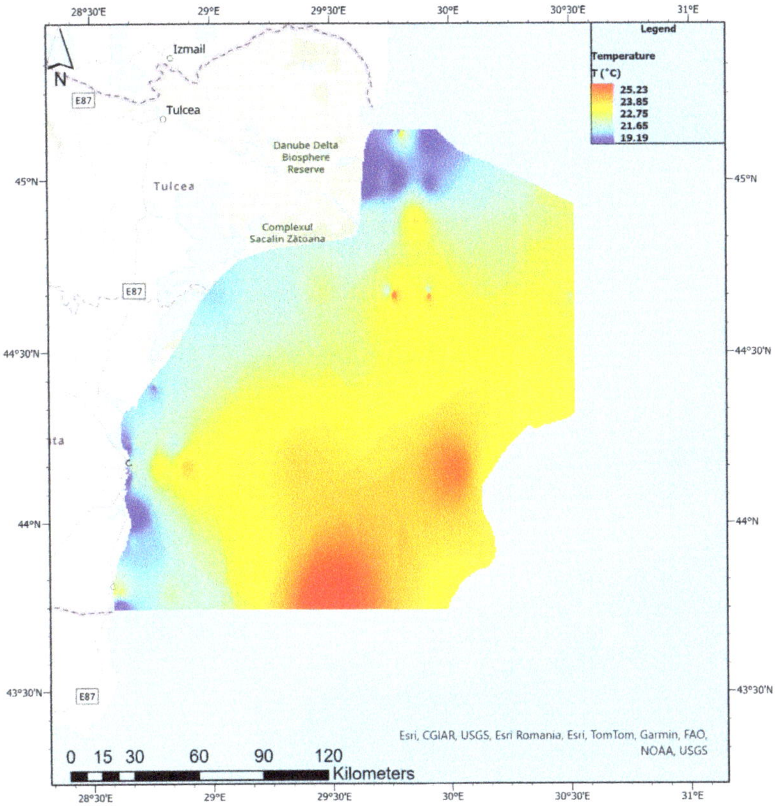

Figure 2. Spatial distribution of surface temperatures along the Romanian Black Sea coast in the warm season of 2013–2020.

The salinity distribution along the Romanian Black Sea coast (Figure 3) showed the lowest levels in the transitional areas, especially near the Danube Delta, which was attributed to the significant freshwater input from the Danube River, leading to a dilution effect. As we move further offshore, the salinity gradually increases, with the highest levels observed in the more southern and southeastern offshore regions.

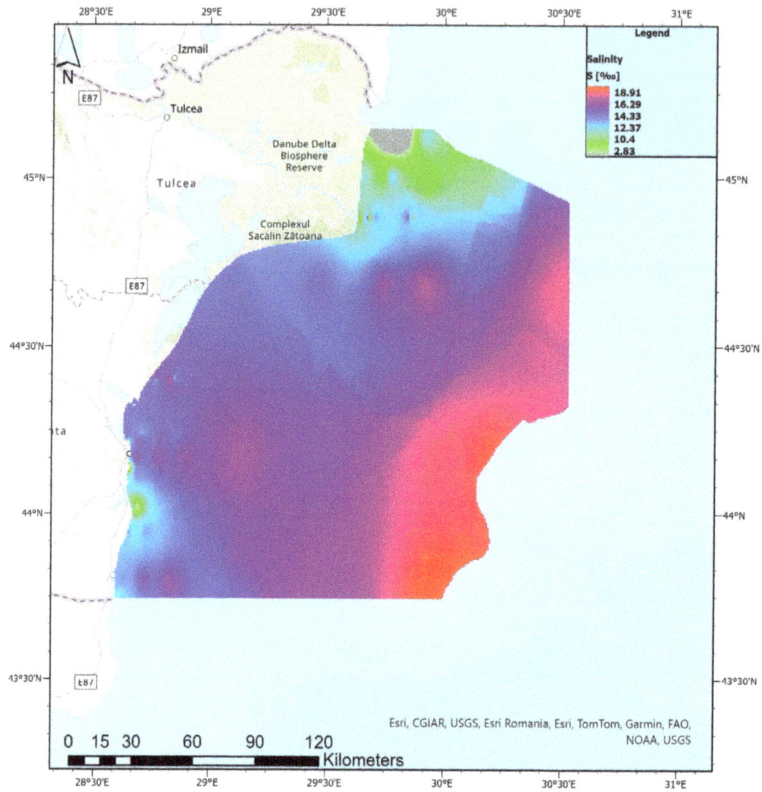

Figure 3. Spatial distribution of salinity along the Romanian Black Sea coast in the warm season of 2013–2020.

The distribution of oxygen saturation levels showed the highest variability in the transitional waters, with a median oxygen saturation slightly above 120% and a broad interquartile range, indicating significant fluctuations likely due to mixing processes and varying biological activity (Figure 4). However, the median O_2% is similar across all three MRUs, hovering around the 120% mark. Still, the Coastal MRU displayed a slightly lower median than the other two. The spread of the data, indicated by the interquartile range (the box), is also quite similar across the MRUs, with the transitional and marine regions showing slightly more variability. Outliers were present in all three groups, particularly in the coastal water, which had one extreme value that was significantly higher than the others. This suggests some variability in oxygen levels within each MRU, but generally consistent central tendencies. Coastal waters, while similar in median to marine waters, displayed slightly less variability and fewer outliers, reflecting more stable oxygen levels, but were still influenced by proximity to land and human activities (Figure 4).

Transitional waters exhibited the widest range and highest variability in SiO_4 (silicate) and NO_3 (nitrate) concentrations, with numerous outliers indicating substantial fluctuations (Figure 5). Similarly, PO_4 (phosphate) and NO_2 (nitrite) levels in transitional waters showed higher medians and significant variability with many outliers. NH_4 (ammonium) concentrations were also highest in transitional waters, characterized by considerable variability and outliers (Figure 5). Coastal waters presented moderate variability across SiO_4, NO_2, NO_3, PO_4 and NH_4 concentrations (Figure 5A–E), with some outliers, influenced by terrestrial runoff, and human activities. In contrast, marine waters generally displayed lower median levels and minimal variability for these parameters, reflecting stable conditions with fewer outliers (Figure 5).

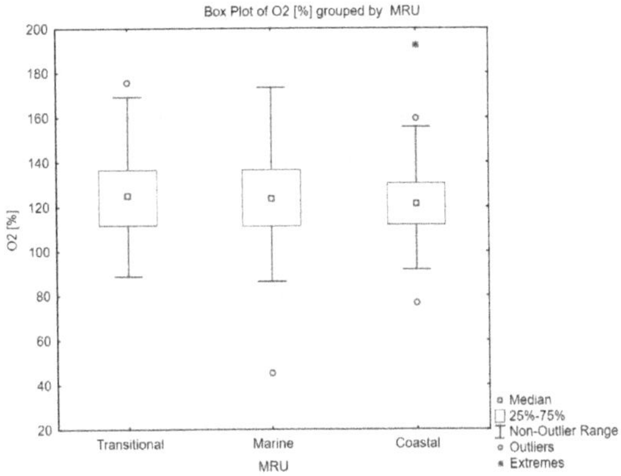

Figure 4. Black Sea oxygen box plot by marine reporting units in the warm seasons of 2013–2020.

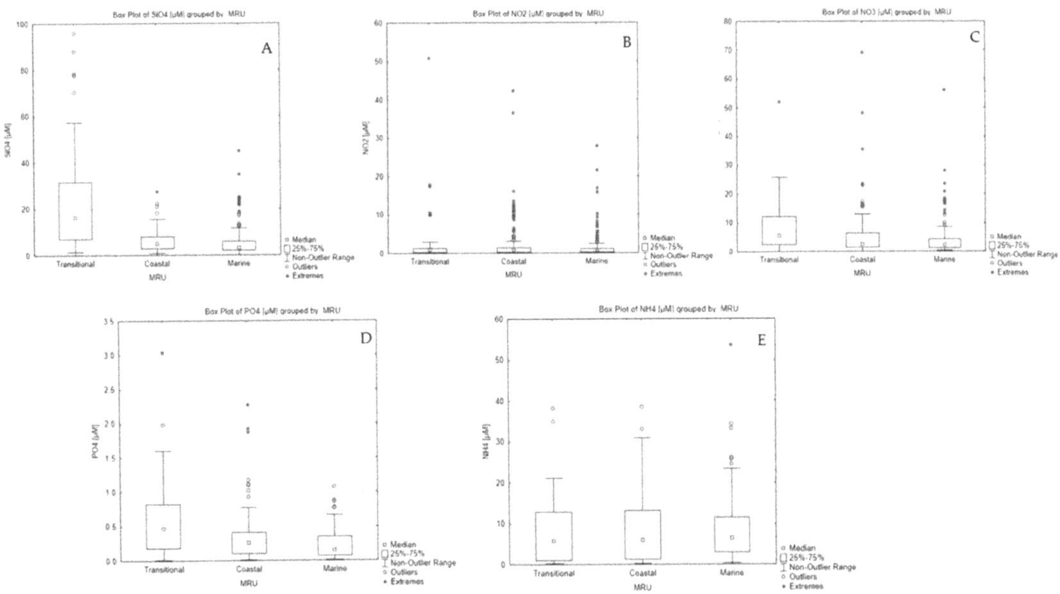

Figure 5. Black Sea nutrient box plots by marine reporting units in the warm seasons of 2013–2020.

3.2. Mesozooplankton's Qualitative Composition and Biomass Structure During Warm Seasons of 2013–2020

To analyze the mesozooplankton in the Romanian Black Sea, we separated the fodder component, consisting of Copepoda, Cladocera, meroplankton, and other groups representing the overall mesozooplankton component, from the non-fodder zooplankton, represented by the dinoflagellate *N. scintillans*. Due to its large cell size (over 200 μm) and phagotrophic feeding behavior, *N. scintillans* is assessed within mesozooplankton [54,67]. With a total number of 27 identified taxa (Table S8) the distribution of mesozooplankton taxa during the warm season exhibits considerable variation across the marine reporting units, with transitional waters typically displaying higher diversity (Table S8). Copepoda are the most numerous in terms of species richness, underscoring their ecological significance (Figure 6). Cladocera and meroplankton exhibited moderate diversity and were evenly distributed across all zones, suggesting they played crucial roles in transitional ecosystems and might be sensitive to changes in environmental conditions (Figure 6). Other groups showed limited diversity but were present across all zones, indicating a generalist nature or less-specialized ecological niches (Figure 6).

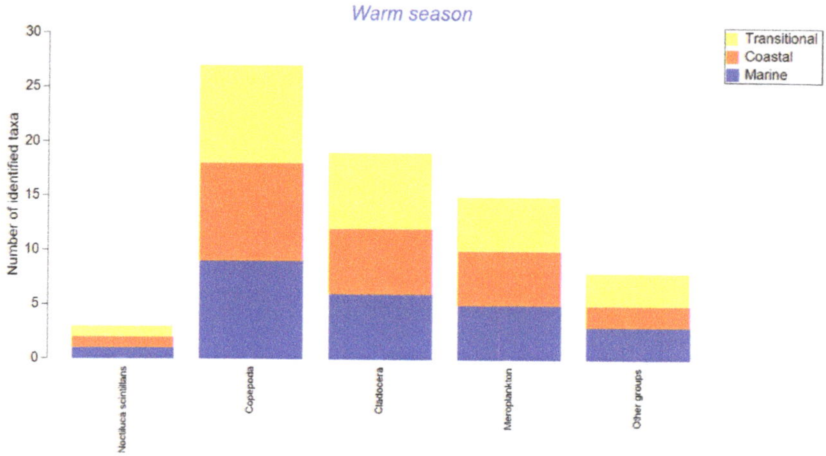

Figure 6. Number of identified mesozooplankton taxa during the warm season of 2013–2020.

The non-fodder species *N. scintillans* was identified throughout the entire period across all three marine reporting units (Figure 7, Table S8), showing a pronounced preference for coastal zones where it reaches its highest biomass. The Copepoda group exhibited variations in species numbers, with a maximum of nine taxa observed in all marine reporting units (Figure 7). The core copepod species consistently identified from 2013 to 2020 included *Acartia clausi*, *Pseudocalanus elongatus*, *Oithona similis*, *Paracalanus parvus*, *Calanus euxinus*, and *Centropages ponticus*. Additionally, seven Cladocera species were identified along the Romanian coast (Figure 7, Table S8). Their presence in large numbers was relatively rare and closely associated with the freshwater input from the Danube River. The high number of Cladocera was attributed to freshwater species such as *Chydorus sphaericus*, *Bosmina longirostris*, *Daphnia longispina*, and *Diaphanosoma brachyurum*, which occasionally appear in transitional waters (Figure 7). The cladocerans *Pleopis polyphemoides*, *Evadne spinifera*, *Penilia avirostris*, and *Pseudevadne tergestina* were consistently identified throughout the analyzed period (Figure 7). The meroplankton showed minimal variation in the number of taxa, with a maximum of five taxa observed (Figure 7): the larval stages of Cirripedia, Decapoda, Gastropoda, Bivalvia, and Polychaeta (Table S8). The other group category reached a maximum of three species (Figure 7), represented by Chaetognatha—*Parasagitta setosa*, Appendicularia- *Oikopleura dioica*, and Mysida- *Mesopodopsis slabberi* (Table S8).

Coastal and transitional waters exhibit a high degree of similarity, with a similarity index close to 90% (Figure 8). This high similarity indicates that the mesozooplankton communities in these zones share many taxa and have comparable biomass distributions. In contrast, the marine waters show a lower similarity to both the coastal and transitional ones, with the similarity index dropping to around 75% (Figure 8).

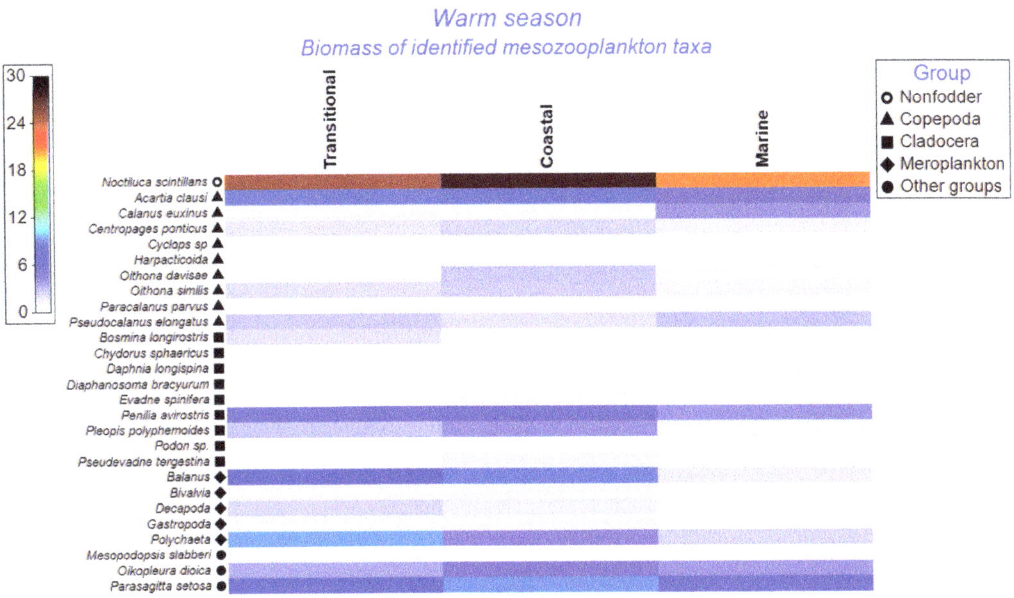

Figure 7. Biomass distribution of identified mesozooplankton taxa during the warm seasons of 2013–2020.

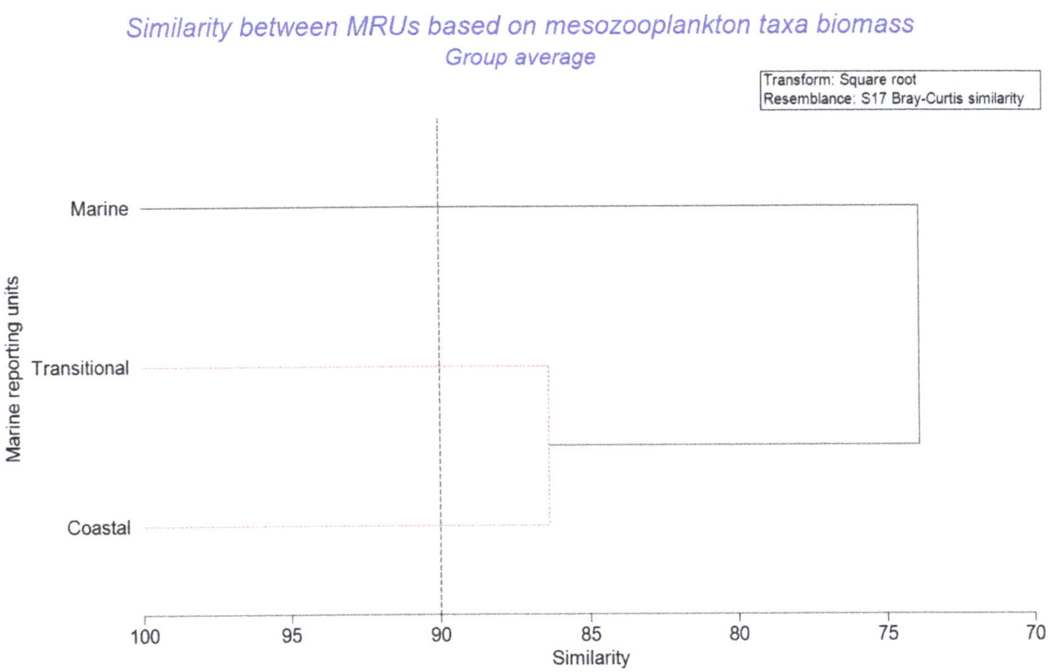

Figure 8. Similarity between marine reporting units (MRUs) based on mesozooplankton taxa biomass during the warm season of 2013–2020.

The analysis of species richness (S), diversity (H′), and evenness (J′) across the transitional, coastal, and marine waters reveals distinct ecological patterns within mesozooplankton (Table 2). The transitional waters, with the highest species richness and diversity, support a highly diverse and relatively balanced mesozooplankton community, as indicated by their high evenness index (Table 2). The coastal waters, although slightly lower in species richness and diversity compared to the transitional ones, reflect an evenly distributed and balanced community (Table 2). In contrast, the marine waters, with intermediate species richness and the lowest diversity and evenness indices, suggest a less balanced community where certain species are likely dominant (Table 2). These findings highlight the influence of environmental conditions and nutrient availability in shaping the mesozooplankton community structure across different marine zones.

Table 2. Diversity indices of mesozooplankton across MRUs during the warm seasons of 2013–2020.

MRU	Taxa Richness (S)	Pielou's Evenness Index (J')	Shannon–Weiner Index H'
Transitional	25	0.7629	2.456
Coastal	23	0.7729	2.423
Marine	24	0.7388	2.348

In transitional waters, *N. scintillans* exhibited a highly skewed distribution, indicating occasional high concentrations and substantial variability. Copepoda biomass showed a more consistent distribution, while Cladocera indicates sparse but occasionally dense populations. Meroplankton and other groups also experienced sporadic high concentrations, contributing to the overall high variability of mesozooplankton biomass in these waters (Table S9). In coastal waters, *N. scintillans* had a higher mean biomass compared to transitional and marine waters, suggesting frequent blooms. Copepoda and Cladocera exhibited similar patterns to those in transitional waters. Meroplankton and other groups reflected a diverse zooplankton community, with moderate variability in total mesozooplankton biomass (Table S9). *N. scintillans* in marine waters recorded the lowest mean biomass compared to transitional and coastal waters, with relatively stable populations of Copepoda. Cladocera and meroplankton had lower mean biomasses, reflecting their lower abundance in marine environments (Table S9). Overall, the mesozooplankton biomass in marine waters is the lowest among the three marine reporting units, indicating less variability compared to transitional and coastal waters (Table S9).

The non-metric multidimensional scaling (nMDS) analysis of Copepoda biomass during the warm seasons demonstrates significant spatial variability across the transitional, coastal, and marine waters (Figure 9). The data revealed that the transitional and coastal waters have higher biomasses, indicative of more nutrient-rich and ecologically favorable conditions for Copepoda. Conversely, the marine waters, characterized by lower biomass (Figure 9), reflect distinct ecological dynamics and nutrient limitations.

Mesozooplankton biomass clusters tightly together within marine environments, indicating a homogeneous composition (Figure 10). In contrast, coastal waters form a distinct cluster separate from the marine samples, suggesting a different biomass composition. Transitional waters are more dispersed across the plot, indicating a higher variability in biomass composition (Figure 10). Distinct clustering patterns for marine, coastal, and transitional zones highlight the ecological differentiation among these regions. The overlap observed with transitional zones underscores their role as intermediary areas with mixed mesozooplankton communities.

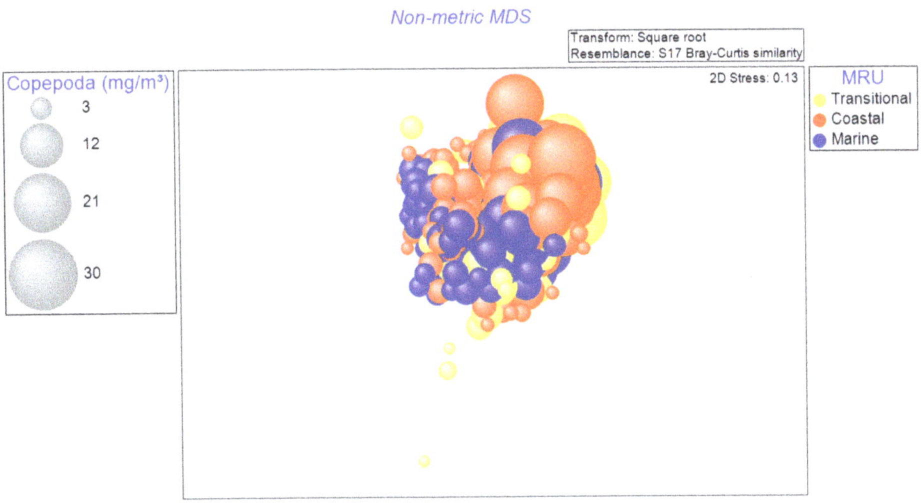

Figure 9. Non-metric multidimensional scaling (n-MDS) plot of Copepoda biomass during the warm seasons of 2013–2020.

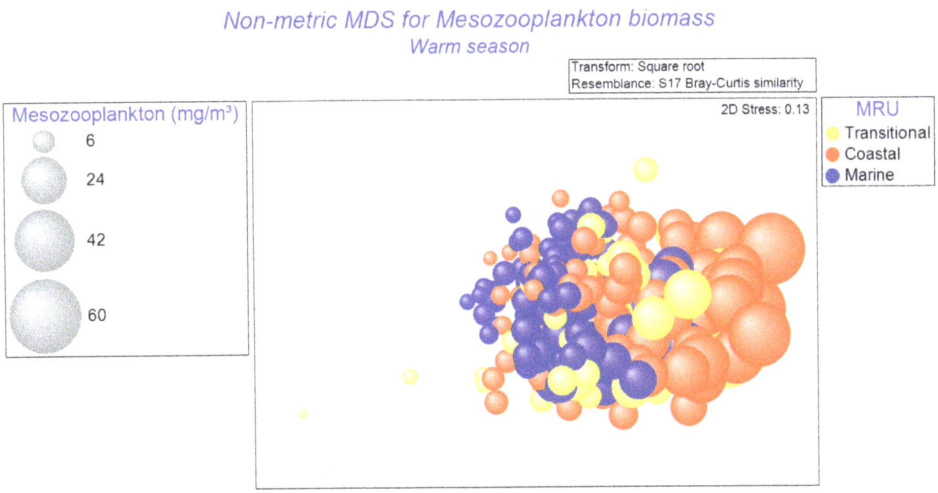

Figure 10. Non-metric multidimensional scaling (n-MDS) plot of Mesozooplankton biomass during the warm seasons of 2013–2020.

The clustering of high amounts of biomass in transitional and coastal areas for *N. scintillans* underscores the influence of nutrient inputs and environmental variability, while the dispersed, lower amount of biomass in marine areas highlights the more stable and oligotrophic nature of this environment (Figure 11). Several studies have examined the role of *N. scintillans* in nutrient-rich environments [53,54]. This dinoflagellate plays a crucial role as both a contributor and source of nutrients, due to its ability to regenerate and release significant amounts of ammonium and inorganic phosphorus, which can fuel primary phytoplankton production [53,54,68]. biomasses of *N. scintillans* may indicate imbalances in the ecosystem, as it competes with other zooplankton for resources and alters predator–prey dynamics [29,66,69,70]. Therefore, the trends in its biomass not only reflect eutrophication but also provide insight into broader ecosystem health and the need for nutrient management strategies [66].

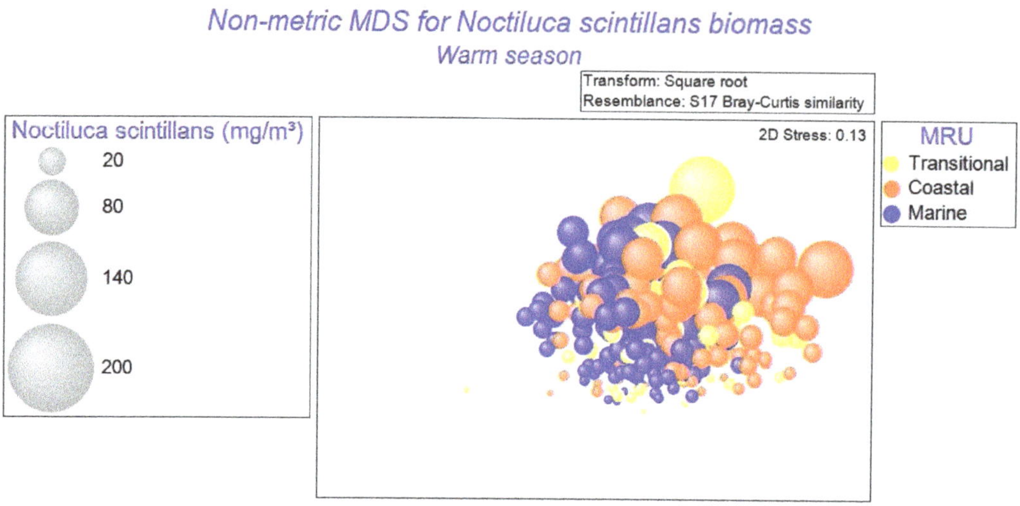

Figure 11. Non-metric multidimensional scaling (n-MDS) plot of *Noctiluca scintillans* biomass during the warm seasons of 2013–2020.

3.3. Mesozooplankton and the Environmental Variables During Warm Seasons of 2013–2020

The first principal component (PC1) in Principal Component Analysis (PCA) accounts for 23.42% of the total variance, capturing the most significant variation within the dataset. The second principal component (PC2) explains 17.56% of the total variance, representing the second most significant variation. Combined, PC1 and PC2 account for 40.98% of the total variance, indicating a substantial representation of the data's variability (Figure 12). The PCA reveals significant correlations among the environmental variables and their interactions with biological factors, providing insights into their interrelationships. The vectors for NO_3 [μM], PO_4 [μM], and SiO_4 [μM] form small angles with each other, indicating a strong positive correlation, particularly for phosphate and silicate which have the same main source—the Danube River's input. This suggests that these nutrients co-vary due to common sources or similar biogeochemical processes, such as agricultural runoff or upwelling events, which can enhance mesozooplankton, Copepoda, and Cladocera populations by increasing food availability. The nearly orthogonal vectors for temperature (T[°C]) and salinity (S[‰]) indicate little to no correlation between these two parameters, suggesting that their variations are influenced by independent factors such as different water masses or seasonal changes. This orthogonality implies that temperature and salinity independently affect the distribution and abundance of mesozooplankton and other marine organisms. Furthermore, the biological groups, including mesozooplankton, Copepoda, and Cladocera, show positive correlations with each other, as indicated by their acute angles. This correlation can be attributed to their similar ecological niches or food web dynamics, where favorable environmental conditions simultaneously boost the populations of these groups. For instance, higher nutrient levels (NO_3, PO_4, SiO_4) can lead to increased primary production, thereby supporting biomasses of these zooplankton groups.

The results also revealed significant correlations between environmental variables and mesozooplankton groups across three marine reporting units (Table S10). Transitional and coastal waters tend to show stronger correlations between nutrients (e.g., nitrate, silicate, phosphate) and zooplankton biomass while marine waters exhibited weaker correlations, suggesting that nutrient limitations might play a more significant role in controlling zooplankton populations.

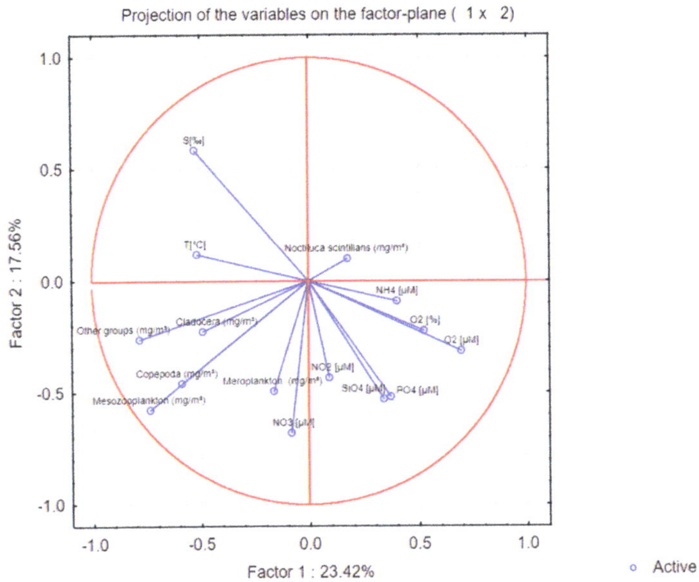

Figure 12. Analysis of the main components (PCA) for mesozooplankton and the environmental factors from the Romanian Black Sea during the warm seasons of 2013–2020.

3.4. Assessment of Black Sea Environmental Status Using Mesozooplankton Indicators for MSFD During the Warm Seasons of 2013–2020

Based on the biomass values of each indicator and by comparing them to the established threshold value, a comprehensive assessment of the Black Sea's environmental status for Descriptor 1 and Descriptor 5 within the MSFD framework was conducted.

The MSFD requires reporting the percentage or square kilometers of surface area that have a Good Environmental Status (GES) for each major habitat within a marine region [9]. Therefore, each marine reporting unit's percentage was established based on the obtained values for each indicator. Based on expert judgment, if more than 50% of the values fall within the established threshold, then the marine reporting unit is considered to have a good environmental status [71].

For Copepoda biomass, Black Sea marine waters predominantly exhibited non-GES, with 60% of the biomass falling into this category, indicating significant ecological stress that needs urgent attention to improve their health (Figure 13). Black Sea coastal waters were in even worse condition, with 62% of the biomass classified as non-GES, highlighting environmental pressure and the necessity for immediate conservation and remediation efforts. In contrast, transitional waters were in relatively better condition, with 56% of the biomass achieving GES, though a substantial portion still requires improvement (Figure 13). The disparity between GESs and non-GES in these regions underscores the need for targeted environmental management strategies. Efforts should prioritize reducing pollution, protecting habitats, and implementing sustainable practices to enhance the ecological health of these waters.

For the Mesozooplankton biomass indicator, Black Sea marine waters showed a nearly balanced distribution between GESs and non-GES, with a slight preference towards GESs for mesozooplankton biomass (Figure 14). This suggests a relatively stable environmental condition, though some areas still need attention. In contrast, Black Sea coastal waters had a higher proportion of non-GES, indicating that over half of these waters are experiencing environmental stress or degradation (Figure 14). This highlights the need for targeted conservation and remediation efforts in coastal regions. Black Sea transitional waters had the highest percentage of non-GES, pointing to significant environmental challenges due to the riverine input.

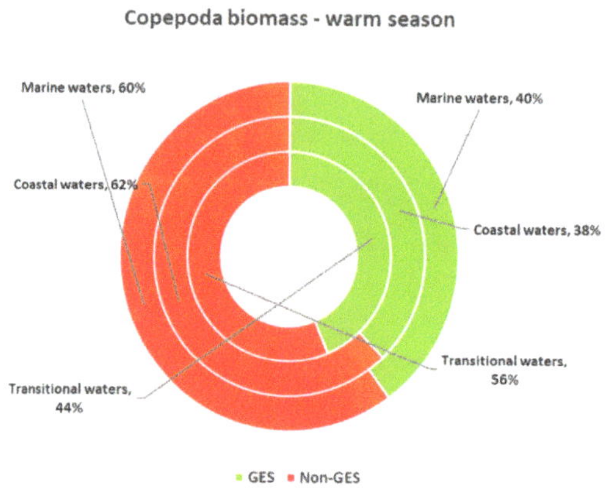

Figure 13. Evaluation of the Black Sea's environmental status based on the Copepoda biomass indicator during the warm seasons of 2013–2020.

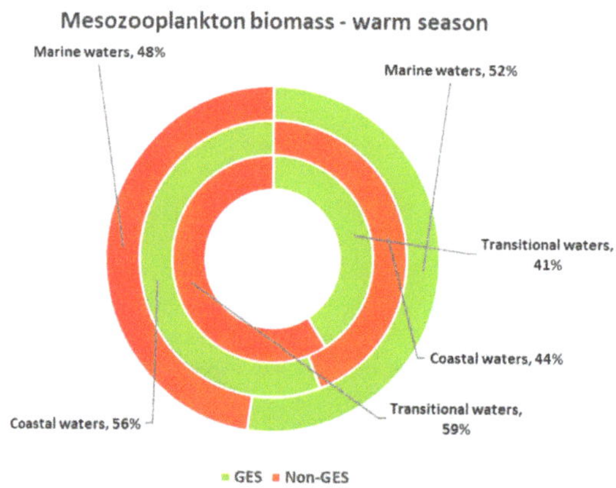

Figure 14. Evaluation of Black Sea environmental status based on the Mesozooplankton biomass indicator during the warm seasons of 2013–2020.

The distribution of *N. scintillans* biomass reveals varied environmental conditions across different water types. Black Sea marine waters exhibited a strong predominance of GES, indicating lower anthropogenic pressures. Black Sea coastal waters presented a mixed scenario, with 41% of the biomass classified as Non-GES, indicating moderate environmental stress that is likely due to pollution, coastal development, and other human activities (Figure 15). The 59% GES shows that while some areas maintain a good environmental status, there is still a need for focused conservation strategies. Black Sea transitional waters showed the highest percentage, with 79% of the biomass achieving GES (Figure 15).

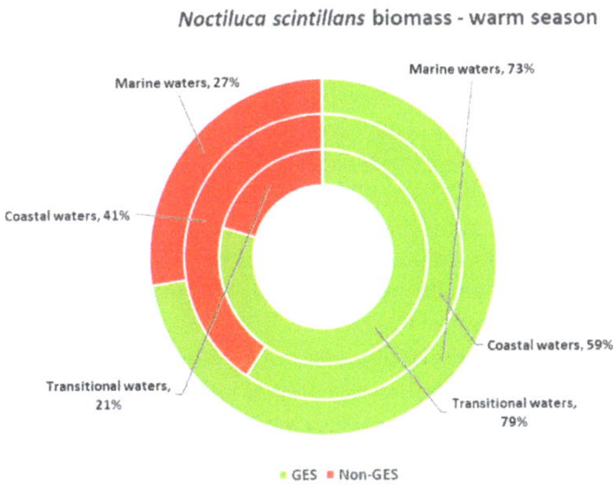

Figure 15. Evaluation of Black Sea environmental statuses based on the *Noctiluca scintillans* biomass indicator during the warm seasons of 2013–2020.

4. Discussion

4.1. Environmental Factors

The findings on temperature, salinity, dissolved oxygen, and nutrient concentrations in the Black Sea's waters reveal a consistent pattern of significant variability during the warm season of 2013–2020. This dynamic nature of the Black Sea ecosystem is influenced by a complex interplay of factors, including geographic location, hydrological processes, anthropogenic activities, and climatic conditions. The observed ranges in temperature and salinity across all three zones underscore the Black Sea's dynamic character, with transitional waters being particularly susceptible to fluctuations due to riverine discharges [43]. The temperature distribution has ecological implications, as different temperature zones can affect the dissolved oxygen content, marine life diversity and productivity, while also impacting local human activities such as fishing and tourism [72,73]. The salinity gradient reflects the decreasing influence of freshwater inputs as one moves away from the coast, with marine waters showing a more typical salinity for open sea conditions [74–76]. The variation in salinity across this region plays a crucial role in shaping the local marine ecosystems, influencing species distribution, water density, and the overall health of the marine environment [77,78].

Coastal waters also exhibit variability that is influenced by human activities, atmospheric conditions, and local circulation patterns, while marine waters display variability attributed to deep-water processes, currents, and seasonal changes [43,79]. The significant variability in dissolved oxygen and oxygen saturation levels is a critical concern, as the Black Sea is known for hypoxia, with fluctuations linked to biological and physical factors [80,81]. This pattern highlights the dynamic nature of oxygen levels in transitional zones compared to the more stable conditions in coastal and marine waters. Marine waters exhibit a similar median oxygen saturation but with less variability, suggesting more stable conditions, although outliers indicate occasional oxygen extremes [66]. Nutrient concentrations exhibit wide ranges, influenced by riverine input, and anthropogenic activities, with implications for eutrophication and ecosystem dynamics [81]. Understanding these differences is crucial for assessing the ecological health and resilience of these environments, as well as for managing the diverse biological communities they support. The variability in nutrient concentrations is a key factor that can influence primary production and the structure of food webs across these environments [66].

4.2. Mesozooplankton's Qualitative Composition and Biomass Structure

The study of the mesozooplankton community in the Romanian Black Sea during the warm seasons of 2013–2020 revealed diverse taxa distributions across different marine reporting units, with a total of 27 identified taxa. Cladocera and meroplankton exhibited moderate diversity and even distributions, underscoring their vital roles in transitional waters [29]. *N. scintillans* w prevalent in all reporting units, particularly in coastal areas, aligning with its preference for nutrient-rich waters [54,67,82]. Copepoda showed variability in species numbers, with a maximum of nine species, with *A. clausi*, and *P. elongatus* consistently observed from 2013 to 2020, as shown in other studies [29,83,84]. Though less diverse, other groups were found across all zones and included *P. setosa*, *O. dioica*, and *M. slabberi*.

The Shannon–Wiener diversity index (H′) for mesozooplankton was found to be more than 2 in all marine reporting units, with the highest values in transitional and coastal waters. The high diversity and evenness in the transitional and coastal waters underscore the importance of nutrient input and environmental variability in maintaining mesozooplankton, while the lower diversity in the marine area reflects stable, nutrient-poor conditions. Coastal zones generally support higher biomass levels, highlighting their productivity and ecological significance [85]. The high similarity index (~90%) between coastal and transitional waters indicates strong ecological connections due to similar environmental factors like nutrient availability and freshwater input [86–88]. This pattern is likely influenced by similar environmental factors, such as nutrient availability and freshwater input from the Danube River, which impact both zones. In contrast, marine waters exhibit a lower similarity index (~75%), indicating distinct mesozooplankton and biomass distributions. This lower similarity suggests a distinct mesozooplankton structure and biomass distribution in the marine zone. These differences can be attributed to the more stable but nutrient-poor conditions in the marine zone compared to the more variable and nutrient-rich environments of the coastal and transitional zones.

Nutrient-rich conditions in coastal and transitional waters support higher primary productivity, influencing larger and more diverse mesozooplankton populations [85]. Freshwater input from rivers like the Danube also impacts salinity and nutrient levels, shaping mesozooplankton [89]. Mesozooplankton in transitional waters exhibit notable variability, characterized by occasional high concentrations of *N. scintillans* and varying Copepoda and Cladocera biomass levels. Meroplankton and other groups also contribute to the overall variability, reflecting dynamic ecological conditions [90]. *N. scintillans* forms frequent blooms in coastal waters, leading to a higher mean biomass than in transitional and marine waters. Copepoda and Cladocera show similar distribution patterns, while a diverse array of meroplankton taxa contributes to moderate variability in total mesozooplankton biomass [29,67,91]. This pattern is likely influenced by similar environmental factors, such as nutrient availability and freshwater input from the Danube River, which impact both zones [92]. Conversely, marine waters exhibit the lowest mean biomass of *N. scintillans*, with relatively stable populations of Copepoda and lower biomasses of Cladocera and meroplankton, reflecting oligotrophic conditions [93–95].

Non-metric multidimensional scaling (nMDS) analysis highlights significant spatial variability in Copepoda biomasses across transitional, coastal, and marine waters during the warm seasons. Transitional and coastal waters consistently exhibit higher biomass, indicative of favorable nutrient conditions for Copepoda, whereas the marine waters show lower biomass due to ecological dynamics and nutrient limitations [79]. This spatial distribution underscores the influence of nutrient availability and environmental variability on mesozooplankton dynamics. Understanding these patterns is crucial for effective marine resource management and conservation efforts [96,97].

4.3. Mesozooplankton and Environmental Variables

The Black Sea is a unique and sensitive marine environment where changes in temperature, salinity, nutrient levels, and oxygen concentration can have pronounced effects on the mesozooplankton community [98,99]. Mesozooplankton are vital components of the marine food web, linking primary producers with higher trophic levels, including commercially important fish species [27]. Therefore, understanding how environmental factors influence these organisms is essential for predicting ecosystem responses to natural variability and human activities [100,101]. Effective management strategies must consider these interactions to mitigate adverse impacts, such as eutrophication and hypoxia, and to preserve the ecological balance and biodiversity of the Black Sea [31,102].

The Principal Component Analysis (PCA) effectively illustrates the interactions between various environmental factors and the mesozooplankton in the Black Sea. Temperature (T[°C]) and salinity (S[‰]) are two fundamental abiotic factors that significantly shape the mesozooplankton structure and are major drivers of the observed data variance. In the Black Sea, temperature stratification and salinity gradients are influenced by riverine inputs, seasonal changes, and water exchange with the Mediterranean Sea through the Bosphorus Strait [79]. Nutrient levels, including those of ammonium (NH_4), nitrate (NO_3), phosphate (PO_4), and silicate (SiO_4), are key factors in the second principal component (PC2), highlighting their importance in the ecosystem. The clustering of a few nutrient vectors (NO_3 [μM], PO_4 [μM], and SiO_4 [μM]) also suggests interconnected nutrient cycling and sources, indicating that changes in one nutrient could predict changes in the others. This interconnectedness has direct implications for the growth and reproduction of biotic communities, as nutrient availability is a key factor in primary and secondary production [103]. Nutrient levels directly drive phytoplankton growth and zooplankton populations by affecting food availability [104]. These interconnected factors create complex trophic interactions, where changes in one can cascade throughout the ecosystem, influencing higher trophic levels and overall ecosystem health [66]. The interactions are often non-linear, with shifts in one parameter triggering cascading effects across the entire system [105,106]. Understanding these processes is essential for predicting and managing planktonic communities' responses to environmental changes, supporting aquatic ecosystem conservation and sustainable management. Additionally, the independence of temperature and salinity variations, as highlighted by their orthogonal vectors, suggests that these parameters should be modeled separately to capture their distinct impacts on aquatic ecosystems accurately. Understanding these correlations between abiotic and biotic factors aids in better ecological modeling and informed environmental management practices, facilitating the maintenance of healthy marine ecosystems.

This study's findings highlight the varied responses of zooplankton groups and species to environmental factors in different marine regions, which is consistent with observations from other studies. For example, in transitional waters, the mixed responses of zooplankton to nitrate and silicate are consistent with previous research indicating that transitional zones often have fluctuating nutrient levels, which can lead to a diverse range of zooplankton communities [32,67,93,107,108]. The negative correlation between meroplankton and temperature, alongside their positive relationship with nitrate, suggests that nutrient availability rather than temperature primarily influences their distribution in these areas, which is a pattern also seen in other studies from the North Sea, Baltic Sea, and Black Sea [109–114].

In coastal regions, the strong positive correlation between Copepoda and nitrate (0.68) reflects nutrient-driven productivity, which is a common finding in coastal ecosystem studies [29,115]. The negative correlation between meroplankton and salinity (−0.29) supports other research that has demonstrated the influence of salinity gradients on zooplankton distributions in coastal waters [75,76,78,116]. Additionally, the positive correlation of other groups with temperature (0.40) and salinity (0.37), along with the negative correlation with oxygen (−0.43), suggests that these factors are critical in shaping coastal zooplankton communities.

In marine regions, the positive correlations of *N. scintillans* with oxygen (0.27) and oxygen saturation (0.37) [53,54,70,117], and the relationships of meroplankton with phosphate (0.23), silicate (0.21), and nitrate (0.35), highlight the influence of well-oxygenated and nutrient-rich conditions that are typical of open marine environments [79,81,95,118]. This aligns with other studies that have shown similar environmental dependencies for marine zooplankton [21,66,119–122]. The negative correlations of other groups with oxygen

(−0.33) and oxygen saturation (−0.27) further emphasize the complexity of factors affecting zooplankton distributions in marine areas [19,21,123–125].

The observed differences emphasize how mesozooplankton adapt to specific environmental conditions across MRUs, with temperature, oxygen, and nutrient availability emerging as key variables that shape the community. For example, increases in nutrient inputs in coastal zones could lead to shifts in the zooplankton community structure, favoring species that thrive in nutrient-rich, low-oxygen conditions, such as Copepoda [19,21,123–125]. In contrast, the transitional and marine zones may respond more strongly to changes in temperature and oxygen availability, reflecting broader climate-driven shifts in oceanography. Changes in mesozooplankton can have cascading effects on higher trophic levels, including commercially important fish species [27,126]. Understanding the interactions between environmental factors and mesozooplankton is essential for predicting ecosystem responses and implementing effective management strategies.

4.4. Assessment of Black Sea Ecological Status Using Mesozooplankton Indicators for MSFD

The Marine Strategy Framework Directive (MSFD) [9] assessment for Descriptor 1 (Biodiversity—Pelagic Habitats) and Descriptor 5 (Eutrophication) in the Black Sea reveals significant insights into the ecological status of the region. Assessing Good Environmental Status (GES) across marine, coastal, and transitional waters reveals distinct patterns that reflect varying environmental conditions and the influence of human activities.

For the Mesozooplankton biomass indicator, marine waters demonstrate a nearly balanced distribution between GESs and non-GES, indicating relatively stable environmental conditions but also areas requiring targeted conservation efforts to maintain their ecological integrity. This observation aligns with studies that have highlighted the mixed impacts of anthropogenic activities on marine environments. In the Black Sea, similar patterns have been observed, with significant human impacts such as eutrophication, overfishing, and pollution affecting marine ecosystems [35,127]. In contrast, coastal waters exhibit a higher proportion of non-GES, highlighting significant environmental stressors such as pollution and coastal development. This underscores the urgent need for focused conservation and remediation actions to mitigate anthropogenic impacts and restore these critical coastal ecosystems. Studies have consistently shown that coastal areas in the Black Sea are more vulnerable to human activities due to their proximity to urban and industrial centers, leading to increased nutrient loading, habitat modification, and contamination [31,34].

The Copepoda biomass indicator underscores considerable environmental challenges across different water types. Marine and coastal waters, in particular, display high percentages of non-GES, pointing to ecological stress that is likely caused by pollution, habitat degradation, overfishing, and climate change. This pattern is consistent with findings that copepod populations are sensitive indicators of environmental change and can reflect broader ecosystem health [15,24–26,52,128]. In the Black Sea, copepod communities have been significantly impacted by environmental changes, including shifts in species composition due to eutrophication and invasive species [129]. The non-GES for transitional waters underscores the complex environmental challenges faced by transitional zones, emphasizing the importance of comprehensive and basin-scale management strategies to address both localized stressors and broader ecological impacts.

The *N. scintillans* biomass provides further insight into these environmental dynamics, with marine waters predominantly achieving a Good Environmental Status (GES), which suggests reduced anthropogenic pressures. This aligns with research indicating that some marine areas benefit from protection measures and lower human impact, resulting in a better ecological status [66]. Coastal waters present a mixed scenario, with a significant proportion classified as non-GES, necessitating targeted efforts to enhance resilience. The findings reveal that the Black Sea continues to face eutrophication issues [66], particularly in coastal areas influenced by agricultural runoff and wastewater discharge. High nutrient loads have resulted in frequent algal blooms, which negatively impact marine life and water quality [66]. The GES in Black Sea transitional waters can be linked to changes in the Danube River due to hydrological constructions, which significantly regulated its flow regime, decreasing nutrient loads in the Black Sea [93].

The Ukraine–Russian conflicts in the Black Sea region have highlighted the substantial effects of military operations on marine biodiversity and conservation efforts. These conflicts has led to both immediate and long-term environmental consequences in the area [130–132]. Direct impacts, such as oil spills, the release of toxic substances, and habitat destruction, have adversely affected marine ecosystems, resulting in the loss of vital breeding grounds [130,132]. Furthermore, the delicate balance of marine life is increasingly threatened by indirect consequences stemming from intensified fishing activities, heightened maritime traffic, and the development of coastal infrastructure driven by the conflict [130,131]. Significant changes in ecosystem components have resulted in declining ecosystem health, characterized by transitions from naturally diverse habitats to areas with reduced biodiversity. This has facilitated the spread of invasive species and increased occurrences of harmful events, such as harmful algal blooms (HABs) and jellyfish outbreaks [132]. These shifts present considerable challenges for the management of human activities within these ecosystems.

Future research and monitoring efforts should prioritize identifying and mitigating the most critical stressors, thereby enhancing the resilience and health of marine ecosystems in the face of ongoing environmental changes. This perspective is supported by calls for integrated coastal zone management and adaptive management strategies that take into account the dynamic and interconnected nature of marine ecosystems. In the Black Sea, integrated management approaches have been emphasized to address the diverse and cumulative impacts of human activities, aiming to restore and protect marine and coastal environments [34].

5. Conclusions

This assessment aims to improve the understanding of ecological dynamics in the Black Sea and to guide future conservation and policy decisions. It highlights the dynamic nature of pelagic habitats and stresses the need for continuous monitoring to inform effective conservation strategies. Changes in zooplankton abundance and distribution serve as key indicators of ecosystem health, making them essential for marine resource management. Coastal areas, such as the Romanian Black Sea, are particularly vulnerable to anthropogenic pressures from industry, tourism, and urbanization, which contribute to habitat degradation and pollution.

The findings of this study reinforce the importance of targeted conservation measures, such as reducing nutrient inflows and promoting sustainable coastal development, to mitigate these impacts. This is especially crucial in the face of climate change, which causes shifts in temperature and salinity, further affecting marine ecosystems. Rising sea temperatures disrupt species' reproductive cycles, while pollution and nutrient runoff fuel eutrophication, leading to harmful algal blooms and oxygen-depleted dead zones. These combined stressors threaten biodiversity, fisheries, and the overall health of marine habitats.

Ongoing research and enhanced monitoring of mesozooplankton and environmental conditions are essential for adaptive management, ensuring the long-term health of marine environments. By providing valuable insights into ecosystem health and the success of conservation efforts, such monitoring helps preserve ecological integrity and build resilience against environmental changes. The intersection of climate change and pollution demands a proactive approach, where adaptive management and informed policy can address the compounded threats to marine biodiversity and ecosystem stability.

Supplementary Materials: The following supporting information can be downloaded at: https://www.mdpi.com/article/10.3390/oceans5040053/s1: Table S1. Descriptive statistics of mesozooplankton biomass values (mg/m^3), on marine reporting units and warm season from 1960 to 2002; Table S2. Descriptive statistics of Copepoda biomass values (mg/m^3) on marine reporting units and warm season from 1960 to 2002; Table S3. Descriptive statistics of *Noctiluca scintillans* biomass values (mg/m^3) on marine reporting units and war seasons from 1960 to 2002;

Table S4. Threshold values for defining the ecological status according to Copepoda biomass (mg/m^3); Table S5. Threshold values for defining the ecological status according to Mesozooplankton biomass (mg/m^3); Table S6. Threshold values for defining the ecological status according to *Noctiluca scintillans* biomass (mg/m^3); Table S7. Descriptive statistics of environmental parameters in the warm season of 2013–2020; Table S8. List of identified mesozooplankton taxa in the Black Sea during the warm season of 2013–2020; Table S9. Descriptive statistics of mesozooplankton biomass in the warm season of 2013–2020; Table S10. Statistically significant correlation ($p < 0.05$) between the mesozooplankton groups and environmental factors in the warm season of 2013–2020.

Author Contributions: Conceptualization, E.B. and L.L.; methodology, E.B.; software, E.B. and L.L.; validation, E.B. and L.L.; formal analysis, E.B. and L.L.; investigation, E.B. and L.L.; resources, E.B. and L.L; data curation, E.B. and L.L.; writing—original draft preparation, E.B. and L.L.; writing—review and editing, E.B. and L.L.; visualization, E.B. and L.L.; supervision, E.B. and L.L.; project administration, E.B. and L.L.; funding acquisition, E.B. and L.L. All authors have read and agreed to the published version of the manuscript.

Funding: This research was funded by the Nucleu Programme INTELMAR 2019–2022 funded by the Ministry of Research, Innovation, and Digitization, grant number 45N/2019 and by the Nucleu Programme SMART-BLUE 2023–2026 funded by the Ministry of Research, Innovation, and Digitization, grant number 33N/2023, PN23230201 and PN23230103.

Institutional Review Board Statement: Not applicable.

Informed Consent Statement: Not applicable.

Data Availability Statement: The data belong to the National Institute for Marine Research and Development "Grigore Antipa" (NIMRD) and can be accessed by upon request to http://www.nodc.ro/data_policy_nimrd.php, accessed on 1 August 2024.

Acknowledgments: We would like to extend our gratitude to all those who contributed to the success of this study: to Timofte Florin, Tabarcea Cristina, and Harcota George, whose expertise was important in developing the mesozooplankton indicators.

Conflicts of Interest: The authors declare no conflicts of interest. The funders had no role in the design of the study; in the collection, analyses, or interpretation of data; in the writing of the manuscript; or in the decision to publish the results.

References

1. Magliozzi, C.; Palma, M.; Druon, J.-N.; Palialexis, A.; Abigail, M.-G.; Ioanna, V.; Rafael, G.-Q.; Elena, G.; Birgit, H.; Laura, B.; et al. Status of Pelagic Habitats within the EU-Marine Strategy Framework Directive: Proposals for Improving Consistency and Representativeness of the Assessment. *Mar. Policy* **2023**, *148*, 105467. [CrossRef]
2. Yao, L.; He, P.; Xia, Z.; Li, J.; Liu, J. Typical Marine Ecological Disasters in China Attributed to Marine Organisms and Their Significant Insights. *Biology* **2024**, *13*, 678. [CrossRef] [PubMed]
3. Kleitou, P.; Crocetta, F.; Giakoumi, S.; Giovos, I.; Hall-Spencer, J.M.; Kalogirou, S.; Kletou, D.; Moutopoulos, D.K.; Rees, S. Fishery Reforms for the Management of Non-Indigenous Species. *J. Environ. Manag.* **2021**, *280*, 111690. [CrossRef]
4. Halpern, B.S.; Frazier, M.; Potapenko, J.; Casey, K.S.; Koenig, K.; Longo, C.; Lowndes, J.S.; Rockwood, R.C.; Selig, E.R.; Selkoe, K.A.; et al. Spatial and Temporal Changes in Cumulative Human Impacts on the World's Ocean. *Nat. Commun.* **2015**, *6*, 7615. [CrossRef] [PubMed]
5. Ducklow, H.; Cimino, M.; Dunton, K.H.; Fraser, W.R.; Hopcroft, R.R.; Ji, R.; Miller, A.J.; Ohman, M.D.; Sosik, H.M. Marine Pelagic Ecosystem Responses to Climate Variability and Change. *Bioscience* **2022**, *72*, 827–850. [CrossRef]
6. Lazar, L.; Boicenco, L.; Pantea, E.; Timofte, F.; Vlas, O.; Bișinicu, E. Modeling Dynamic Processes in the Black Sea Pelagic Habitat—Causal Connections between Abiotic and Biotic Factors in Two Climate Change Scenarios. *Sustainability* **2024**, *16*, 1849. [CrossRef]
7. Lazar, L.; Spanu, A.; Boicenco, L.; Oros, A.; Damir, N.; Bisinicu, E.; Abaza, V.; Filimon, A.; Harcota, G.; Marin, O.; et al. Methodology for Prioritizing Marine Environmental Pressures under Various Management Scenarios in the Black Sea. *Front. Mar. Sci.* **2024**, *11*, 1388877. [CrossRef]
8. Magliozzi, C.; Druon, J.; Palialexis, A.; Aguzzi, L.; Antoniadis, K.; Artigas, L.F.; Azzellino, A.; Bisinicu, E.; Boicenco, L.; Bojanić, N.; et al. *Pelagic Habitats Under MSFD D1 Scientific Advice of Policy Relevance: Recommendations to Frame Problems and Solutions for the Pelagic Habitats' Assessment*; Publications Office of the European Union: Luxembourg, 2021; ISBN 9789276359586.
9. European Commission. Commission Decision (EU) 2017/848 of 17 May 2017 Laying down Criteria and Methodological Standards on Good Environmental Status of Marine Waters and Specifications and Standardised Methods for Monitoring and Assessment, and Repealing Decision 2010/477/EU. *Off. J. Eur. Union* **2017**, *125*, 43–74.

10. European Commission. *Report on the Implementation of the Marine Strategy Framework Directive*; European Commission: Brussels, Belgium, 2020.
11. Borja, Á.; Galparsoro, I.; Irigoien, X.; Iriondo, A.; Menchaca, I.; Muxika, I.; Pascual, M.; Quincoces, I.; Revilla, M.; Germán Rodríguez, J.; et al. Implementation of the European Marine Strategy Framework Directive: A Methodological Approach for the Assessment of Environmental Status, from the Basque Country (Bay of Biscay). *Mar. Pollut. Bull.* **2011**, *62*, 889–904. [CrossRef]
12. Steinberg, D.K.; Landry, M.R. Zooplankton and the Ocean Carbon Cycle. *Ann. Rev. Mar. Sci.* **2017**, *9*, 413–444. [CrossRef]
13. Bucklin, A.; Peijnenburg, K.T.C.A.; Kosobokova, K.N.; O'Brien, T.D.; Blanco-Bercial, L.; Cornils, A.; Falkenhaug, T.; Hopcroft, R.R.; Hosia, A.; Laakmann, S.; et al. Toward a Global Reference Database of COI Barcodes for Marine Zooplankton. *Mar. Biol.* **2021**, *168*, 78. [CrossRef]
14. Turner, J.T. The Importance of Small Planktonic Copepods and Their Roles in Pelagic Marine Food Webs. *Zool. Stud.* **2004**, *43*, 255–266.
15. Jeppesen, E.; Nõges, P.; Davidson, T.A.; Haberman, J.; Nõges, T.; Blank, K.; Lauridsen, T.L.; Søndergaard, M.; Sayer, C.; Laugaste, R.; et al. Zooplankton as Indicators in Lakes: A Scientific-Based Plea for Including Zooplankton in the Ecological Quality Assessment of Lakes According to the European Water Framework Directive (WFD). *Hydrobiologia* **2011**, *676*, 279–297. [CrossRef]
16. Kehayias, G.; Chalkia, E.; Doulka, E. Zooplankton Variation in Five Greek Lakes. In *Zooplankton*; Nova Science Publishers: New York, NY, USA, 2014.
17. Preston, N.D.; Rusak, J.A. Homage to Hutchinson: Does Inter-Annual Climate Variability Affect Zooplankton Density and Diversity? *Hydrobiologia* **2010**, *653*, 165–177. [CrossRef]
18. Caroni, R.; Irvine, K. The Potential Of Zooplankton Communities For Ecological Assessment Of Lakes: Redundant Concept Or Political Oversight? *Biol. Environ. Proc. R. Ir. Acad.* **2010**, *110*, 35–53. [CrossRef]
19. Richardson, A.J. In Hot Water: Zooplankton and Climate Change. *ICES J. Mar. Sci.* **2008**, *65*, 279–295. [CrossRef]
20. Irigoien, X.; Huisman, J.; Harris, R.P. Global Biodiversity Patterns of Marine Phytoplankton and Zooplankton. *Nature* **2004**, *429*, 863–867. [CrossRef]
21. Dam, H.G. Evolutionary Adaptation of Marine Zooplankton to Global Change. *Ann. Rev. Mar. Sci.* **2013**, *5*, 349–370. [CrossRef]
22. Ratnarajah, L.; Abu-Alhaija, R.; Atkinson, A.; Batten, S.; Bax, N.J.; Bernard, K.S.; Canonico, G.; Cornils, A.; Everett, J.D.; Grigoratou, M.; et al. Monitoring and Modelling Marine Zooplankton in a Changing Climate. *Nat. Commun.* **2023**, *14*, 564. [CrossRef]
23. Steinberg, D.K.; Lomas, M.W.; Cope, J.S. Long-term Increase in Mesozooplankton Biomass in the Sargasso Sea: Linkage to Climate and Implications for Food Web Dynamics and Biogeochemical Cycling. *Global Biogeochem. Cycles* **2012**, *26*, GB1004. [CrossRef]
24. Jakhar, P. Role of Phytoplankton and Zooplankton as Health Indicators of Aquatic Ecosystem: A Review. *Int. J. Innov. Res. Stud.* **2013**, *2*, 489–500.
25. Caroppo, C.; Buttino, I.; Camatti, E.; Caruso, G.; De Angelis, R.; Facca, C.; Giovanardi, F.; Lazzara, L.; Mangoni, O.; Magaletti, E. State of the Art and Perspectives on the Use of Planktonic Communities as Indicators of Environmental Status in Relation to the EU Marine Strategy Framework Directive. *Biol. Mar. Mediterr.* **2013**, *20*, 65–73.
26. Giordani, G.; Zaldívar, J.M.; Viaroli, P. Simple Tools for Assessing Water Quality and Trophic Status in Transitional Water Ecosystems. *Ecol. Indic.* **2009**, *9*, 982–991. [CrossRef]
27. Lomartire, S.; Marques, J.C.; Gonçalves, A.M.M. The Key Role of Zooplankton in Ecosystem Services: A Perspective of Interaction between Zooplankton and Fish Recruitment. *Ecol. Indic.* **2021**, *129*, 107867. [CrossRef]
28. Damir, N.A.; Coatu, V.; Pantea, E.D.; Galațchi, M.; Botez, E.; Birghilă, S. Assessment of Polycyclic Aromatic Hydrocarbons Content in Marine Organisms of Commercial Interest from the Romanian Black Sea Coast. *Polycycl. Aromat. Compd.* **2022**, *42*, 7595–7606. [CrossRef]
29. Bișinicu, E.; Lazăr, L.; Timofte, F. Dynamics of Zooplankton along the Romanian Black Sea Coastline: Temporal Variation, Community Structure, and Environmental Drivers. *Diversity* **2023**, *15*, 1024. [CrossRef]
30. Lazar, L.; Boicenco, L.; Moncheva, S. *Impact of the Rivers on the Black Sea Ecosystem*; CD Press: Constanta, Romania, 2021.
31. Kideys, A.E. Ecology: Fall and Rise of the Black Sea Ecosystem. *Science (1979)* **2002**, *297*, 1482–1484. [CrossRef]
32. Kharytonova, Y.V.; Nabokin, M.V.; Mgeladze, M.M.; Vadachkoria, P.A.; Dyadichko, V.G. Current State and Long-Term Changes in the Mesozooplankton Community of the Ukrainian and Georgian Parts of the Black Sea as Indicators of Its Ecological Status. *Biosyst. Divers.* **2021**, *29*, 47–58. [CrossRef]
33. Shiganova, T.; Musaeva, E.; Arashkevich, E.; Shirshov, P.P.; Kamburska, L.; Stefanova, K.; Mihneva, V.; Polishchuk, L.; Timofte, F.; Ustun, F.; et al. Chapter 6 The State Of Zooplankton (T. Shiganova et al.). Available online: https://fishermenassociation.wordpress.com/wp-content/uploads/2014/10/biological-diversity-in-the-black-sea-1.pdf (accessed on 1 August 2024).
34. BSC. *State of the Environment of the Black Sea (2009-2014/5)*; BSC: Istanbul, Turkey, 2019.
35. Oguz, T. *State of the Environment of the Black Sea (2001–2006/7)*; Oguz, T., Ed.; Publications of the Commission on the Protection of the Black Sea Against Pollution (BSC): Istanbul, Turkey, 2008.
36. Zaitsev, Y.P. Recent Changes in the Trophic Structure of the Black Sea. *Fish. Oceanogr.* **1992**, *1*, 180–189. [CrossRef]
37. Zaitsev, Y.; Mamaev, V. *Biological Diversity in the Black Sea: A Study of Change and Decline*; United Nations: New York, NY, USA, 1997; ISBN 9211260426.

38. Shiganova, T.A.; Shirshov, P.P. Invasion of the Black Sea by the Ctenophore *Mnemiopsis leidyi* and Recent Changes in Pelagic Community Structure. *Fish. Oceanogr.* **1998**, *7*, 305–310. [CrossRef]
39. Shiganova, T.A.; Mikaelyan, A.S.; Moncheva, S.; Stefanova, K.; Chasovnikov, V.K.; Mosharov, S.A.; Mosharova, I.N.; M Mavrodieva, R.; Stefanova, E.; et al. Effect of Invasive Ctenophores *Mnemiopsis leidyi* and *Beroe ovata* on Low Trophic Webs of the Black Sea Ecosystem. *Mar. Pollut. Bull.* **2019**, *141*, 434–447. [CrossRef] [PubMed]
40. Shiganova, T.A.; Sommer, U.; Javidpour, J.; Molinero, J.C.; Malej, A.; Kazmin, A.S.; Isinibilir, M.; Christou, E.; Siokou-Frangou, I.; Marambio, M.; et al. Patterns of Invasive Ctenophore *Mnemiopsis leidyi* Distribution and Variability in Different Recipient Environments of the Eurasian Seas: A Review. *Mar. Environ. Res.* **2019**, *152*, 104791. [CrossRef] [PubMed]
41. Konsutov, A.S.; Kamburska, L.T. Ecological Determination of the New Ctenophora-*Beroe ovata* invasion in the Black Sea. *Oceanology* **1998**, *2*, 195–198.
42. Shiganova, T.A.; Dumont, H.J.; Mikaelyan, A.; Glazov, D.M.; Bulgakova, Y.V.; Musaeva, E.I.; Studenikina, E. Interaction between the Invading Ctenophores *Mnemiopsis leidyi* (A. Agassiz) and *Beroe ovata* Mayer 1912, and Their Influence on the Pelagic Ecosystem of the Northeastern Black Sea. In *Aquatic Invasions in the Black, Caspian, and Mediterranean Seas*; Niermann, U., Ed.; Kluwer Academic Publishers: Dordrecht, The Netherlands, 2004; Volume 35, pp. 33–70.
43. Bakan, G.; Büyükgüngör, H. The Black Sea. *Mar. Pollut. Bull.* **2000**, *41*, 24–43. [CrossRef]
44. Mihailov, M.E.; Nicolaev, S.; Buga, L.; Jelescu, S.; Boicenco, L.; Spinu, A.; Ganea, G. Identification of the Romanian Black Sea Waters Types—Assessment Related to the Marine Strategy Framework Directive Implementation. In Proceedings of the 14th International Multidisciplinary Scientific Geoconference SGEM, Albena, Bulgaria, 17–26 June 2014; pp. 623–630.
45. Alexandrov, B.; Arashkevich, E.; Gubanova, A.; Korshenko, A. *Manual for Mesozooplankton Sampling and Analysis in the Black Sea Monitoring*; Black Sea Commission: Istanbul, Turkey, 2014.
46. Mordukhay-Boltovskoy, F.D. (Ed.) *Identification Manual on the Fauna of the Black and Azov Seas*; Naukova Dumka: Kiev, Ukraine, 1968.
47. Mordukhay-Boltovskoy, F.D. (Ed.) *Guide of the Black Sea and the Sea of Azov Fauna*; Naukova Dumka: Kiev, Ukraine, 1972.
48. WoRMS Editorial Board World Register of Marine Species. Available online: https://www.marinespecies.org/ (accessed on 1 August 2024).
49. Petipa, T.S. On the Mean Weight of the Principle Forms of Zooplankton in the Black Sea. *Sevast. Biol. Station* **1957**, *9*, 39–57.
50. Grasshoff, K.; Kremling, K.; Ehrhardt, M. *Methods of Seawater Analysis*, 3rd ed.; Grasshoff, K., Kremling, K., Ehrhardt, M., Eds.; Willey-VCH: Weinheim, Germany, 1999.
51. Mullin, J.B.; Riley, J.P. The Spectrophotometric Determination of Nitrate in Natural Waters, with Particular Reference to Sea-Water. *Anal. Chim. Acta* **1955**, *12*, 464–480. [CrossRef]
52. Muñoz-Colmenares, M.E.; Soria, J.M.; Vicente, E. Can Zooplankton Species Be Used as Indicators of Trophic Status and Ecological Potential of Reservoirs? *Aquat. Ecol.* **2021**, *55*, 1143–1156. [CrossRef]
53. Elbrächter, M.; Qi, Z. *Aspects of Noctiluca (Dinophyceae) Population Dynamics, Physiological Ecology of Harmful Algal Blooms*; Anderson, D.M., Cembella, A.D., Hallegraeff, G.M., Eds.; Springer: Berlin/Heidelberg, Germany, 1998.
54. Aytan, U.; Senturk, Y. Dynamics of *Noctiluca scintillans* (Macartney) Kofoid & Swezy and Its Contribution to Mesozooplankton in the Southeastern Black Sea. *Aquat. Sci. Eng.* **2018**, *33*, 84–89. [CrossRef]
55. Clarke, K.R.; Gorley, R.N.; Somerfield, P.J. *Change in Marine Communities: An Approach to Statistical Analysis*, 3rd ed.; PRIMER-E: Plymouth, UK, 2014.
56. TIBCO Software, Inc. *TIBCO Statistica, version 14.0.1.25*; TIBCO Software, Inc.: Palo Alto, CA, USA, 2023.
57. Clarke, K.R. Nonmetric multivariate analysis in community-level ecotoxicology. *Environ. Toxicol. Chem. An. Int. J.* **1999**, *18*, 118–127.
58. Clarke, K.R.; Somerfield, P.J.; Gorley, R.N. Clustering in Non-Parametric Multivariate Analyses. *J. Exp. Mar. Biol. Ecol.* **2016**, *483*, 147–155. [CrossRef]
59. Clarke, K.R.; Warwick, R.M. Similarity Based Testing for Community Pattern the two-way layout with no replication. *Mar. Biol.* **1994**, *118*, 167–176. [CrossRef]
60. Clarke, K.R.; Somerfield, P.J.; Chapman, M.G. On Resemblance Measures for Ecological Studies, Including Taxonomic Dissimilarities and a Zero-Adjusted Bray-Curtis Coefficient for Denuded Assemblages. *J. Exp. Mar. Biol. Ecol.* **2006**, *330*, 55–80. [CrossRef]
61. Islam, M.S.; Azadi, M.A.; Nasiruddin, M.; Sarker, M.M. Plankton Species Composition, Abundance and Diversity Indices in Three Ponds of Chittagong University Campus, Bangladesh Water Quality and Plankton Diversity of the River Halda and Its Boalia Khal (Canal) Tributary View Project. *J. Biol. Pharm. Chem. Res.* **2022**, *9*, 1–14.
62. Clarke, K.R.; Tweedley, J.R.; Valesini, F.J. Simple Shade Plots Aid Better Long-Term Choices of Data Pre-Treatment in Multivariate Assemblage Studies. *J. Mar. Biol. Assoc. United Kingd.* **2014**, *94*, 1–16. [CrossRef]
63. ESRI. *ArcGIS Desktop, version 10.7*; Environmental Systems Research Institute: Redlands, CA, USA, 2019.
64. Al-Mamoori, S.K.; Al-Maliki, L.A.; Al-Sulttani, A.H.; El-Tawil, K.; Al-Ansari, N. Statistical Analysis of the Best GIS Interpolation Method for Bearing Capacity Estimation in An-Najaf City, Iraq. *Environ. Earth Sci.* **2021**, *80*, 683. [CrossRef]
65. Putra, M.; Lewis, S.; Kurniasih, E.; Prabuning, D.; Faiqoh, E. Plankton Biomass Models Based on GIS and Remote Sensing Technique for Predicting Marine Megafauna Hotspots in the Solor Waters. *IOP Conf. Ser. Earth Environ. Sci.* **2016**, *47*, 012015. [CrossRef]

66. Lazar, L.; Vlas, O.; Pantea, E.; Boicenco, L.; Marin, O.; Abaza, V.; Filimon, A.; Bisinicu, E. Black Sea Eutrophication Comparative Analysis of Intensity between Coastal and Offshore Waters. *Sustainability* **2024**, *16*, 5146. [CrossRef]
67. Bișinicu, E.; Boicenco, L.; Pantea, E.; Timofte, F.; Lazăr, L.; Vlas, O. Qualitative Model of the Causal Interactions between Phytoplankton, Zooplankton, and Environmental Factors in the Romanian Black Sea. *Phycology* **2024**, *4*, 168–189. [CrossRef]
68. Turkoglu, M. Red Tides of the Dinoflagellate *Noctiluca scintillans* Associated with Eutrophication in the Sea of Marmara (the Dardanelles, Turkey). *Oceanologia* **2013**, *55*, 709–732. [CrossRef]
69. Zhang, S.; Liu, H.; Glibert, P.M.; Guo, C.; Ke, Y. Effects of Prey of Different Nutrient Quality on Elemental Nutrient Budgets in Noctiluca Scintillans. *Nature* **2017**, *7*, 1–12. [CrossRef]
70. Zhang, S.; Submitted, T. Ecological Roles of *Noctiluca scintillans* in Marine Food Web—As a Predator, "Prey" and Nutrient Regenerator. Ph. D. Thesis, Hong Kong University of Science and Technology, Hong kong, China, 2016.
71. Boicenco, L.; Abaza, V.; Anton, E.; Bișinicu, E.; Buga, L.; Coatu, V.; Damir, N.; Diaconeasa, D.; Dumitrache, C.; Filimon, A.; et al. Studiu Privind Elaborarea Raportului Privind Starea Ecologică a Eco-943 Sistemului Marin Marea Neagră Conform Cerinţelor Art. 17 Ale Directivei Cadru Strategia 944 Pentru Mediul Marin (2008/56/EC). 2018. Available online: https://www.mmediu.gov.ro/app/webroot/uploads/files/STUDIU%20MSFD%20V1.9.pdf (accessed on 1 August 2024).
72. Zhao, Q.; Liu, S.; Niu, X. Effect of Water Temperature on the Dynamic Behavior of Phytoplankton–Zooplankton Model. *Appl. Math. Comput.* **2020**, *378*, 125211. [CrossRef]
73. Montagnes, D.J.S.; Franklin, D.J. Erratum: Effect of Temperature on Diatom Volume, Growth Rate, and Carbon and Nitrogen Content: Reconsidering Some Paradigms (Limnology and Oceanography 46 (2008–2018)). *Limnol. Oceanogr.* **2002**, *47*, 1272. [CrossRef]
74. Telesh, I.; Schubert, H.; Skarlato, S. Life in the Salinity Gradient: Discovering Mechanisms behind a New Biodiversity Pattern. *Estuar. Coast. Shelf Sci.* **2013**, *135*, 317–327. [CrossRef]
75. Anton-Pardo, M.; Armengol, X. Effects of Salinity and Water Temporality on Zooplankton Community in Coastal Mediterranean Ponds. *Estuar. Coast. Shelf Sci.* **2012**, *114*, 93–99. [CrossRef]
76. Hall, C.A.M.; Lewandowska, A.M. Zooplankton Dominance Shift in Response to Climate-Driven Salinity Change: A Mesocosm Study. *Front. Mar. Sci.* **2022**, *9*, 861297. [CrossRef]
77. Akimova, A.; Núñez-Riboni, I.; Kempf, A.; Taylor, M.H. Spatially-Resolved Influence of Temperature and Salinity on Stock and Recruitment Variability of Commercially Important Fishes in the North Sea. *PLoS ONE* **2016**, *11*, e0161917. [CrossRef]
78. Svetlichny, L.; Hubareva, E.; Kideys, A.; İşinibilir, M.; Shmeleva, A. Zooplankton Community State in the Northeastern Marmara Sea during Early Autumn with Comments on Mass Mortality of the Black Sea Species Due to the Salinity Gradient. *J. Black Sea/Mediterr. Environ.* **2006**, *12*, 213–231.
79. Stanev, E.V. Understanding Black Sea Dynamics. *Oceanography* **2005**, *18*, 56–75. [CrossRef]
80. Zaitsev, Y.P.; Alexandrov, B.G.; Berlinsky, N.A.; Zenetos, A. *Seas Around Europe: The Black Sea: An Oxygen-Poor Sea. Europe's Biodiversity: Biogeographical Regions and Seas*; European Environment Agency: Copenhagen, Denmark, 2002.
81. He, Y.; Stanev, E.V.; Yakushev, E.; Staneva, J. Black Sea Biogeochemistry: Response to Decadal Atmospheric Variability during 1960-2000 Inferred from Numerical Modeling. *Mar. Environ. Res.* **2012**, *77*, 90–102. [CrossRef]
82. Stefanova, E.S.; Stefanova, K.B.; Kozuharov, D.S. Mesozooplankton Assemblages in Northwestern Part of Black Sea. *Acta Zool. Bulg.* **2012**, *64*, 403–412.
83. Gubanova, A.; Altukhov, D.; Stefanova, K.; Arashkevich, E.; Kamburska, L.; Prusova, I.; Svetlichny, L.; Timofte, F.; Uysal, Z. Species Composition of Black Sea Marine Planktonic Copepods. *J. Mar. Syst.* **2014**, *135*, 44–52. [CrossRef]
84. Tabarcea, C.; Bisinicu, E.; Harcota, G.E.; Timofte, F.; Gomoiu, M.T. Zooplankton Community Structure And Dynamics Along The Romanian Black Sea Area In 2017. *J. Environ. Prot. Ecol.* **2019**, *20*, 742–752.
85. Nixon, S.W.; Buckley, B.A. "A Strikingly Rich Zone"—Nutrient Enrichment and Secondary Production in Coastal Marine Ecosystems. *Estuaries* **2002**, *25*, 782–796. [CrossRef]
86. Lønborg, C.; Müller, M.; Butler, E.C.V.; Jiang, S.; Ooi, S.K.; Trinh, D.H.; Wong, P.Y.; Ali, S.M.; Cui, C.; Siong, W.B.; et al. Nutrient Cycling in Tropical and Temperate Coastal Waters: Is Latitude Making a Difference? *Estuar. Coast. Shelf Sci.* **2021**, *262*, 107571. [CrossRef]
87. Zessner, M.; Van Gils, J. Nutrient Fluxes from the Danube Basin to the Black Sea. *Water Sci. Technol.* **2002**, *46*, 9–17. [CrossRef]
88. Panin, N.; Jipa, D. Danube River Sediment Input and Its Interaction with the North-Western Black Sea. *Estuar. Coast. Shelf Sci.* **2002**, *54*, 551–562. [CrossRef]
89. Parr, W.; Volovik, Y.; Nixon, S.; Lipan, I. *Improving the Understanding of the Danube River Impact on the Status of the Black Sea*; UNDP: Istanbul, Turkey, 2005.
90. Kovalev, A.V.; Mazzocchi, M.G.; Siokou-Frangou, I.; Kideys, A.E. Zooplankton of the Black Sea and the Eastern Mediterranean: Similarities and Dissimilarities. *Mediterr. Mar. Sci.* **2001**, *2*, 69–77. [CrossRef]
91. Bologa, A.S.; Sava, D. Present State and Evolution Trends of Biodiversity in the Black Sea: Decline and Restoration. *J. Black Sea/Mediterr. Environ.* **2012**, *18*, 144–154.
92. Bănaru, D.; Harmelin-Vivien, M.; Gomoiu, M.-T.; Onciu, T.-M. Influence of the Danube River Inputs on C and N Stable Isotope Ratios of the Romanian Coastal Waters and Sediment (Black Sea). *Mar. Pollut. Bull.* **2007**, *54*, 1385–1394. [CrossRef]

93. Cozzi, S.; Ibáñez, C.; Lazar, L.; Raimbault, P.; Giani, M. Flow Regime and Nutrient-Loading Trends from the Largest South European Watersheds: Implications for the Productivity of Mediterranean and Black Sea's Coastal Areas. *Water* **2019**, *11*, 1. [CrossRef]
94. Üstün, F.; Bat, L.; Mutlu, E. Seasonal Variation and Taxonomic Composition of Mesozooplankton in the Southern Black Sea (off Sinop) between 2005 and 2009. *Turk. J. Zool.* **2018**, *42*, 541–556. [CrossRef]
95. Arashkevich, E.G.; Stefanova, K.; Bandelj, V.; Siokou, I.; Terbiyik Kurt, T.; Ak Orek, Y.; Timofte, F.; Timonin, A.; Solidoro, C. Mesozooplankton in the Open Black Sea: Regional and Seasonal Characteristics. *J. Mar. Syst.* **2014**, *135*, 81–96. [CrossRef]
96. Ward, D.; Melbourne-Thomas, J.; Pecl, G.T.; Evans, K.; Green, M.; McCormack, P.C.; Novaglio, C.; Trebilco, R.; Bax, N.; Brasier, M.J.; et al. Safeguarding Marine Life: Conservation of Biodiversity and Ecosystems. *Rev. Fish. Biol. Fish.* **2022**, *32*, 65–100. [CrossRef] [PubMed]
97. Corcoran, E.; Skuja, M.; Wild, C.; Brown, C.; Harvey, A.; Press, T.L.; Centre, U.W.C.M. *Global Marine Assessments: A Survey of Global and Regional Assessments and Related Scientific Activities of the Marine Environment*; UNEP: Nairobi, Kenya, 2007; ISBN 9789280728002.
98. Oguz, T.; Dippner, J.W.; Kaymaz, Z. Climatic Regulation of the Black Sea Hydro-Meteorological and Ecological Properties at Interannual-to-Decadal Time Scales. *J. Mar. Syst.* **2006**, *60*, 235–254. [CrossRef]
99. Murphy, G.E.P.; Romanuk, T.N.; Worm, B. Cascading Effects of Climate Change on Plankton Community Structure. *Ecol. Evol.* **2020**, *10*, 2170–2181. [CrossRef]
100. Halpern, B.S.; Walbridge, S.; Selkoe, K.A.; Kappel, C.V.; Micheli, F.; D'Agrosa, C.; Bruno, J.F.; Casey, K.S.; Ebert, C.; Fox, H.E.; et al. A Global Map of Human Impact on Marine Ecosystems. *Science (1979)* **2008**, *319*, 948–952. [CrossRef]
101. Crain, C.M.; Halpern, B.S.; Beck, M.W.; Kappel, C.V. Understanding and Managing Human Threats to the Coastal Marine Environment. *Ann. N. Y. Acad. Sci.* **2009**, *1162*, 39–62. [CrossRef]
102. Mee, L.; Friedrich, J.; Gomoiu, M. Restoring the Black Sea in Times of Uncertainty. *Oceanography* **2012**, *18*, 100–111. [CrossRef]
103. Moschonas, G.; Gowen, R.J.; Paterson, R.F.; Mitchell, E.; Stewart, B.M.; McNeill, S.; Glibert, P.M.; Davidson, K. Nitrogen Dynamics and Phytoplankton Community Structure: The Role of Organic Nutrients. *Biogeochemistry* **2017**, *134*, 125–145. [CrossRef]
104. Kudela, R.M.; Seeyave, S.; Cochlan, W.P. The Role of Nutrients in Regulation and Promotion of Harmful Algal Blooms in Upwelling Systems. *Prog. Oceanogr.* **2010**, *85*, 122–135. [CrossRef]
105. Bisinicu, E.; Harcota, G.E.; Lazar, L. Interactions between Environmental Factors and the Mesozooplankton Community from the Romanian Black Sea Waters. *Turk. J. Zool.* **2023**, *47*, 202–215. [CrossRef]
106. Bișinicu, E.; Abaza, V.; Cristea, V.; Harcotă, G.E.; Lazar, L.; Tabarcea, C.; Timofte, F. The Assessment of the Mesozooplankton Community from the Romanian Black Sea Waters and the Relationship to Environmental Factors. *Cercet. Mar.—Rech. Mar.* **2021**, *51*, 108–128. [CrossRef]
107. Oviatt, C.A. Effects of Different Mixing Schedules on Phytoplankton. Zooplankton and Nutrients in Marine Microcosms. *Mar. Ecol. Prog. Ser.* **1981**, *4*, 57–67. [CrossRef]
108. Yang, L.-J.; Tao, Y.; Jiang, X.; Wang, Y.; Li, Y.-H.; Zhou, L.; Wang, P.-Z.; Li, Y.-Y.; Zhao, X.; Wang, H.-J.; et al. Interactive Effects of Nutrients and Salinity on Zooplankton in Subtropical Plateau Lakes with Contrasting Water Depth. *Front. Environ. Sci.* **2023**, *11*, 1110746. [CrossRef]
109. Cade-Menun, B.J.; Paytan, A. Nutrient Temperature and Light Stress Alter Phosphorus and Carbon Forms in Culture-Grown Algae. *Mar. Chem.* **2010**, *121*, 27–36. [CrossRef]
110. van Beusekom, J.E.E.; Loebl, M.; Martens, P. Distant Riverine Nutrient Supply and Local Temperature Drive the Long-Term Phytoplankton Development in a Temperate Coastal Basin. *J. Sea Res.* **2009**, *61*, 26–33. [CrossRef]
111. Almroth, E.; Tengberg, A.; Andersson, J.H.; Pakhomova, S.; Hall, P.O.J. Effects of Resuspension on Benthic Fluxes of Oxygen, Nutrients, Dissolved Inorganic Carbon, Iron and Manganese in the Gulf of Finland, Baltic Sea. *Cont. Shelf Res.* **2009**, *29*, 807–818. [CrossRef]
112. EEA. Nutrients in Transitional, Coastal and Marine Waters (CSI 021). 2011. Available online: https://www.eea.europa.eu/en/analysis/indicators/nutrients-in-transitional-coastal-and?activeAccordion= (accessed on 1 August 2024).
113. Friedl, G.; Dinkel, C.; Wehrli, B. Benthic Fluxes of Nutrients in the Northwestern Black Sea. *Mar. Chem.* **1998**, *62*, 77–88. [CrossRef]
114. Kalnejais, L.H.; Martin, W.R.; Bothner, M.H. The Release of Dissolved Nutrients and Metals from Coastal Sediments Due to Resuspension. *Mar. Chem.* **2010**, *121*, 224–235. [CrossRef]
115. Wei, Y.; Ding, D.; Gu, T.; Jiang, T.; Qu, K.; Sun, J.; Cui, Z. Different Responses of Phytoplankton and Zooplankton Communities to Current Changing Coastal Environments. *Environ. Res.* **2022**, *215*, 114426. [CrossRef]
116. Sun, X.; Zhang, H.; Wang, Z.; Huang, T.; Tian, W.; Huang, H. Responses of Zooplankton Community Pattern to Environmental Factors along the Salinity Gradient in a Seagoing River in Tianjin, China. *Microorganisms* **2023**, *11*, 1638. [CrossRef] [PubMed]
117. Miyaguchi, H.; Fujiki, T.; Kikuchi, T.; Kuwahara, V.S.; Toda, T. Relationship between the Bloom of *Noctiluca scintillans* and Environmental Factors in the Coastal Waters of Sagami Bay, Japan. *J. Plankton Res.* **2006**, *28*, 313–324. [CrossRef]
118. Stanev, E.V.; Peneva, E.; Chtirkova, B. Climate Change and Regional Ocean Water Mass Disappearance: Case of the Black Sea. *J. Geophys. Res. Oceans* **2019**, *124*, 4803–4819. [CrossRef]
119. Vollenweider, R.A. *Marine Coastal Eutrophication*; Elsevier Science: Amsterdam, The Netherlands, 1992; ISBN 0444899901.
120. Barbier, E.B. Marine Ecosystem Services. *Curr. Biol.* **2017**, *27*, R507–R510. [CrossRef] [PubMed]

121. Stoecker, D.K.; Hansen, P.J.; Caron, D.A.; Mitra, A. Mixotrophy in the Marine Plankton. *Ann. Rev. Mar. Sci.* **2016**, *9*, 311–335. [CrossRef]
122. Legendre, L.; Rassoulzadegan, F. Plankton and Nutrient Dynamics in Marine Waters. *Ophelia* **1995**, *41*, 153–172. [CrossRef]
123. Porumb, F. Le Zooplancton de La Mer Noire. Biologie Des Eaux Saumâtres de La Mer Noire. *IRCM Constanta* **1977**, *1*, 99–108.
124. Litchman, E.; Ohman, M.D.; Kiørboe, T. Trait-Based Approaches to Zooplankton Communities. *J. Plankton Res.* **2013**, *35*, 473–484. [CrossRef]
125. Brandão, M.C.; Benedetti, F.; Martini, S.; Soviadan, Y.D.; Irisson, J.-O.; Romagnan, J.-B.; Elineau, A.; Desnos, C.; Jalabert, L.; Freire, A.S.; et al. Macroscale Patterns of Oceanic Zooplankton Composition and Size Structure. *Sci. Rep.* **2021**, *11*, 15714. [CrossRef]
126. Heneghan, R.F.; Everett, J.D.; Blanchard, J.L.; Sykes, P.; Richardson, A.J. Climate-Driven Zooplankton Shifts Cause Large-Scale Declines in Food Quality for Fish. *Nat. Clim. Change* **2023**, *13*, 470–477. [CrossRef]
127. Oguz, T. Long-Term Impacts of Anthropogenic Forcing on the Black Sea Ecosystem. *Oceanography* **2012**, *18*, 112–121. [CrossRef]
128. Butchart, S.H.M.; Walpole, M.; Collen, B.; van Strien, A.; Scharlemann, J.P.W.; Almond, R.E.A.; Baillie, J.E.M.; Bomhard, B.; Brown, C.; Bruno, J.; et al. Global Biodiversity: Indicators of Recent Declines. *Science (1979)* **2010**, *328*, 1164–1168. [CrossRef] [PubMed]
129. Kovalev, A.V.; Mazzocchi, M.G. Seasonal Changes in the Composition and Abundance of Zooplankton in the Seas of the Mediterranean Basin. *Turk. J. Zool.* **2003**, *27*, 205–219.
130. Childs, N. The Black Sea in the Shadow of War. *Survival* **2023**, *65*, 25–36. [CrossRef]
131. Renolafitri, H.; Yolandika, C. Impact of the Russia-Ukraine War on the Environmental, Social and Economic Conditions of the Black Sea. *Econ. Manag. Soc. Sci. J.* **2022**, *1*, 120–128. [CrossRef]
132. Bisinicu, E.; Abaza, V.; Boicenco, L.; Adrian, F.; Harcota, G.-E.; Marin, O.; Oros, A.; Pantea, E.; Spinu, A.; Timofte, F.; et al. Spatial Cumulative Assessment of Impact Risk-Implementing Ecosystem-Based Management for Enhanced Sustainability and Biodiversity in the Black Sea. *Sustainability* **2024**, *16*, 4449. [CrossRef]

oceans

Article

Locally Adapted Coral Species Withstand a 2-Week Hypoxic Event

Noelle Lucey [1,2,*], Carolina César-Ávila [1], Alaina Eckert [1], Paul Veintimilla [1] and Rachel Collin [1,*]

[1] Smithsonian Tropical Research Institute, Balboa, Ancon 0843-03092, Panama
[2] High Meadows Environmental Institute, Princeton University, Princeton, NJ 08544, USA
* Correspondence: nlucey@princeton.edu (N.L.); collinr@si.edu (R.C.)

Abstract: One approach to improve long-term coral restoration success utilizes naturally stress-tolerant corals from the wild. While the focus has primarily been on thermal stress, low oxygen is a growing threat to coral reefs and restoration efforts should also consider hypoxia tolerance. Here we determine if *Siderastrea siderea* and *Agaricia tenuifolia* populations from a reef with a historical record of low oxygen exhibit evidence of local adaptation to hypoxic events, compared to populations from a reference reef. We employed a laboratory-based reciprocal transplant experiment mimicking a severe 14-night hypoxic event and monitored bleaching responses, photo-physiology, metabolic rates, and survival of all four populations during, and for two weeks following the event. In both species, we found the populations from the hypoxic reef either fully persisted or recovered within 3 days of the event. In contrast, the conspecific naïve populations from the well-oxygenated reference reef experienced bleaching and death. This showcases the vulnerability of naïve corals exposed to low oxygen but also suggests that corals from the hypoxic reef locally adapted to survive severe episodic hypoxia. Other reefs with past episodic low oxygen may also be home to corals with adaptation signatures to hypoxia and may be useful for restoration efforts.

Keywords: Caribbean; low oxygen; hypoxia; bleaching; *Siderastrea siderea*; *Agaricia tenuifolia*; reef restoration

1. Introduction

The world's corals are under threat from global change, and there is almost unanimous agreement that stopping CO_2 emissions is the most direct and best solution to help save them [1]. However, even if the most optimistic CO_2 emission targets are reached, corals are expected to decline by 70–90% relative to their current abundance by 2050 [2,3]. Coral restoration efforts aimed at saving corals must be considered through this lens.

One subset of coral restoration efforts is based on harnessing the innate stress tolerance in certain coral species and populations [2,4]. Active interventions use stress-tolerant corals, primarily thermally-tolerant corals, by (1) transplanting corals from their native habitat to either the edge of their range to promote assisted gene flow or by moving them to a location outside of their current range (assisted evolution/migration), (2) selectively breeding corals to promote heat tolerant adaptations and outplanting their offspring, or (3) growing asexual fragments or asexual propagation in a nursery and outplanting them to reefs, reviewed in [5]. Despite numerous potential risks that demand careful consideration [6], these approaches have shown promise by increasing the presence of resilient genotypes and/or more optimal symbiont communities [7]. This raises the question: where are these climate-ready stress-tolerant corals that can thrive despite the multiple climate stressors facing them?

It is possible to find climate-ready stress-tolerant corals in naturally stressful environments such as semi-enclosed lagoons, mangrove lagoons, highly variable reef flats or back reef habitats [7,8]. Importantly, many of these systems are often not only warmer, but also have regularly low pH and low oxygen conditions, compared to nearby reference reef locations [9]. Therefore, corals from these stressful habitats may be tolerant not only to thermal stress but also to acidification and deoxygenation.

Reciprocal transplant experiments that test populations from stressful and less stressful environments provide a relatively simple way to identify stress-tolerant traits [10,11]. This approach has been used to find evidence of local adaptation in corals, including within symbionts and associated microbial communities [12]. For example, in a semi-enclosed lagoon in New Caledonia, *Acropora tenuis*, *Montipora digitata*, and *Porites* sp. corals were found to be locally adapted to grow and calcify faster in low pH conditions, compared to conspecifics from a cooler, less acidified reef [13]. Similarly, in the semi-enclosed Nikko Bay, in Palau, *Porites cylindrica* living under naturally acidified and warm conditions showed signs of local adaptation as they maintained high calcification rates when transplanted to a non-stressful reference reef, compared to the reduced calcification rates of counterparts from the reference reef transplanted to the naturally warm, acidified reef [14]. In the Great Barrier Reef Woody Isle mangrove lagoon, *Porites lutea* is both morphologically and energetically adapted to low pH, increased temperature, and low oxygen, compared to outer reference reef corals [15]. While most of these studies have tested coral responses to in situ diel environmental cycles, adaptive responses to laboratory-regulated environmental extremes have not been widely examined [16,17].

Laboratory-based reciprocal transplant experiments can constrain the specific environmental variable driving organismal stress and improve the predictive power of responses to future conditions. An important and often underappreciated variable we consider here is episodic low oxygen, which are often referred to as hypoxia events because they cause respiratory hypoxic stress in marine animals unable to meet their oxygen demands. This is in part because oxygen saturation decreases with increasing temperature [18,19]. Additionally, with increased nutrients, bacterial respiration can disproportionately use up available oxygen in the water column [20]. Through a combination of drivers, episodic hypoxic events have had severe consequences on marine ecosystems throughout the last few decades. Importantly, event occurrences are now increasing at an exponential rate around the world, leading to more frequent mass mortalities and ecosystem collapse [21,22].

There are a growing number of reefs where the aftermath of hypoxic events has been documented [23–27]. Across these systems, the impacts have included widespread coral bleaching and mass mortality. In a global synthesis of hypoxia-driven mass mortality events, we found they reoccur in the same location 54% of the time [22]. This is true in Bahía Almirante, Panama, where coral reefs experienced a hypoxic event in 2010 [28], and again in 2017, resulting in the death of 50% of the living coral community [23]. This suggests corals living on this reef system today have experienced previous hypoxia and may be adapted to survive future hypoxic events. However, no studies have tested corals from historically deoxygenated reefs for signs of local adaptation to hypoxia.

Here we show that two species of coral native to this frequently oxygen-depleted site display signs of local adaptation and can survive severe hypoxic events. We first parameterized hypoxic events on a reef in Bahía Almirante near the mainland of Panama with a long-term record of hourly oxygen data and found these events manifest as a sequence of consecutive low oxygen nights. We replicated this in situ hypoxic event in a laboratory reciprocal transplant experiment. We found severe bleaching and mortality in naïve *A. tenuifolia* and *S. siderea* populations, while the hypoxia-native populations showed signs of local adaptation that included bleaching recovery and overall survival. These findings identify reef habitats and corals that can withstand severe episodic hypoxia. Exploring other historically deoxygenated reefs as potential sources of hypoxia-tolerant corals could aid coral restoration projects contending with probable future oxygen loss.

2. Materials and Methods

2.1. Reef Sites, Environmental History, and Hypoxic Events

Bahía Almirante is a semi-enclosed tropical lagoon that consists primarily of shallow coral reefs, mangrove islands, and restricted water flow [29]. It also has a well-studied temperature, oxygen, and pH gradient that becomes increasingly more extreme with increased proximity to the mainland [30,31]. Oxygen is seasonally low at depths > 10 m from September to November and on shallower reefs the severity of the diel O_2 cycle increases during these months [32]. Based on these past findings, we selected two reef sites at contrasting ends of this gradient with similar reef communities: Cayo Roldan (9.21416, −82.32422) with a history of low oxygen, including the nearby mortality event documented in [23], and Hospital Point (9.33329, −82.21741), a reference reef with no known oxygen-driven mortality (Figure 1).

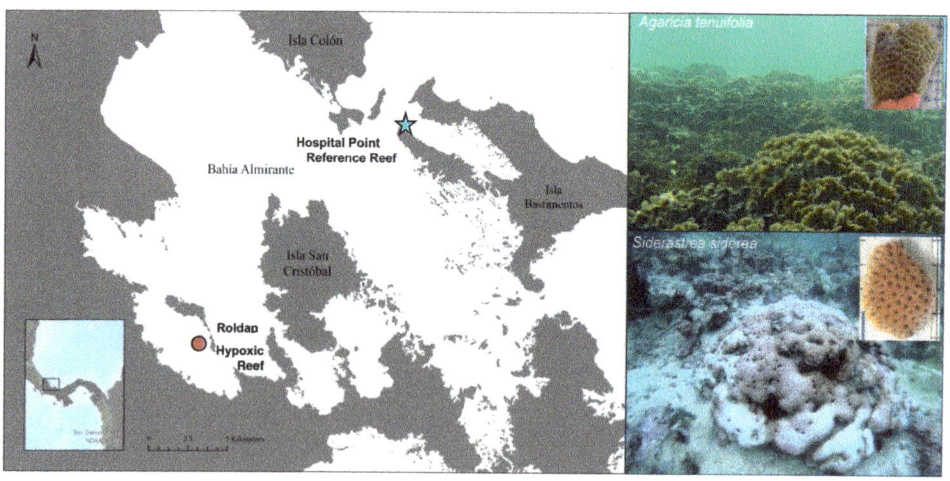

Figure 1. Reef sites in Bahía Almirante, Panama, and study species. Five colonies of both *A. tenuifolia* and *S. siderea* were collected from the historically deoxygenated reef on the south side of Cayo Roldan (red circle), and from a reference reef on the north side of Hospital Point (blue star). Sites are approximately 15 km apart.

Our goal was to replicate the severity and duration of a naturally occurring hypoxic event in the laboratory, which first required we identify these parameters. To identify the severity and duration of low oxygen naturally experienced in the field we analyzed a 2-year dataset of hourly dissolved O_2 from both sites, from September 2021 to November 2023 [24]. The minimum O_2 at Hospital Point reef was 0.91 mg L^{-1} and lasted for a single hour at 4:30 on the morning of 30 August 2023. This episode occurred in the middle of a 15-night stretch of O_2 conditions consecutively falling between 1–2 mg L^{-1}. In contrast, Roldan reef experienced severe low O_2, with a minimum concentration of 0.33 mg L^{-1}, remaining

<1 mg L^{-1} for 3 h on one day in Apr. 2023 (Figure 2A). During this period, nightly low O_2 < 2 mg L^{-1} reoccurred for ~18 consecutive nights (Figure 2B). We define hypoxic events using these parameters, where event duration is 14 consecutive nights of oxygen depletion, with a severity of 0.3 mg L^{-1} for 6 h periods every night (Figure 2C). The O_2 severity chosen matches the minimum O_2 measured on Roldan reef but also represents a level known to cause physiological hypoxic stress for some coral species, including *Agaricia tenuifolia* [24].

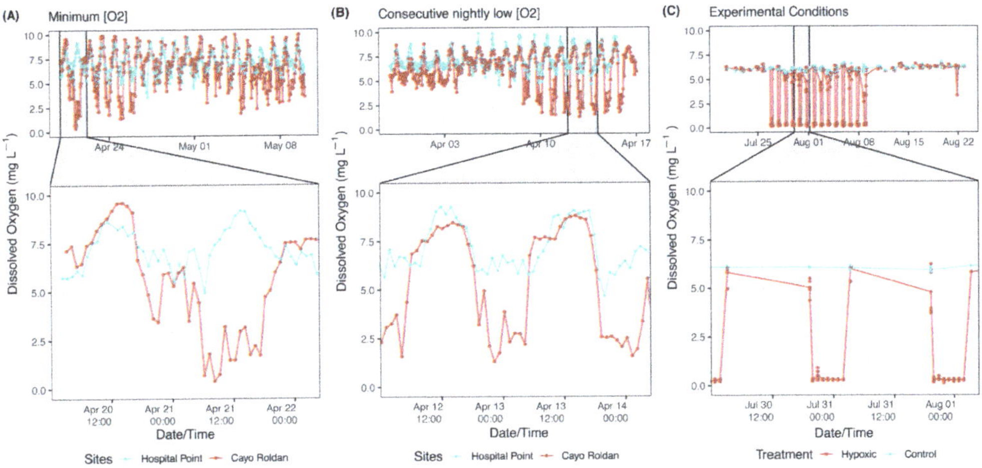

Figure 2. Hypoxic events manifest as a sequence of consecutive extreme low oxygen conditions occurring at night. (**A**) A two-month window between 2020–2023 highlighting that the minimum oxygen at Roldan Reef was 0.3 mg L^{-1} (red), while the minimum on the reference reef, Hospital Point, was 1.0 mg L^{-1} (blue). (**B**) A three-week window highlighting the pattern of low O_2 conditions, which reoccurs consecutively for multiple nights at a time. We define these multi-night low O_2 periods as a single hypoxic event. (**C**) In the laboratory we mimicked an in situ 14 night hypoxic event: red points show measured O_2 in all five of the hypoxic treatment tanks during the experiment and blue points show O_2 in the 5 control tanks. Tank conditions during the one-week laboratory acclimation period prior to low O_2 exposure, and the 2-week post-exposure period are also shown.

2.2. Experimental Design and Maintenance

Using a hammer and chisel we collected five colonies of *A. tenuifolia* and five colonies of *S. siderea* from the hypoxic reef and from the reference reef in July 2022. *Agaricia tenuifolia* was collected from ~3 m, while *S. siderea* was from 8–10 m. All colonies were collected with a ~10 m distance between each to minimize the chance they were clonal fragments of the same colony. We brought the 20 colonies to the Bocas del Toro Research Station and placed them in outdoor water tables with flowing seawater. We cut each colony into 20 fragments, approximately 6 cm long within 3 days of collection using pliers for *A. tenuifolia* and a wet saw for *S. siderea* (Gryphon C-40 Diamond Band Saw, Gryphon Corporation, San Diego, CA, USA) (Figure 3). *Siderastrea siderea* colonies from the reference reef were dense, and it took 2 days to fragment these, whereas colonies from the hypoxic reef took ~3 h in total to fragment. Skeletal quality was visibly reduced, and each colony was riddled with boring bivalves both alive and dead (Figures 1 and 3C). During this fragmentation period, we observed necrotic tissue spreading within the hypoxic colonies. We attributed this to bacteria and treated colonies from each site by dipping them into a dilute iodine water bath for 20 min and then returning them to an outdoor water table with high-flow seawater. All visibly affected fragments were removed prior to the experimental setup.

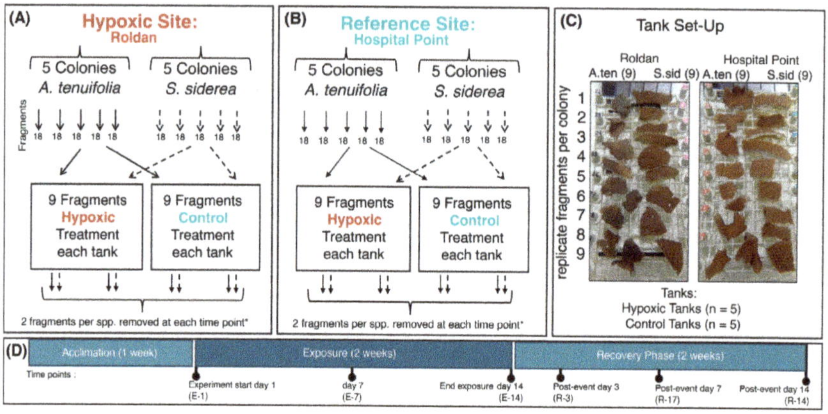

Figure 3. (**A–C**) Experimental set-up and design. Five colonies of each species were collected from a hypoxic (**A**) and a reference (**B**) site to determine local hypoxia adaptation potential. These colonies were all fragmented to produce 20 replicate fragments of a given genotype. Replicate colony fragments of each species were distributed into both control and hypoxic experimental treatments to attain 5 tank replicates per treatment (**C**). In the hypoxic treatment, corals were exposed to diel cycling where the O_2 was lowered to 0.3 mg L^{-1} for 6 h at night, and reoxygenated during the day, for 14 nights. All tanks were then fully oxygenated for an additional two weeks to assess recovery potential. Two coral fragments from each tank were destructively sampled at each time point to assess metabolic rate changes through time. (**D**) Experimental timeline, showing the one-week acclimation period post-fragmentation, exposure period, and recovery phase with associated time points indicating when measurements were made.

After fragmentation, we positioned 18 fragment replicates (i.e., ramets) from each colony onto two plastic egg crate racks, which were then put in two of the 10 experimental indoor tanks. The other two fragments were taken and used to measure baseline metabolic rates. Experimental tanks each held 55 L and were equipped with an air pump, submersible water pump, and submersible heater, with a fresh seawater inflow at a rate of 110 L h^{-1}. Two aquarium lights were positioned over each tank to maintain a 12:12 day/night cycle (Aqua Illumination Hydra 64 HD LED Reef Lights, Bethlehem, PA, USA). We adjusted the day-night cycle one week prior to starting the experiment so dark O_2 manipulation started at 9:00 h. During the 12 h day period, lights were ramped up for 6 h to reach full light intensity, matching midday light measured on the inner bay reef at 10 m (Li-COR LI-250A Light Meter, Lincoln, NE, USA), and then they were ramped down for the following 6 h until completely dark.

During this one-week laboratory acclimation (Figure 3), conditions remained fully oxygenated at 6.1 ± 0.1 mg L^{-1}, and at an ambient temperature of 29.0 ± 0.3 °C (mean ± SD), with a salinity of 30.3 ± 0.7. After acclimation, we initiated two experimental treatments for 2 weeks: hypoxic and control. In the 5 tanks of the control treatment, no changes were made from the acclimation period (i.e., control tanks). In the 5 tanks of the hypoxic treatment, O_2 was lowered to 0.3 mg L^{-1} in each of the 5 tanks for 6 h at the beginning of the night (dark phase) and allowed to return to 6.1 mg L^{-1} for the rest of the 24 h cycle (Figure 2C). O_2 was regulated in hypoxic tanks by bubbling pure nitrogen gas into each tank with a diffusion stone and was monitored by hand every 15 min using a PreSens Dipping Probe (PreSens Precision Sensing, Regensburg, Germany). Adjustments to gas flows were made to maintain the concentration at 0.33 ± 0.01 mg L^{-1}. Temperatures were maintained at 29.99 ± 0.07 °C. Full water changes were made daily during the oxygenated light phase. We also siphoned excess debris from the bottom of tanks daily and replaced the racks holding the corals every four days, to avoid algal growth.

After two weeks of experimental treatments, we kept all corals in their tanks for an additional two weeks. During this time, all tanks were constantly well-oxygenated by replacing all N_2 gas lines with regular air lines and maintaining the same aquarium conditions described above for the acclimation period (i.e., same temperature, lighting, tank cleaning protocol). We monitored corals daily during this recovery phase (Figures 2C and 3).

2.3. *Coral Physiological Assessments*

2.3.1. Bleaching and Mortality

Each day after the hypoxic night phase, we scored corals for bleaching and/or mortality. We did this by placing a white photographic standard in each tank and changing the lights to a set-white intensity before making visual observations. We recorded if color loss or tissue loss occurred in each fragment (y/n binary data). Tissue loss impacting more than 50% of the fragment's surface area was considered a mortality response and determined by eye from the same person.

2.3.2. Fluorescence Analysis

Directly after the dark phase, we assessed the photosynthetic capacity of all coral fragments. With pulse-amplitude-modulated fluorometry (PAM), we measured the maximum quantum yield of PSII (F_v:F_m), a photosynthetic parameter used to describe the condition of coral's Symbiodiniaceae algae [33]. F_v:F_m values range from 0–1 with values closer to 1 indicating higher photosynthetic performance. Measurements were taken in the dark at the end of the night cycle, i.e., dark acclimation of ~12 h, in the middle of each coral fragment with the following settings: gain: 1, saturation: 10, spacing: 10 mm, and damping: 2 for *A. tenuifolia*, 1 for *S. siderea*. The different parameters for each species are based on preliminary testing and chosen to optimize the signal-to-noise ratio.

2.3.3. Metabolic Rates

After each experimental period, at the end of the day phase, one fragment from each colony in each treatment was taken out of the experiment, and its metabolic rate was measured using closed-chamber respirometry. Corals were placed in 0.2 L glass respirometry chambers equipped with a magnetic stir bar and filled with 0.5 μm filtered seawater. Air bubbling from an air pump kept seawater O_2 fully saturated for 1 h while corals acclimated to the new conditions. After 1 h, chambers were sealed with care ensuring no air bubbles remained inside. Sealed chambers were then placed in a 40 L dark water bath standing upon two stir plates and filled with fresh water at the designated 29 °C temperature. The temperature was kept constant with a precision cooker at 29 °C to match experimental tank conditions (Sous Vide Culinary precision cooker PRO, ANOVA applied electronics; San Francisco, CA, USA). Chambers inside the water bath were equipped with O_2 sensors and associated fiber optic cables (PreSens OXY—4 SMA G3; Regensburg, Germany), set to record the partial pressure of oxygen, pO_2, every minute. Each cohort of coral trials included one chamber filled with seawater from the acclimation bath to determine background respiration rates. Trials were stopped when pO_2 within the chamber was depleted to approximately 80%, which took ~45 ± 15 min for each fragment. Immediately following the trial, fragment volume was measured via the water displacement method and surface area via the tin-foil method [34]. Salinity was measured to correct pO_2 values. By recording pO_2 within the chambers every minute we produced curves displaying oxygen consumption over time for each fragment. From each individual time series dataset, we calculated the standard metabolic rate (SMR) by fitting a linear model of oxygen against time (~45 timepoints per fragment), where the slope of each regression defined the rate. Final rates were adjusted to account for salinity and microbial respiration, as well as standardized to the living tissue surface area of the fragment (μmol $O_2{}^{-1}$ h^{-1} cm^2). These calculations were performed using the R package respR [35]. In total, 96 fragments per species were processed over the course of the experiment for SMR (Table 1).

Table 1. Total number of fragments for each species at each timepoint, and for each site and treatment. In the final column, fragment designation, the number of fragments removed from the tanks for metabolic rate (MR) measurements and symbiont density counts are listed (zoox).

			Reference Site (Hosp. Pt.)		Hypoxic Site (Roldan)		Fragment
	Ttimepoint	Date	Hypoxic	Control	Hypoxic	Control	Designation
A. tenuifolia	E-1 Experiment start	25-Jul	25	25	25	25	
		26-Jul	5	5	5	5	MR/zoox
	E-7 Mid-exposure	1-Aug	20	20	20	20	
		2-Aug	5	5	5	5	MR/zoox
	E-14 End exposure	8-Aug	14	15	15	15	
		9-Aug	5	4	5	5	MR/zoox
	R-3 Post-exposure (3 day)	11-Aug	8	10	10	10	
		12-Aug	4	5	5	5	MR/zoox
	R-7 Post-exposure (7 day)	15-Aug	3	5	5	5	
	R-14 Post-exposure (14 day)	22-Aug	3	5	5	5	
		23-Aug	3	5	5	5	MR/zoox
S. siderea	E1 Experiment start	25-Jul	25	25	25	25	
		26-Jul	5	5	5	4	MR/zoox
	E7 Mid-exposure	1-Aug	20	20	20	20	
		2-Aug	5	5	5	5	MR/zoox
	E14 End exposure	8-Aug	15	14	15	15	
		9-Aug	5	4	5	5	MR/zoox
	R3 Post-exposure (3 day)	11-Aug	10	10	10	10	
		12-Aug	4	5	5	5	MR/zoox
	R7 Post-exposure (7 day)	15-Aug	4	4	5	5	
	R14 Post-exposure (14 day)	22-Aug	5	5	5	5	
		23-Aug	5	4	5	4	MR/zoox

Samples were frozen after respirometry trials and later processed for symbiont cell counts. The live tissue was initially separated from the coral skeleton using a water pick with filtered seawater, and then zooxanthellae algal cells were separated from animal cells. After cleaning the algal fraction, samples were diluted and counted with a hemocytometer. Counts were normalized to the coral's living tissue surface area (measured above) and are reported in cells per cm^2.

2.3.4. Statistical Analysis of Traits Through Time

All statistical analyses and data visualizations were performed with R software v. 4.3.1. We first identified any significant differences in the physiological traits between the treatments (hypoxic and control) by building separate models for each species and site, at every timepoint during the full 4-week experiment (see Figure 3 for experimental timepoints). We constructed generalized linear models (GLMs) with treatment and parent colony as factors, setting family to binomial for binary data. Residual diagnostics for all models were determined using the DHARMa package, i.e., tests for Kolmogorov–Smirnov tests, dispersion, outliers, overdispersion, and zero-inflation. Significant differences between treatments at each timepoint, for each species' trait, are identified with asterisks (e.g., Section 3.1).

2.3.5. Statistical Analysis of Traits Across Populations

To assess local adaptation potential, we employed the framework described by Kawecki and Ebert, 2004 [11] that considers the performance of populations in their native and naïve habitats. Adapted populations are expected to have higher fitness in their respective source habitats (native site) than in their transplant habitats (naïve site). This performance is explored with reaction norms and analysis of variance, where signs of adaptation are indicated if there is a significant interaction between the genotype (i.e., population) and the environment (i.e., treatment). Such tests of local adaptation expect population sizes to include large numbers of individuals (e.g., >500), but to avoid oversampling the reef and to maintain low density within tanks, we kept our population sizes small (n = 5). We note these limitations below and maintain this terminology, defining our coral 'population' response as the average response of five distinct genetic colonies collected from one site. We considered bleaching, survival, metabolic rate, and photosynthetic rate to each be a component of fitness.

To determine if either species showed signs of adaptation for hypoxia tolerance, we built reaction norms for each population and identified statistical differences between the O_2 treatment, population, and their interaction, for each trait, with 2-way ANOVA using GLMs. These factors are equivalent to the environment (treatment) and genotype (population) terms in the reaction norm examples shown in Figure 4A. Two sets of models were built to assess each trait (1) immediately after the experiment (timepoint E-14) and (2) 3-days post-experiment (R-3). Models assessing traits measured on the R-7 and R-14 timepoints were also built to identify recovery potential but were not reported because they did not differ from R-3 models and were not as robust due to decreased sample sizes from the destructive sampling protocol (Figure 3, Table 1).

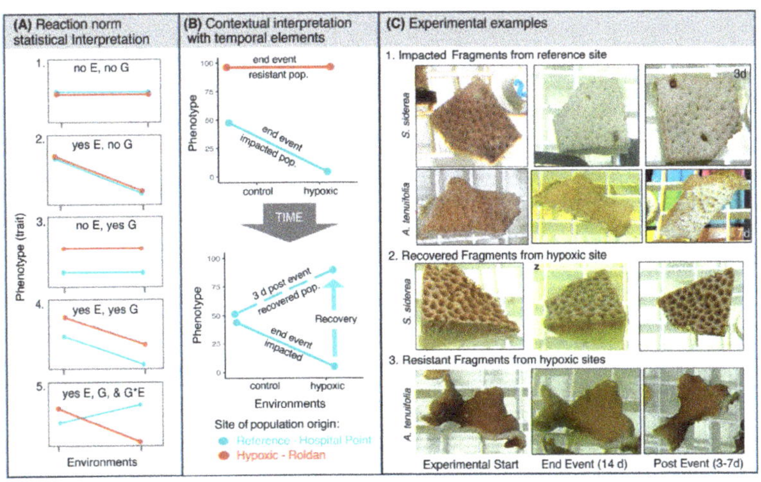

Figure 4. (**A**) In the interpretation of a reaction norm, each line represents a response from one genotype (i.e., population) when tested in different environments (i.e., treatments). When lines are perfectly horizontal and overlapping, there is no effect of the environment (E) or genotype (G), i.e., no plasticity or adaptation (1); when the lines are not horizontal but still overlapping, there is an environmental effect but not a genotype effect (2). When the lines are horizontal but not overlapping, there is a genotype effect that is not influenced by the environment (3). If lines are not horizontal but

are parallel, there is an effect of environment and genotype, but no genotype X environment interaction (4). If lines intersect, there is a genotype X environment interaction indicating local adaptation (5) [36]. (**B**) A contextual interpretation of reaction norms as they pertain to this study's experimental results. Solid lines indicate trait responses measured at the end of the experimental event and dashed reaction norms indicate the responses 3 days later. Any significant changes between the solid and dashed reaction norm lines of the same color indicate recovery potential in that trait (bottom plot). (**C**) Coral responses to the lab-replicated hypoxic event included bleaching and mortality (1), recovery (2), and no change/stress resistance (3).

To determine if the traits measured at the end of the experiment (i.e., E-14) significantly differed from those measured after a recovery period (i.e., R-3), we pooled response data from both timepoints and tested the effect of the interaction of population X treatment and included date as a main factor. To account for repeated measures, we initially added colony ID as a random effect to track the same individuals through time but these were not used in most final models due to overfitting. Note, no tank effect was tested as colonies were not randomly dispersed between tanks (Figure 3). In all models, significant population X treatment interactions indicate potential local adaptation, whereas a significant treatment effect alone indicates environmental effects. Residual diagnostics for all models were determined using the DHARMa package, i.e., tests for Kolmogorov–Smirnov tests, dispersion, outliers, overdispersion, and zero-inflation. These results are reported in Figure 6, Section 3.2.

3. Results

We replicated a single hypoxic event in the laboratory and then identified the local adaptation potential of two *A. tenuifolia* and *S. siderea* populations to this stress. We monitored bleaching, photo-physiology, metabolic rate, and survival of all colonies throughout the 2-week hypoxic event, and for 2 weeks following the event. Physiological responses of both species were most pronounced at the end of the event, or 3 days after, with negative impacts largely affecting corals from the reference site and not the hypoxic site. These responses provide evidence that both species from historically hypoxic reefs may be adapted to withstand severe 2-week hypoxic events.

3.1. Responses to Hypoxia Vary Through Time, Impacting A. tenuifolia Most

The number of bleached fragments in the reference population increased over time, reaching 56 ± 10% at the end of the two-week event (Figure 5A). This response was consistent with a 9.4% decrease in photosynthetic capacity three days after the event, where F_v:F_m was reduced from 0.64 in the control to 0.58 in the hypoxic treatment (Figure 5B). However, no changes in symbiont density between treatments or populations were measured in the fragments processed at each timepoint (Figure 5C, Table 1). Likewise, metabolic rates varied between 0.022–0.038 μmol h^{-1} cm^2, but rates never significantly differed between treatment or population at any timepoint (Figure 5D, Table 1). Mortality responses closely mirrored bleaching responses, and by the end of the event 28 ± 9.2% S.E. of the fragments from the reference population had died. This increased to 40 ± 16.3% (mean ± S.E.) in the following week, despite the fully reoxygenated conditions (Figure 5E). In contrast, no colonies from the hypoxic population bleached, experienced reduced photosynthetic capacity, or died at any point (Figure 5A–E). This indicates *A. tenuifolia* colonies without prior hypoxia conditioning are highly vulnerable to future low O_2 compared to colonies that have experienced hypoxia in the past.

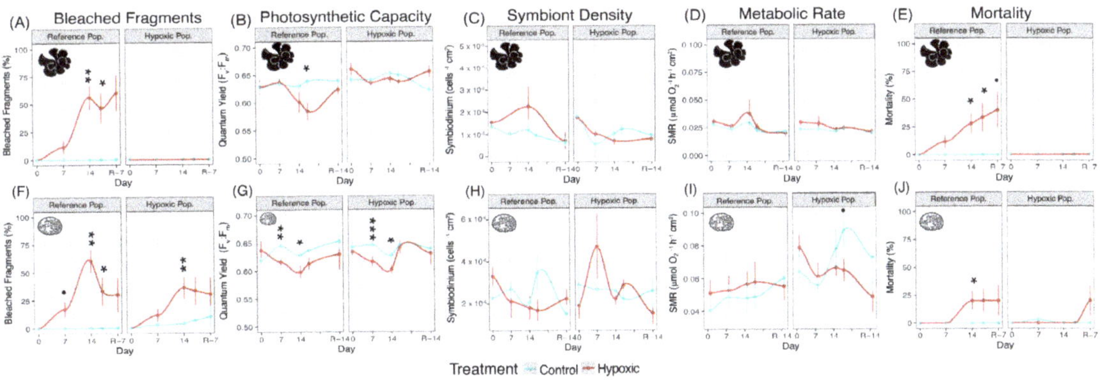

Figure 5. Stress responses accrued in both species as the hypoxic event persisted, but the hypoxic population tolerated more stress than the reference population. (**A–E**) Top row shows the measured responses of *A. tenuifolia* colonies collected from both a reference and hypoxic site (paired plots) and exposed to 14 nights of severe low oxygen in the laboratory (grey shaded area) followed by 14 days of full reoxygenation in non-shaded area, i.e., recovery period. (**F–J**) Bottom row shows the same responses in *S. siderea*. In both species, we counted the number of fragments that bleached at each timepoint (left plots), the photosynthetic capacity, symbiont cell density, metabolic rate, and mortality (right plots). Responses in the control treatment are shown in blue and the hypoxic treatment in red (mean ± S.E.). Stars show significant differences between treatments at each timepoint, for each species and population; $p < 0.001$ ***, 0.01 **, 0.05 *, 0.1 •.

We performed the same comparison in the reference and hypoxic populations of *S. siderea* (Figure 5F–J). We found bleaching peaked at the end of the event in both populations but was nearly twice as high in the reference population (60 vs. 36%; Figure 5F). While visibly observable bleaching occurred at the end of the event, both populations experienced significant reductions in their photosynthetic capacity after the first week (Figure 5G). There was a 4.6% decrease in yield after the first 7 nights, which worsened into the second week. Three days after the event, yield from the hypoxic population increased to near pre-exposure levels (~0.64), and their color began returning (Figure 5G). The color and photosynthetic yield of the reference population marginally increased post-event so that treatment effects were statistically undetectable, but they never reached their respective pre-event levels (Figure 5G). Symbiont density was not impacted by treatment at any point (Figure 5H). Metabolic rates were slightly elevated in all reference corals under hypoxia, but no treatment effects were statistically significant. This differed from responses of the hypoxic population, where rates were marginally lower under hypoxia compared to controls 3 d after the event (Figure 5I). In mortality assessments, only one of the five colonies from the reference population died, with all its replicate fragments impacted at the end of the two-week event. While this resulted in 20% mortality for the population (Figure 5J), it should be noted that the colony was a significant factor here, and every other fragment from the reference populations' other four colonies survived. In the hypoxic population, one fragment died in the first week post-event, but this was attributed to infection. This indicates that *S. siderea* colonies—regardless of prior hypoxia history—bleach quickly under hypoxia but can recover with minimal mortality. This recovery response appears strongest in the hypoxic population.

3.2. Hypoxic Population Locally Adapted to Withstand Severe Hypoxic Events

To determine if either species showed signs of adaptation for hypoxia tolerance, we built reaction norms for each trait immediately after the 2-week event (Figure 6 solid lines). In *A. tenuifolia*, we found the hypoxic population from Roldan reef was resistant to hypoxia in all traits regardless of timepoint, i.e., marginal slopes (Figure 6A–E, red lines). In contrast, the reference population had a 5% reduction in photosynthetic capacity resulting in 56% of its fragments bleaching and dying in response to hypoxia (significant treatment effects, shown by sloped blue lines). The different population responses to the same hypoxic stress are demonstrated most clearly by the significant interactions between population and treatment in photosynthetic capacity and mortality, suggesting local adaptation in these traits (Figure 6B and E, respectively). Symbiont density and metabolic rate were not significantly different across either population or treatment (Figure 6C,D).

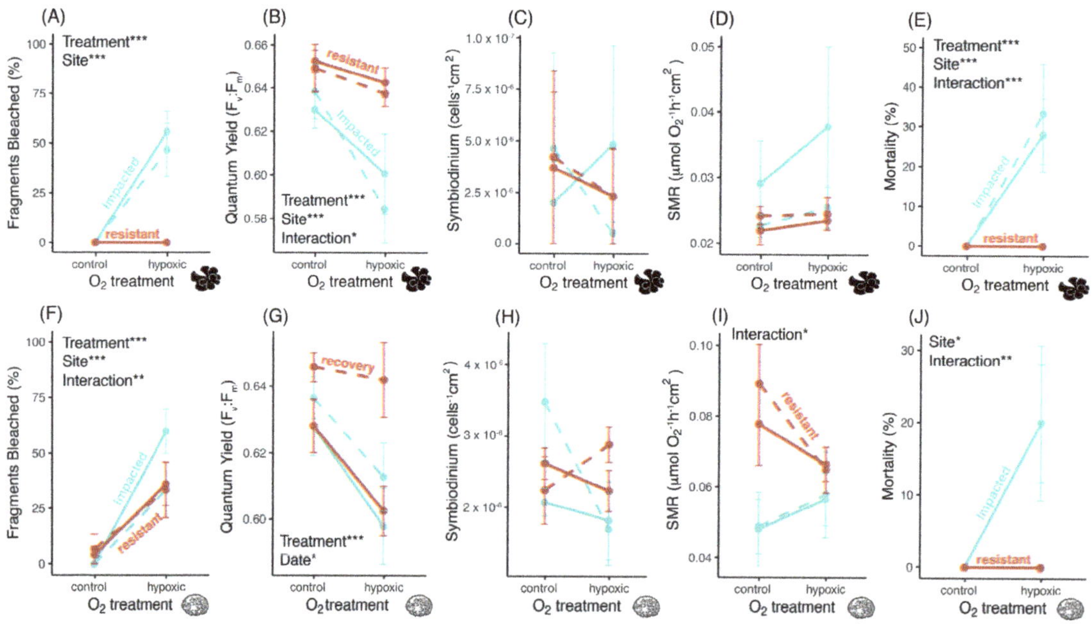

Figure 6. Populations of *A. tenuifolia* and *S. siderea* from hypoxic reefs show strong signs of local adaptation to hypoxic events compared to populations from the reference reef. Top row shows reaction norms for *A. tenuifolia* (**A–E**) and bottom row shows them for *S. siderea* (**F–J**). Points and error bars on each plot indicate the mean and S.E. Responses from the reference populations are shown in blue, while responses from the hypoxic populations are in red. The number of bleached fragments, photosynthetic capacity, symbiont cell density, metabolic rate, and mortality are shown from left to right. Solid lines show responses directly after the 14-night hypoxic event, while dashed lines show responses 3 days post-event when conditions were reoxygenated. Statistically significant terms in the 'treatment X population + date' models for each trait are indicated above each plot, with the number of asterisks next to each term identifying the significance: $p = 0.001$ ***, 0.01 **, 0.05 *. Colored labeling on lines aids in the interpretation of population responses and highlights significant effects for each trait, i.e., resistance or negatively impacted (see Figure 4). A significant effect of 'date' is indicated only in (**G**), which shows a recovery response with contrasting dashed and solid red lines. Unlabeled reaction norms have no corresponding significant effects.

When considering recovery potential for *A. tenuifolia*, reaction norms for traits measured 3 d post-event showed no evidence that recovery time influenced any response in either population. Responses 3-d and 2-weeks post-event were no different from those measured immediately after the event in all traits (all dashed lines overlapped solid lines, and there was no significant date effect in any trait). Overall, these results indicate that reference *A. tenuifolia* population is highly vulnerable to hypoxia and cannot recovery, but that the hypoxic population just 15 km away is well adapted to cope with the same stress.

In *S. siderea*, the hypoxic population also showed signs of local adaptation (Figure 5F–J). Bleaching was noticeably more severe in the reference population, affecting 60% of the fragments compared to 36% of those from the hypoxic population (Figure 6F, significant interaction). Photosynthetic yield decreased by 4% after the event in both populations but during the 3-d post-event, the hypoxic population was able to recover all but 0.3% of its lost yield (Figure 6G, significant treatment and date). No differences in symbiont density were found between populations (Figure 6H), but the hypoxic population did show signs of a metabolic adaptation (Figure 6I). Metabolic rate was significantly higher when conditions were well oxygenated (0.08 μmol h^{-1} cm^2) and was suppressed by 15% to 0.06 μmol h^{-1} cm^2 after the hypoxic event. In the reference population, metabolic rates remained low at ~0.05 μmol h^{-1} cm^2 regardless of changing oxygen conditions (Figure 6I, significant interaction with post-hoc comparisons identifying treatment differences in the hypoxic population response, $p < 0.05$).

These physiological responses culminate with mortality, where we again find differences between the two populations (Figure 6J, significant interaction). None of the hypoxic population fragments died, but 20% of the reference populations did. This response was strongly driven by colony identity, with all the fragments that succumbed to hypoxia coming from the same parent colony. Besides mortality, the parent colony was a significant factor in the other traits, highlighting the high within-population variance. Together, these results indicate the hypoxia *S. siderea* population has local adaptation signatures promoting both metabolic downregulation, survival of, and recovery from, bleaching after hypoxic events.

4. Discussion

Coral colonies locally adapted to the short-term acute hypoxia on coral reefs have until now not been identified. We found evidence suggesting local adaptation in native populations of both *A. tenuifolia* and *S. siderea* from a reef that experienced episodic hypoxia in the past. The traits associated with these adaptations include bleaching resistance and recovery, metabolic suppression, and survival following a simulated hypoxic event where low oxygen persisted nightly for 2 weeks. In contrast, a reference population with hypoxia-naïve colonies suffered from bleaching, tissue loss, and mortality.

This study employed a laboratory-based reciprocal transplant experiment, for which there are several inherent limitations to consider. First, the traits being assessed are based on the average response from an experimentally selected sub-population composed of assumed genetically distinct individuals [11]. In corals, their clonal nature makes these experiments simpler than with other metazoans because genetically distinct colonies are thought of as individuals and are split into different testing environments, instead of being raised over multiple generations [37]. However, research has shown that there can be high intra-colonial genetic variability in stress-related traits that can confound results, such that one part of the colony will perform better than another even though they are genetically identical [38]. Secondly, the assumed genetic responses cannot be differentiated from carry-over of previously expressed plastic responses, e.g., inter- or trans-generational plasticity or epigenetic effects, not genetic adaptation [39,40]. The third limitation involves the fact that corals are symbiotic animals, and one colony can harbor many different lineages of

Symbiodiniaceae, even from multiple species. Different *Symbiodinium* populations can have differential physiologies, some of which have distinct thermal tolerances causally linked to coral acclimation and resilience [41,42]. Identifying if certain lineages can also offer hypoxia tolerance will be a useful next step to disentangle host and symbiont responses to hypoxia. Likewise, *Symbiodinium* populations have independently evolved, or perhaps even co-evolved, in ways that can impart stress tolerance on their coral hosts [43,44] and may obscure host-specific responses. Lastly, the small coral sub-populations we used in this study likely do not capture true population-level responses. Given these issues, our results are suggestive but not conclusive of evolutionary adaptation. Additional studies will be necessary to identify if there is a genetic basis to the hypoxia tolerance observed, in both the corals and in their symbionts. We discuss our results with these limitations in mind.

We found *S. siderea* was generally more hypoxia tolerant than *A. tenuifolia*, and that it showed evidence of local adaptation potential. It is broadly known as a stress-tolerant generalist that has historically been one of the primary reef builders in the tropical Atlantic [45]. On reefs in Belize, *S. siderea* colonies transplanted between reefs with contrasting environmental variability for multiple years showed high transcriptome plasticity that facilitated acclimatization to environmental perturbation [46]. Such gene expression plasticity could be a mechanism of coral adaptation to variable environments [47] and may be the underlying reason for the phenotypic differences we found between the reference and hypoxic *S. siderea* populations in this study. There is also new evidence that these local populations could represent different independent genetic lineages [48]. Preliminary morphology analyses support this possibility that the two populations tested in this study belong to distinct lineages.

Hypoxia tolerance in *S. siderea* may come at a cost. The skeletal densities of hypoxic population colonies were markedly less dense than those from the reference site. Each colony was filled with boring bivalve holes and sustained severe skeletal erosion in areas without living tissue (Figures 1 and 4C). Similar signs of reduced skeletal density, including thickness and high porosity, were found in *Porites* from a mangrove lagoon in Australia [15]. The authors of that study attributed the high porosity to metabolic energy redirection driven by low oxygen and low pH. Co-occurring low pH at Roldan may influence *S. siderea* morphological differences [30], but an energetic trade-off from oxygen alone is unlikely given the hypoxic *S. siderea* populations' strong metabolic plasticity (Figure 6I). Another concern besides low skeletal density is high susceptibility to disease. After fragmentation, before the experiment, we noted signs of rapid infection only impacting colonies from the hypoxic site. Despite these observed impacts, *S. siderea* still dominates these hypoxic reefs. In 2023, *S. siderea* accounted for 94% of the total living coral cover at 10 m in Roldan [49].

On the Roldan reef slope 2–5 m directly above the *S. siderea* zone, *A. tenuifolia* is also thriving and showing signs of local adaptation. *A. tenuifolia* makes up the second most abundant species at this depth after *Porites furcata* [24]. Like *S. siderea*, *A. tenuifolia* is known as an opportunistic species; however, it differs in that it is a fast-growing weedy species that can quickly colonize shallow reef sites [50]. It is also a brooding species, where fertilization is internal and colonies release planulae larvae that settle quickly, often near the parent colony and with limited dispersal, a life history strategy that can promote local adaptation [51]. Unlike *S. siderea*, which demonstrated a greater overall stress resistance, the two populations of *A. tenuifolia* in this study had distinct divergent responses to hypoxia. The colonies native to the hypoxic reef were able to persist, while the naïve colonies from the reference site died. The observable impact occurred rapidly, with bleaching so tightly coupled to tissue loss and mortality that it was difficult to differentiate between the responses (Figure 5A,E). Given this limitation and the lack of symbiont cell densities between treatments (Figure 6), host necrosis may precede symbiont expulsion under acute hypoxia stress in this species [52]. The combination of life history strategies, high abundance on hypoxic reefs, and the experimental evidence presented here, all indicate *A. tenuifolia* has locally adapted to flourish regardless of severe hypoxia.

The physical and ecological attributes in Bahía Almirante appear to promote hypoxia adaptation in ways that may also apply to other tropical semi-enclosed lagoon reef systems. Often the first serious hypoxic event in an area has a disproportionately large response [53], rapidly culling intolerant species and populations [11], e.g., *Acropora cervicornis* [24]. This may be why the reference populations of both coral species in this study suffered so severely. The corals from the Hospital Point reference reef were not impacted by past extreme hypoxic events in 2010 or 2017 [23,28] despite being less than 15 km from the Roldan hypoxic reef. Such differences between nearby sites also illustrate how hypoxia can impart selective pressure on relatively small spatial scales. Additionally, the spatial gradient was well defined by multiple environmental conditions that not only included oxygen, but also pH, temperature, and nutrients [30,31]. This implies corals from these systems, and along these gradients, may also be tolerant to a suite of stressful conditions. Testing corals' capacity to cope with these co-occurring stressors will provide insight into fitness trade-offs and limitations, and mechanisms of long-term realistic survival [54,55]. This has been studied in reef-associated brittle stars with tests of critical oxygen pressures (pO_2^{crit}) over a range of temperatures which identified species-specific thresholds accounting for thermal and oxygen conditions changing together [31]. Given many tropical semi-enclosed lagoons share these characteristics [8], further investigations will be needed to understand the breadth of coral hypoxia tolerance at larger scales and the generality of these responses over thermal and pH gradients.

There is growing recognition that coral reef managers and practitioners should embrace coral reefs that are not "pristine" to develop effective strategies for coral reef protection [56,57]. As episodic low oxygen is expected to become more frequent and severe on coral reefs with continued warming [31,58], coral restoration and conservation efforts should identify, protect, and explore the potential applications of corals that are also hypoxia-tolerant. Here we identified populations of two coral species from a Caribbean reef that are tolerant to low oxygen. Our findings present possibilities to explore their potential in assisted migration and assisted evolution efforts, or in activities like ecological replacement, a concept based on finding corals that have adapted to stress and relocating them to new reefs to serve the same functional roles as the local species that have disappeared [59]. With all approaches, protecting hypoxia-tolerant corals from reefs that have undergone past hypoxic events will likely be a fruitful endeavor. This is especially relevant as the compounding threats of extreme ocean weather continue to add pressure to coral reefs.

Author Contributions: Conceptualization, N.L. and R.C.; methodology, N.L. and R.C.; formal analysis, N.L.; investigation, N.L., C.C.-Á., A.E. and P.V.; resources, N.L. and R.C.; data curation, N.L., C.C.-Á. and A.E.; writing—original draft preparation, N.L.; writing—review and editing, C.C.-Á. and R.C.; visualization, N.L.; supervision, N.L. and R.C.; project administration, R.C.; funding acquisition, N.L. and R.C. All authors have read and agreed to the published version of the manuscript.

Funding: This work was supported by the National Science Foundation grant BIO-OCE 2048955 and the High Meadows Environmental Institute at Princeton University.

Institutional Review Board Statement: Not applicable.

Informed Consent Statement: Not applicable.

Data Availability Statement: Raw data from experiments can be accessed here: https://figshare.com/s/a700a76da2d7e3e1bf94 (accessed on 20 August 2024). Upon publication, access to this data will be updated and open to the public.

Acknowledgments: We thank Andrew Altieri and Julie Meyer for their comments on the experimental design and securing funding for this research. We also thank Adriana M. Cruz Olmo for laboratory assistance, and Plinio Gondola with all the STRI Bocas del Toro Research Station staff for technical support. Finally, we thank the Panamanian government for permission to conduct this research (Permit #4149).

Conflicts of Interest: The authors declare no conflicts of interest.

References

1. Calvin, K.; Dasgupta, D.; Krinner, G.; Mukherji, A.; Thorne, P.W.; Trisos, C.; Romero, J.; Aldunce, P.; Barrett, K.; Blanco, G.; et al. *Climate Change 2023: Synthesis Report. Contribution of Working Groups I, II and III to the Sixth Assessment Report of the Intergovernmental Panel on Climate Change*; Core Writing Team, Lee, H., Romero, J., Eds.; Intergovernmental Panel on Climate Change (IPCC): Geneva, Switzerland, 2023.
2. Hoegh-Guldberg, O.; Kennedy, E.V.; Beyer, H.L.; McClennen, C.; Possingham, H.P. Securing a Long-Term Future for Coral Reefs. *Trends Ecol. Evol.* **2018**, *33*, 936–944. [CrossRef]
3. Mellin, C.; Brown, S.; Cantin, N.; Klein-Salas, E.; Mouillot, D.; Heron, S.F.; Fordham, D.A. Cumulative Risk of Future Bleaching for the World's Coral Reefs. *Sci. Adv.* **2024**, *10*, eadn9660. [CrossRef] [PubMed]
4. Camp, E.F.; Schoepf, V.; Suggett, D.J. How Can "Super Corals" Facilitate Global Coral Reef Survival under Rapid Environmental and Climatic Change? *Glob. Change Biol.* **2018**, *24*, 2755–2757. [CrossRef] [PubMed]
5. Camp, E.F. Contingency Planning for Coral Reefs in the Anthropocene; The Potential of Reef Safe Havens. *Emerg. Top. Life Sci.* **2022**, *6*, 107–124. [CrossRef]
6. Boström-Einarsson, L.; Babcock, R.C.; Bayraktarov, E.; Ceccarelli, D.; Cook, N.; Ferse, S.C.A.; Hancock, B.; Harrison, P.; Hein, M.; Shaver, E.; et al. Coral Restoration—A Systematic Review of Current Methods, Successes, Failures and Future Directions. *PLoS ONE* **2020**, *15*, e0226631. [CrossRef]
7. Camp, E.F.; Schoepf, V.; Mumby, P.J.; Hardtke, L.A.; Rodolfo-Metalpa, R.; Smith, D.J.; Suggett, D.J. The Future of Coral Reefs Subject to Rapid Climate Change: Lessons from Natural Extreme Environments. *Front. Mar. Sci.* **2018**, *5*, 4. [CrossRef]
8. Schoepf, V.; Baumann, J.H.; Barshis, D.J.; Browne, N.K.; Camp, E.F.; Comeau, S.; Cornwall, C.E.; Guzmán, H.M.; Riegl, B.; Rodolfo-Metalpa, R.; et al. Corals at the Edge of Environmental Limits: A New Conceptual Framework to Re-Define Marginal and Extreme Coral Communities. *Sci. Total Environ.* **2023**, *884*, 163688. [CrossRef] [PubMed]
9. Camp, E.F.; Edmondson, J.; Doheny, A.; Rumney, J.; Grima, A.J.; Huete, A.; Suggett, D.J. Mangrove Lagoons of the Great Barrier Reef Support Coral Populations Persisting under Extreme Environmental Conditions. *Mar. Ecol. Prog. Ser.* **2019**, *625*, 1–14. [CrossRef]
10. Caruso, C.; Hughes, K.; Drury, C. Selecting Heat-Tolerant Corals for Proactive Reef Restoration. *Front. Mar. Sci.* **2021**, *8*, 632027. [CrossRef]
11. Kawecki, T.J.; Ebert, D. Conceptual Issues in Local Adaptation. *Ecol. Lett.* **2004**, *7*, 1225–1241. [CrossRef]
12. Chan, Y.-F.; Chen, Y.-H.; Yu, S.-P.; Chen, H.-J.; Nozawa, Y.; Tang, S.-L. Reciprocal Transplant Experiment Reveals Multiple Factors Influencing Changes in Coral Microbial Communities across Climate Zones. *Sci. Total Environ.* **2024**, *907*, 167929. [CrossRef]
13. Tanvet, C.; Camp, E.F.; Sutton, J.; Houlbrèque, F.; Thouzeau, G.; Rodolfo-Metalpa, R. Corals Adapted to Extreme and Fluctuating Seawater pH Increase Calcification Rates and Have Unique Symbiont Communities. *Ecol. Evol.* **2023**, *13*, e10099. [CrossRef]
14. Kurihara, H.; Watanabe, A.; Tsugi, A.; Mimura, I.; Hongo, C.; Kawai, T.; Reimer, J.D.; Kimoto, K.; Gouezo, M.; Golbuu, Y. Potential Local Adaptation of Corals at Acidified and Warmed Nikko Bay, Palau. *Sci. Rep.* **2021**, *11*, 11192. [CrossRef]
15. Scucchia, F.; Zaslansky, P.; Boote, C.; Doheny, A.; Mass, T.; Camp, E.F. The Role and Risks of Selective Adaptation in Extreme Coral Habitats. *Nat. Commun.* **2023**, *14*, 4475. [CrossRef]
16. Mayfield, A.B.; Chan, P.-H.; Putnam, H.M.; Chen, C.-S.; Fan, T.-Y. The Effects of a Variable Temperature Regime on the Physiology of the Reef-Building Coral Seriatopora Hystrix: Results from a Laboratory-Based Reciprocal Transplant. *J. Exp. Biol.* **2012**, *215*, 4183–4195. [CrossRef]
17. Jury, C.P.; Delano, M.N.; Toonen, R.J. High Heritability of Coral Calcification Rates and Evolutionary Potential under Ocean Acidification. *Sci. Rep.* **2019**, *9*, 20419. [CrossRef]
18. Ito, T.; Minobe, S.; Long, M.C.; Deutsch, C. Upper Ocean O_2 Trends: 1958–2015. *Geophys. Res. Lett.* **2017**, *44*, 4214–4223. [CrossRef]

19. Schmidtko, S.; Stramma, L.; Visbeck, M. Decline in Global Oceanic Oxygen Content during the Past Five Decades. *Nature* **2017**, *542*, 335–351. [CrossRef] [PubMed]
20. Rabalais, N.N.; Díaz, R.J.; Levin, L.A.; Turner, R.E.; Gilbert, D.; Zhang, J. Dynamics and Distribution of Natural and Human-Caused Hypoxia. *Biogeosciences* **2010**, *7*, 585–619. [CrossRef]
21. Diaz, R.J. Overview of Hypoxia around the World. *J. Environ. Qual.* **2001**, *30*, 275–281. [CrossRef] [PubMed]
22. Deutsch, C.; Penn, J.L.; Lucey, N. Climate, Oxygen, and the Future of Marine Biodiversity. *Annu. Rev. Mar. Sci.* **2024**, *16*, 217–245. [CrossRef] [PubMed]
23. Johnson, M.D.; Scott, J.J.; Leray, M.; Lucey, N.M.; Bravo, L.M.R.; Wied, W.L.; Altieri, A.H. Rapid Ecosystem-Scale Consequences of Acute Deoxygenation on a Caribbean Coral Reef. *Nat. Commun.* **2021**, *12*, 4522. [CrossRef] [PubMed]
24. Lucey, N.; Cesar, C.; Eckert, A.; Rajagopalan, A.; Brister, W.; Altieri, A.; Deutsch, C.; Collin, R. Coral Community Composition Linked to Hypoxia Exposure. *Glob. Change Biol.* **2024**, *30*, e17545. [CrossRef] [PubMed]
25. Laboy-Nieves, E.N.; Klein, E.; Conde, J.E.; Losada, F.; Cruz, J.J.; Bone, D. Mass Mortality of Tropical Marine Communities in Morrocoy, Venezuela. *Bull. Mar. Sci.* **2001**, *68*, 163–179.
26. Johnston, M.A.; Nuttall, M.F.; Eckert, R.J.; Blakeway, R.D.; Sterne, T.K.; Hickerson, E.L.; Schmahl, G.P.; Lee, M.T.; Macmillan, J. Localized Coral Reef Mortality Event at East Flower Garden Bank, Gulf of Mexico. *Bull. Mar. Sci.* **2019**, *95*, 239–250. [CrossRef]
27. Rodríguez-martínez, R.E.; Medina-valmaseda, A.E.; Blanchon, P.; Monroy-velázquez, L.V.; Ciencias, I.D.; Nacional, U.; México, A.D.; Avenida, P.; Héroes, N.; Morelos, P.; et al. Faunal Mortality Associated with Massive Beaching and Decomposition of Pelagic Sargassum. *Mar. Pollut. Bull.* **2019**, *146*, 201–205. [CrossRef]
28. Altieri, A.H.; Harrison, S.B.; Seemann, J.; Collin, R.; Diaz, R.J.; Knowlton, N. Tropical Dead Zones and Mass Mortalities on Coral Reefs. *Proc. Natl. Acad. Sci. USA* **2017**, *114*, 3660–3665. [CrossRef] [PubMed]
29. Adelson, A.E.; Altieri, A.H.; Boza, X.; Collin, R.; Davis, K.A.; Gaul, A.; Giddings, S.N.; Reed, V.; Pawlak, G. Seasonal Hypoxia and Temperature Inversions in a Tropical Bay. *Limnol. Oceanogr.* **2022**, *67*, 2174–2189. [CrossRef]
30. Lucey, N.M.; Haskett, E.; Collin, R. Multi-Stressor Extremes Found on a Tropical Coral Reef Impair Performance. *Front. Mar. Sci.* **2020**, *7*, 588764. [CrossRef]
31. Lucey, N.M.; Deutsch, C.A.; Carignan, M.-H.; Vermandele, F.; Collins, M.; Johnson, M.D.; Collin, R.; Calosi, P. Climate Warming Erodes Tropical Reef Habitat through Frequency and Intensity of Episodic Hypoxia. *PLoS Clim.* **2023**, *2*, e0000095. [CrossRef]
32. Lucey, N.M.; Collins, M.; Collin, R. Oxygen-Mediated Plasticity Confers Hypoxia Tolerance in a Corallivorous Polychaete. *Ecol. Evol.* **2020**, *10*, 1145–1157. [CrossRef]
33. Warner, M.E.; Fitt, W.K.; Schmidt, G.W. Damage to Photosystem II in Symbiotic Dinoflagellates: A Determinant of Coral Bleaching. *Proc. Natl. Acad. Sci. USA* **1999**, *96*, 8007–8012. [CrossRef] [PubMed]
34. Veal, C.J.; Carmi, M.; Fine, M.; Hoegh-Guldberg, O. Increasing the Accuracy of Surface Area Estimation Using Single Wax Dipping of Coral Fragments. *Coral Reefs* **2010**, *29*, 893–897. [CrossRef]
35. Harianto, J.; Carey, N.; Byrne, M. respR—An R Package for the Manipulation and Analysis of Respirometry Data. *Methods Ecol. Evol.* **2019**, *10*, 912–920. [CrossRef]
36. Pigliucci, M. *Phenotypic Plasticity: Beyond Nature and Nurture*; Johns Hopkins University Press: Baltimore, MA, USA, 2001.
37. Reznick, D.N.; Ghalambor, C.K. The Population Ecology of Contemporary Adaptations: What Empirical Studies Reveal about the Conditions That Promote Adaptive Evolution. *Genetica* **2001**, *112*, 183–198. [CrossRef]
38. Schweinsberg, M.; Weiss, L.C.; Striewski, S.; Tollrian, R.; Lampert, K.P. More than One Genotype: How Common Is Intracolonial Genetic Variability in Scleractinian Corals? *Mol. Ecol.* **2015**, *24*, 2673–2685. [CrossRef] [PubMed]
39. Ghalambor, C.K.; McKay, J.K.; Carroll, S.P.; Reznick, D.N. Adaptive versus Non-Adaptive Phenotypic Plasticity and the Potential for Contemporary Adaptation in New Environments. *Funct. Ecol.* **2007**, *21*, 394–407. [CrossRef]
40. Chevin, L.-M.; Hoffmann, A.A. Evolution of Phenotypic Plasticity in Extreme Environments. *Philos. Trans. R. Soc. B Biol. Sci.* **2017**, *372*, 20160138. [CrossRef] [PubMed]
41. Berkelmans, R.; van Oppen, M.J.H. The Role of Zooxanthellae in the Thermal Tolerance of Corals: A 'Nugget of Hope' for Coral Reefs in an Era of Climate Change. *Proc. R. Soc. B Biol. Sci.* **2006**, *273*, 2305–2312. [CrossRef] [PubMed]
42. Manzello, D.P.; Matz, M.V.; Enochs, I.C.; Valentino, L.; Carlton, R.D.; Kolodziej, G.; Serrano, X.; Towle, E.K.; Jankulak, M. Role of Host Genetics and Heat-Tolerant Algal Symbionts in Sustaining Populations of the Endangered Coral Orbicella Faveolata in the Florida Keys with Ocean Warming. *Glob. Change Biol.* **2019**, *25*, 1016–1031. [CrossRef] [PubMed]
43. Rowan, R. Thermal Adaptation in Reef Coral Symbionts. *Nature* **2004**, *430*, 742. [CrossRef] [PubMed]
44. Fabina, N.S.; Putnam, H.M.; Franklin, E.C.; Stat, M.; Gates, R.D. Transmission Mode Predicts Specificity and Interaction Patterns in Coral-Symbiodinium Networks. *PLoS ONE* **2012**, *7*, e44970. [CrossRef] [PubMed]

45. Gelais, A.T.S.; Chaves-Fonnegra, A.; Brownlee, A.S.; Kosmynin, V.N.; Moulding, A.L.; Gilliam, D.S. Fecundity and Sexual Maturity of the Coral Siderastrea Siderea at High Latitude along the Florida Reef Tract, USA. *Invertebr. Biol.* **2016**, *135*, 46–57. [CrossRef]
46. Castillo, K.D.; Bove, C.B.; Hughes, A.M.; Powell, M.E.; Ries, J.B.; Davies, S.W. Gene Expression Plasticity Facilitates Acclimatization of a Long-Lived Caribbean Coral across Divergent Reef Environments. *Sci. Rep.* **2024**, *14*, 7859. [CrossRef] [PubMed]
47. Kenkel, C.D.; Matz, M.V. Gene Expression Plasticity as a Mechanism of Coral Adaptation to a Variable Environment. *Nat. Ecol. Evol.* **2016**, *1*, 14. [CrossRef] [PubMed]
48. Grupstra, C.G.B.; Gómez-Corrales, M.; Fifer, J.E.; Aichelman, H.E.; Meyer-Kaiser, K.S.; Prada, C.; Davies, S.W. Integrating Cryptic Diversity into Coral Evolution, Symbiosis and Conservation. *Nat. Ecol. Evol.* **2024**, *8*, 622–636. [CrossRef]
49. Lucey, N.M.; Cesar, C.; Collin, R.; Deutsch, C.A. Rates of Community Metabolism on Tropical Coral Reefs Severely Underestimated. Unpublished Manuscript.
50. Aronson, R.B.; Macintyre, I.G.; Precht, W.F.; Murdoch, T.J.T.; Wapnick, C.M. The Expanding Scale of Species Turnover Events on Coral Reefs in Belize. *Ecol. Monogr.* **2002**, *72*, 233–249. [CrossRef]
51. Sanford, E.; Kelly, M.W. Local Adaptation in Marine Invertebrates. *Annu. Rev. Mar. Sci.* **2011**, *3*, 509–535. [CrossRef] [PubMed]
52. Helgoe, J.; Davy, S.K.; Weis, V.M.; Rodriguez-Lanetty, M. Triggers, Cascades, and Endpoints: Connecting the Dots of Coral Bleaching Mechanisms. *Biol. Rev.* **2024**, *99*, 715–752. [CrossRef] [PubMed]
53. Camillo, C.G.D.; Cerrano, C. Mass Mortality Events in the NW Adriatic Sea: Phase Shift from Slow- to Fast-Growing Organisms. *PLoS ONE* **2015**, *10*, e0126689. [CrossRef]
54. Boyd, P.W.; Lennartz, S.T.; Glover, D.M.; Doney, S.C. Biological Ramifications of Climate-Change-Mediated Oceanic Multi-Stressors. *Nat. Clim. Change* **2014**, *5*, 71–79. [CrossRef]
55. Sampaio, E.; Santos, C.; Rosa, I.C.; Ferreira, V.; Pörtner, H.; Duarte, C.M.; Levin, L.A.; Rosa, R. Impacts of Hypoxic Events Surpass Those of Future Ocean Warming and Acidification. *Nat. Ecol. Evol.* **2021**, *5*, 311–321. [CrossRef] [PubMed]
56. McClanahan, T.R. Reconsidering and Rescaling Climate Change Predictions for Coral Reefs. *Nat. Clim. Change* **2024**, *14*, 779–781. [CrossRef]
57. Streit, R.P.; Morrison, T.H.; Bellwood, D.R. Coral Reefs Deserve Evidence-Based Management Not Heroic Interference. *Nat. Clim. Chang.* **2024**, *14*, 773–775. [CrossRef]
58. Pezner, A.K.; Courtney, T.A.; Barkley, H.C.; Chou, W.-C.; Chu, H.-C.; Clements, S.M.; Cyronak, T.; DeGrandpre, M.D.; Kekuewa, S.A.H.; Kline, D.I.; et al. Increasing Hypoxia on Global Coral Reefs under Ocean Warming. *Nat. Clim. Change* **2023**, *13*, 403–409. [CrossRef]
59. Webster, M.M.; Schindler, D.E. Ecological Replacement for Reef-Building Corals. *Nat. Clim. Change* **2024**, *14*, 776–778. [CrossRef]

oceans

Article

An Unexpected Small Biodiversity Oasis of Sea Slugs (Mollusca, Gastropoda, Heterobranchia) in the Largest Petrochemical Hub of Italy (Central Mediterranean)

Andrea Lombardo [1] and Giuliana Marletta [2,*]

1 Department of Biological, Geological and Environmental Sciences, University of Catania, 95124 Catania, Italy; andylombardo94@gmail.com
2 Independent Researcher, 95124 Catania, Italy
* Correspondence: giulianamarletta93@gmail.com

Abstract: The Magnisi peninsula is a small portion of land located near the largest Italian petrol-chemical pole of Augusta–Priolo–Melilli (40 km^2), which, since the 1950s, devastated the local environment and landscape and unloaded directly into the sea an impressive quantity of pollutants. Unlike the terrestrial part of the area, where a Natural Oriented Reserve (NOR) called "Saline di Priolo" was established in the 2000s, no concrete legislative action has been implemented or proposed for the marine environment. At the same time, the fauna of the marine environment has not been studied in the same way as that of the terrestrial environment. As concerns the molluscan fauna, most of the information dates back to the 1800s. These studies were exclusively focused on the shells of some mollusks in the area. Instead, no study related to this area has ever been carried out on the group of sea slugs. This study conducted between 2022 and 2023, through snorkeling activities, allowed to provide the first faunistic list of the sea slugs of this area, together with information on the biology and ecology of these gastropods, highlighting the potential biodiversity present in this small stretch of coastline affected by high industrial pollution.

Keywords: industrial pollution; marine heterobranchia; Sicily; Mediterranean sea

1. Introduction

The Magnisi peninsula (also called Thapso, Thapsos, or Thapsus in ancient times [1,2]) is a small tear-shaped land portion located between the Augusta and Priolo–Melilli gulfs along the southern–eastern coast of Sicily (Italy), connected to the mainland by a thin and low sandy isthmus [1,3] (Figure 1A–C). The name Magnisi derives from the denomination "Magna Insula" (i.e., large island), a term formerly given to this peninsula by the Norsemen [1,2]. This calcareous peninsula, presenting a length of 2000–2300 m and a maximum width of 700–800 m in its central axis, is characterized by a generally flat surface (its maximum height is 16–20 m above the sea level) and steep coasts [1,4]. These latter are generally high along almost the entire peninsula border, except on the southeast coasts [1].

This small peninsula is located close to the largest Italian petrol–chemical pole of Augusta–Priolo–Melilli (40 km^2), which since the 1950s has not only devastated and heavily disfigured the local environment and landscape (as well as people's health) but has also dumped an impressive amount of pollutants such as benzene, chlorobenzenes, and mercury directly into the sea [3,5,6]. In the area that connects the peninsula with the sandy isthmus, there was an industrial area (nowadays abandoned) where bromide was synthesized through the acquisition and subsequent discharge of seawater, causing extensive damage to the marine environment [3].

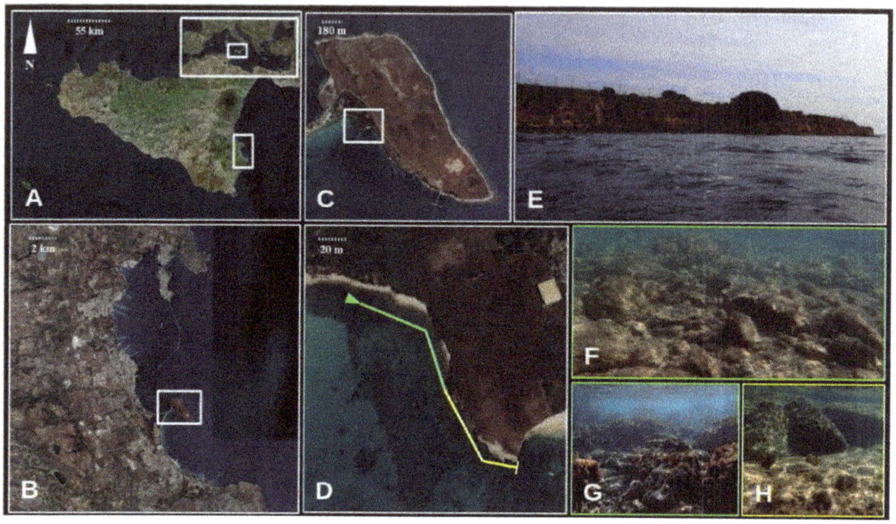

Figure 1. Study area (**A**) Sicily and the location of the gulfs of Augusta and Priolo–Melilli; these latter are indicated by the white box in the lower right corner (the white box in the top right corner shows the position of Sicily (Italy) in the Mediterranean Sea); (**B**) Gulfs of Augusta and Priolo–Melilli, the Magnisi peninsula is indicated by the white box; (**C**) the Magnisi peninsula, the study area is indicated by the white box; (**D**) in the examined coastal stretch, the location of the two different bottom types is indicated by the different colorations. The arrow indicates the start point and the followed direction of the snorkeling path; (**E**) a general view of the examined coastal tract; (**F**,**G**) the appearance of the bottom of the first area, edge colored in green; (**H**) aspect of the bottom of the second area, edge colored in yellow.

These remarkable anthropogenic impacts have caused several damages to the local marine environment, such as a high quantity of non-living particles suspended in the water column, seawater turbidity, sedimentation, contamination of sediments and marine organisms, the presence of alien species, degradation of natural phytobenthic communities, and almost total destruction of *Posidonia oceanica* (Linnaeus) Delile meadows located in the area [7].

Although the anthropogenic impacts caused by the petrochemical hub have not yet disappeared, since the 2000s several environmental actions in the area have led to the restoration of some important wetlands and the associated birdlife through the establishment of a Natural Oriented Reserve (NOR) called "Saline di Priolo" and the homonymous Natura 2000 site (ITA 090013), into which the Magnisi peninsula was incorporated in 2013, due to its remarkable importance as both an ornithological and archaeological site [8].

However, these important environmental achievements only concerned the terrestrial domain of the area, almost completely neglecting the marine realm. Indeed, the only restoration action of the local marine environment involved the replanting of *P. oceanica* meadows with considerable success [7,9]. Nevertheless, unlike the terrestrial part of the area, no concrete legislative action has been implemented or proposed for the marine environment. At the same time, unlike the terrestrial part of the Magnisi peninsula, whose fauna is very well known, the marine fauna of this area is generally unstudied and almost completely lacking in specific studies.

As concerns the marine fauna, some studies regarding the phylum Mollusca were carried out in this area during the 1800s [10–16]. As highlighted by De Martino [17], the Magnisi peninsula (especially its south side) represents one of the most shell-rich coastal stretches of the entire Sicily. However, although these conchological studies provide an important knowledge base on the mollusks of the Magnisi peninsula, these are not specific to the area and only contain a list of mollusks with shells (a few with brief shell descriptions), without biological or ecological data of the living organisms. To date, the only description of the molluscan fauna of the Magnisi peninsula is that reported by De Martino [17], albeit very concise and concerning only stranded shells.

Within the Gastropod Mollusca, the subclass Heterobranchia includes the sea slugs (formerly belonging to the subclass "Opisthobranchia" and no longer in use) in the broad sense, with the taxonomic groups Rhodopoidea, Acteonoidea, Cephalaspidea, Runcinida, Aplysiida, Pteropoda, Umbraculida, Ringiculimorpha, Pleurobranchida, Nudibranchia, Acochlidiimorpha, and Sacoglossa [18,19]. Thanks to their showy and colorful body shapes, these marine critters have become one of the most favorite subjects of underwater photographers worldwide. This can be easily noted by the large number of books and websites on them, e.g., [20–24].

For the Magnisi peninsula area, there are no data on the sea slugs. The only information about this group of mollusks is regarded as isolated or old findings of a few stranded sea slug shells (Table 1). Consequently, given the presumed malacological importance of this area and the almost total absence of data on this group of gastropods, this study aims to provide the first faunistic list (together with biological and ecological information) of the sea slugs of this Sicilian coastal stretch.

Table 1. Sea slug species documented in the literature of the Magnisi peninsula.

Taxa	References
Superorder Ringiculimorpha	
Family Ringiculidae R. A. Philippi, 1853	
Ringicula auriculata (Ménard de la Groye, 1811)	[10,11]
Ringicula gianninii F. Nordsieck, 1974	[12,14]
Order Pleurobranchida	
Family Pleurobranchidae Gray, 1827	
Berthella aurantiaca (Risso, 1818)	[16]
Order Cephalaspidea	
Family Bullidae Gray, 1827	
Bulla striata Bruguière, 1792	[17]
Family Retusidae Thiele, 1925	
Retusa laevisculpta (Granata Grillo, 1877)	[16]
Retusa mammillata (R. A. Philippi, 1836)	[10,11]
Retusa umbilicata (Montagu, 1803)	[12]
Family Rhizoridae Dell, 1952	
Volvulella acuminata (Bruguière, 1792)	[10–12]
Family Haminoeidae Pilsbry, 1895	
Haminoea navicula (da Costa, 1778)	[12,17]
Roxaniella jeffreysi (Weinkauff, 1866)	[16]
Weinkauffia turgidula (Forbes, 1844)	[16]
Family Philinidae Gray, 1850 (1815)	
Philine quadripartite Ascanius, 1772	[17]

2. Materials and Methods

This study was carried out from January 2022 to December 2023 along a small coastal stretch of the Magnisi peninsula located between the dismissed ESPESI industrial complex and the left side of Cala Magnisi (Figure 1C–E). The examined coastal stretch runs from 37°09′09.1″ N 15°13′45.7″ E to 37°09′05.0″ N 15°13′51.2″ E (Figure 1D). This area is very shallow and characterized by two different types of bottoms. The first one is present in the corresponding area of the thin white pebble beach that delimits the north and northeast parts of the shoreline and is characterized by a large number of scattered pebbles, boulders, and rocks, which almost cover a sandy gravel bed (Figure 1F,G). The second one, which characterizes the area at the base of the small southeast cliff, has very shallow, large submarine rock terraces interspersed with large oblique rock blocks (Figure 1H).

Data collection, performed through snorkeling (within 0.50–0.70 m of depth), was carried out once a month during the study period, following the same underwater path (Figure 1) from 09:30 to 11:30 a.m. in the morning. During each snorkeling session, all the sea slug specimens found during the path were photographed by an Olympus TG-4 underwater camera and counted. Moreover, the seawater temperature (in Celsius degrees) was measured and recorded for each snorkeling session with a SUUNTO D6i dive computer and included in a line chart (Figure 2). In the case of two or more temperatures recorded in the same session, only their average value was entered in the chart.

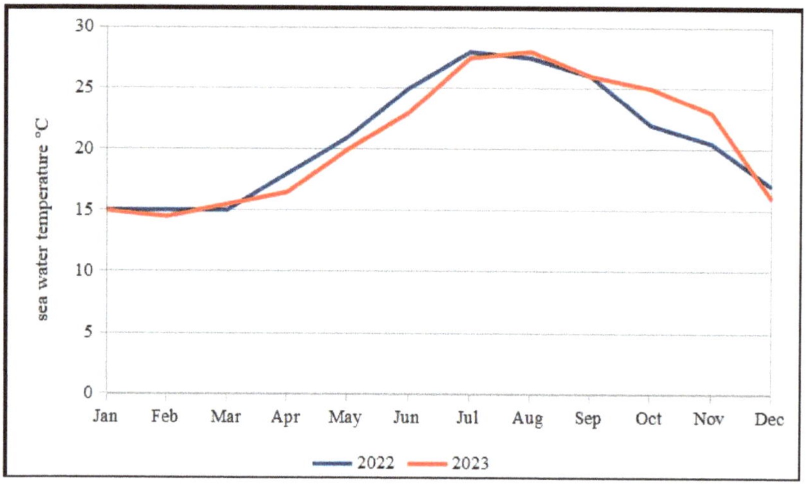

Figure 2. Monthly temperature values in the study area.

Through subsequent examination of the photographs, it was possible to make the specific identification of each sea slug and to note the different substrates on which the specimens were found. The sources consulted for the identification of the sea slugs were Prkić et al. [19], Rudman [20], Trainito and Doneddu [21], Valdés et al. [25], Galià-Camps et al. [26], and OPK Opistobranquis [24]. The identification of zoo- and phytobenthic components was performed following Trainito and Baldacconi [27] and Rodríguez-Prieto et al. [28], respectively. The used nomenclature is that reported on WoRMS [29].

For each found sea slug species, the following information was reported: abundance, temperature range, seasonality, substrates/habitats, and any remarks. The abundance data of the different species during the months of the two years of study can be seen in Table 2.

Table 2. List of all sea slug species found during this study. The numbers indicate the number of specimens documented for each month. The symbols + and ‡ indicate the finding of egg masses and the observation of mating, respectively.

Taxa	Year	January	February	March	April	May	June	July	August	September	October	November	December
Order Pleurobranchida													
Family Pleurobranchidae Gray, 1827													
Berthellina cf. *edwardsii* (Vayssière, 1897)	**2022**	2	0	0	1	0	0	0	2	2	1	0	1
	2023	0	0	0	0	0	0	2	4	6	3	2	0
Berthella stellata (Risso, 1826)	**2022**	0	0	0	0	0 +	0	0	0	0 +	0 +	0	0
	2023	0	0	0	2	0	0 +	0	0	0	0	0	0
Order Nudibranchia													
Suborder Cladobranchia													
Family Aeolidiidae Gray, 1827													
Berghia verrucicornis (A. Costa, 1867)	**2022**	0	0	0	3 ‡	1	0	0	0	0	0	0	0
	2023	0	0	0	0	0	0	0	0	0	0	0	0
Spurilla neapolitana (Delle Chiaje, 1841)	**2022**	0	2	0	0	0	0	0	0	0	0	0	0
	2023	0	0	1	1	0	0	2	0	1	1	0	1

Table 2. *Cont.*

Taxa	Year	January	February	March	April	May	June	July	August	September	October	November	December
Family Facelinidae Bergh, 1889													
Cratena peregrina (Gmelin, 1791)	2022	8	1	0	0	0	0	0	0	0	0	0	1
	2023	3	0	1	0	0	0	0	0	0	0	0	2
Facelina annulicornis (Chamisso and Eysenhardt, 1821)	2022	0	0	0	0	0	0	0	0	0	0	0	0
	2023	0	0	0	2	0	0	0	0	0	0	0	0
Facelina rubrovittata (A. Costa, 1866)	2022	6	13	0	6	2	0	0	0	0	0	0	0
	2023	0	0	0	0	0	0	0	0	1	0	0	1
Favorinus branchialis (Rathke, 1806)	2022	0	0	0	0	0	0	0	0	0	0	0	0
	2023	0	0	0	0	0	0	0	0	0	0	0	1
Family Flabellinidae Bergh, 1889													
Calmella cavolini (Vérany, 1846)	2022	0	0	0	3	0	0	1	0	0	0	1	0
	2023	0	0	0	0	0	0	0	1	1	0	0	1
Calmella gaditana (Cervera, Arcega-Gómez and F. J. García, 1987)	2022	0	1	0	0	2 ‡	0	0	0	0	0	0	0
	2023	0	0	2	0	0	0	0	0	0	0	0	0
Edmundsella pedata (Montagu, 1816)	2022	15 +	1	0	0	0	0	0	0	0	0	0	0
	2023	6	8 +	3	0	0	0	0	0	0	0	0	22
Family Dotidae Gray, 1853													
Doto paulinae Trinchese, 1881	2022	0	1 +	0	0	0	0	0	0	0	0	0	0
	2023	1 +	0	0	0	0	0	0	0	0	0	0	0 +
Family Trinchesiidae F. Nordsieck, 1972													
Trinchesia foliata (Forbes and Goodsir, 1839)	2022	0	2	0	0	0	0	0	0	0	0	0	0
	2023	0	0	0	0	0	0	0	0	0	0	0	0
Trinchesia morrowae Korshunova, Picton, Furfaro et al., 2019	2022	0	0	0	1	0	0	0	0	0	0	0	0
	2023	0	0	0	0	0	0	0	0	0	0	0	0
Trinchesia sp.	2022	1	0	0	0	0	0	0	0	0	0	0	0
	2023	0	0	0	0	0	0	0	0	0	0	0	0
Aeolid sp.	2022	0	0	0	1	0	0	0	0	0	0	0	0
	2023	0	0	0	0	0	0	0	0	0	0	0	0
Family Tritoniidae Lamarck, 1809													
Candiella manicata (Deshayes, 1853)	2022	0	0	0	0	0	0	0	0	0	0	0	0
	2023	0	0	0	0	0	0	0	0	1	0	0	0
Suborder Doridina													
Family Chromodorididae Bergh, 1891													
Felimare gasconi (Ortea, 1996)	2022	0	2	0	0	0	0	0	0	0	0	0	0
	2023	0	5	1	0	1	0	0	6	4	4	2	5
Felimare picta (R. A. Philippi, 1836)	2022	1	0	0	0	0	0	0	1	0	2	0	0
	2023	0	0	0	1	0	0	0	0	0	0	0	0
Felimare villafranca (Risso, 1818)	2022	0	0	0	0	0	0	0	1	0	0	1	0
	2023	2	1	0	0	1	1	0	0	0	0	0	0
Family Discodorididae Bergh, 1891													
Aporodoris cf. *millegrana* (Alder and Hancock, 1854)	2022	0	0	0	0	0	0	0	1	0	0	0	0
	2023	0	0	0	0	0	0	0	0	0	0	0	0
Baptodoris cinnabarina Bergh, 1884	2022	0	0	0	0	0	0	0	0	0	0	0	0
	2023	0	0	0	0	0	1	0	0	0	0	0	0
Jorunna tomentosa (Cuvier, 1804)	2022	0	0	0	0	0	0	0	0	0	0	0	0
	2023	0	0	0	1 +	0	0	0	0	0	0	0	0
Taringa sp.	2022	0	0	0	0	0	2 ‡	0	0	0	0	0	0
	2023	0	0	0	0	0	0	0	0	0	0	0	0
Tayuva lilacina (A. Gould, 1852)	2022	0	0	0	0	0	0	0	0	1	0	0	0
	2023	0	0	0	0	0	0	2 ‡	0	0	0	0	0
Doridina sp.	2022	0	0	0	0	0	0	0	0	0	0	0	0
	2023	0	0	0	0	0	0	0	1	0	0	0	0
Family Dendrodorididae O'Donoghue, 1924 (1864)													
Dendrodoris grandiflora (Rapp, 1827)	2022	0	0	0	0	0	0	1	0	0	0	0	0
	2023	0	0	0	0	0	0	0	0	0	0	0	0
Dendrodoris limbata (Cuvier, 1804)	2022	0	1	0	0	0	1	2	1	1	1	0 +	2
	2023	1	0	0	0	0	0	2	1	0	0	0	0 +
Dendrodoris temarana Pruvot-Fol, 1953	2022	0	0	0	0	0	1	1	0	1	0	0	0
	2023	0	0	0	0	0	1	1	0	0	0	0	0
Family Polyceridae Alder and Hancock, 1845													
Polycera quadrilineata (O. F. Müller, 1776)	2022	0	0	0	0	0	0	0	0	0	0	0	0
	2023	0	1	0	0	0	0	0	0	0	0	0	0
Order Cephalaspidea													
Family Bullidae Gray, 1827													
Bulla striata Bruguière, 1792 (shell only)	2022	0	0	0	0	0	1	0	0	0	0	0	1
	2023	0	0	0	0	1	0	0	2	0	1	0	1
Order Aplysiida													
Family Aplysiidae Lamarck, 1809													
Aplysia depilans Gmelin, 1791	2022	0	0	0	1	0	0	0	0	0	0	0	0
	2023	0	0	0	0	1	0 +	0	0	0	0	0	0
Superorder Sacoglossa													
Family Plakobranchidae Gray, 1840													
Elysia gordanae T. E. Thompson and Jaklin, 1988	2022	0	0	0	1	1	0	0	0	0	0	0	0
	2023	0	0	0	0	0	0	0	0	0	0	0	0

Table 2. *Cont.*

Taxa	Year	January	February	March	April	May	June	July	August	September	October	November	December
Elysia timida (Risso, 1818)	**2022**	53 ‡	75 ‡	10	70 ‡	108 ‡	184 ‡	88 ‡	94 ‡	42 ‡	17	37	51
	2023	96	38 ‡	65 ‡	15	18	109	98 ‡	146	43	83	73 ‡	168 ‡
Elysia viridis (Montagu, 1804)	**2022**	1	0	1	0	0	0	0	0	0	0	0	0
	2023	0	0	0	0	0	0	0	0	0	0	0	0
Thuridilla hopei (Vérany, 1853)	**2022**	1	0	0	0	0	0	0	0	0	0	0	0
	2023	0	0	0	0	0	0	0	0	1	0	0	0
Family Hermaeidae H. Adams et A. Adams, 1854 *Cyerce cristallina* (Trinchese, 1881)	**2022**	0	0	0	0	0	0	2	0	0	0	0	0
	2023	0	0	0	0	0	0	0	0	0	0	0	0

The following graphs were created from the collected data: a bar chart on the number of species for each taxonomic group, a line chart on the variation in the number of nudibranch species per month, a bar chart on the variation in the number of nudibranch species per season, and a bar chart on the total abundance for each species. In this case, the seasons were considered as follows: winter (January + February + March), spring (April + May + June), summer (July + August + September), and autumn (October + November + December). Monthly water temperature values can be seen in Figure 2.

3. Results

3.1. General Section

Through the present study, a total of 37 sea slugs were found, divided into five taxonomic groups. The most species-rich group was the order Nudibranchia, with 28 species. The two suborders of this group were similarly represented in the examined area: 15 Cladobranchia and 13 Doridina. The other four taxonomic groups were represented by a lower number of species compared to the abovementioned order: Sacoglossa (5 species), Pleurobranchida (2), Cephalaspidea (1), and Aplysiida (1).

Considering the temporal variation in species number of the group Nudibranchia, several differences during months, years, and seasons were observed (Figure 3).

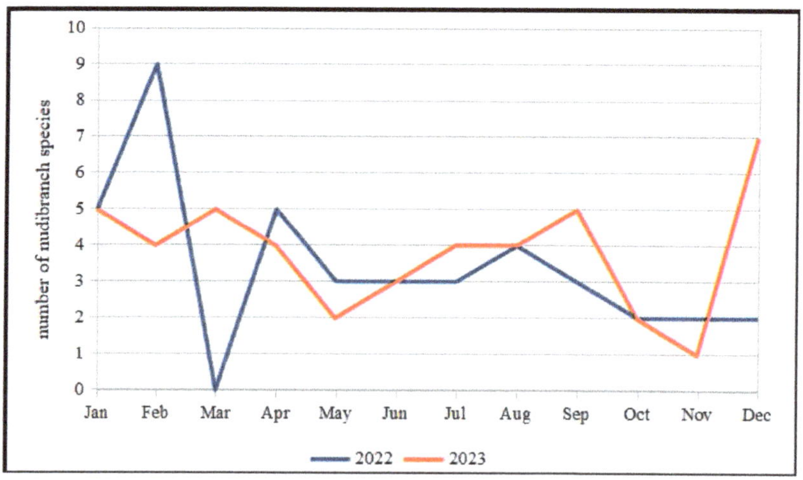

Figure 3. Monthly variation in observed nudibranch species number.

In 2022, after a rapid rise in species number between January (5 species) and February (9), there was a total decline in nudibranch species number in March (0). After this month, the nudibranch species number rose in April (5), then it fell slightly in the subsequent months until the stable lowering of October, November, and December (2) (Figure 3). In 2023, the nudibranch species number remained more or less constant from January (5) to April (4) and then decreased in May (2). After this month, there was an increase in the species number until September (5). Later, October (2) and November (1) recorded the lowest nudibranch species number of 2023. After this decline, there was a sharp increase in December (7).

Considering the variation in nudibranch species number per season (Figure 4), it can be seen that in 2022 there was an almost constant decline in species number across seasons [Winter (14 species), Spring (11), Summer (10), and Autumn (6)]. In 2023, on the other hand, the trend was more constant between seasons [Winter (14), Spring (9), Summer (13) and Autumn (10).

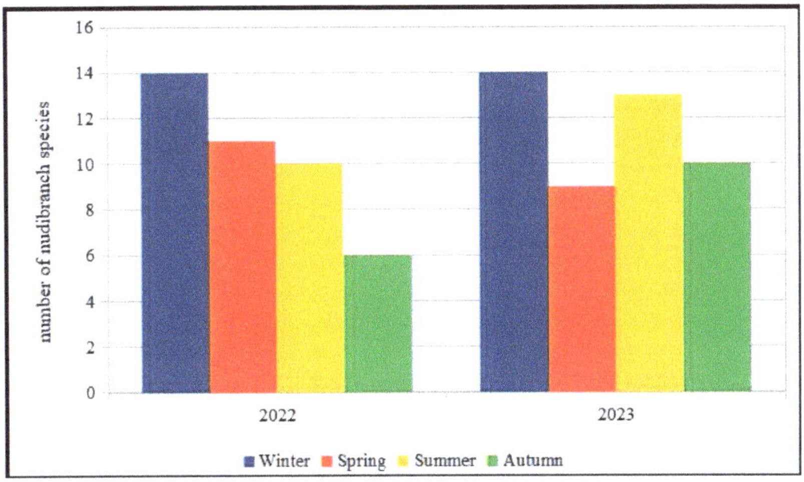

Figure 4. Seasonal variation in observed nudibranch species number.

3.2. Faunistic Section

The list of the species found and their abundances throughout the months of study can be seen in Table 2.

Order Pleurobranchida

Family Pleurobranchidae Gray, 1827

Berthellina cf. *edwardsii* (Vayssière, 1897) (about 25–30 mm in length) (Figure 5A–G).

Abundance: 26 specimens.

Temperature range: from 15 to 28 °C.

Seasonality: this species was found in the area from July to November. However, it could also be found in December, January, and April.

Substrates/habitats: *B.* cf. *edwardsii* specimens were always found on the underside of rocks. These latter could be almost totally bare, colonized with scattered calcareous red algae, serpulids, and sponges, or covered with detritus.

Remarks: When found under the rocks, *B.* cf. *edwardsii* specimens had a roughly spherical body shape in which the rhinophores and oral veil were retracted under the notum (Figure 5A). As soon as they were exposed to light (natural or produced by the underwater camera flashes), the animals started to slightly expose the distal part of the rhinophores (Figure 5B) and, if they were still disturbed by the light, began to crawl toward a darker area with fully extended rhinophores, oral veil, and body (Figure 5C). Almost all of the specimens had a general orange body coloration; however, a few animals were pale yellow (Figure 5E).

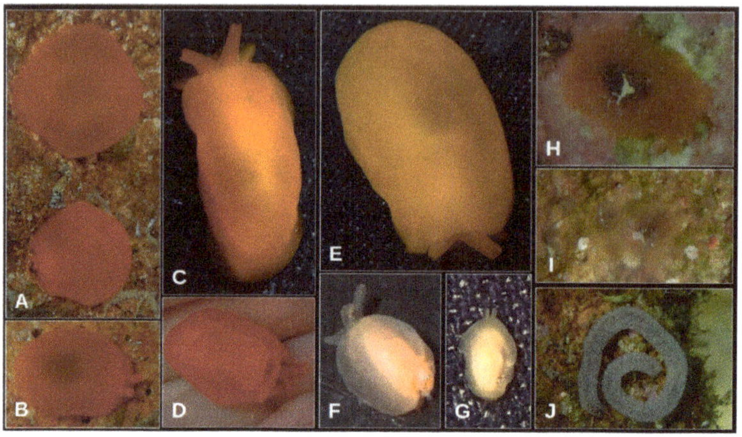

Figure 5. *Berthellina* cf. *edwardsii* (**A**) two specimens with the resting/contracted spherical body shape; (**B**) a specimen with its rhinophores slightly exposed; (**C**) a fully extended individual; (**D**) ventral view; (**E**) a yellowish specimen; (**F**) the teratological individual with the peculiar separation of the posterior part of the notum; (**G**) the specimen that lacked some part of the mantle margin. *Berthella stellata* (**H**) a specimen partially extended; (**I**) an individual with a very flattened appearance; (**J**) egg mass.

During this study, a small teratological specimen of *B.* cf. *edwardsii* (about 4–5 mm in length) was found with a peculiar separation of the posterior part of the notum in two lobes, which, rejoining posterior-dorsally, formed a strange cavity from which a grey tegumentary structure (with white tips) leaked out (Figure 5F).

Another small specimen (4–5 mm in length) had part of the notum edge missing (Figure 5G). Both the strange posterior teratology and the lack of part of the mantle margin were probably caused by some hypothetical predator (such as a small crustacean) or by the movement of the rocks under which these two specimens were found.

It is important to highlight that *Berthellina edwardsii* is indistinguishable from *Berthella aurantiaca* (Risso, 1818) by examining only the external morphology of these animals [20,21]. Consequently, the information reported here for the examined area is associated with an entity whose determination was not possible with absolute certainty. However, the specimens we found were identified as *B.* cf. *edwardsii* because, in the Mediterranean, *B. aurantiaca* is not a common species, unlike *B. edwardsii* [20]. As proof of this, Cervera [20] pointed out that it is likely that most Mediterranean finds of *B. aurantiaca* are actually to be considered *B. edwardsii*.

Berthella stellata (Risso, 1826) (about 10–12 mm in length) (Figure 5H–J).

Abundance: 2 specimens.

Temperature range: the animals were found at 16.5 °C, while the egg masses of this species were documented between 21 and 26 °C.

Seasonality: *B. stellata* specimens were documented in April. The egg masses (Figure 5J) were found in May, June, September, and October.

Substrates/habitats: this species was found on the underside of rocks covered by calcareous red algae, serpulids, remains of *Posidonia oceanica* rhizomes, and detritus. The egg masses were documented on the underside of rocks covered with a very low substrate of filamentous algae.

Remarks: in contrast to previous species, *B. stellata* individuals were always found with a very flattened body strongly compressed against the substrate (Figure 5I). The animals stretched their bodies and only began to crawl away after being touched several times (Figure 5H).

Order Nudibranchia Cuvier, 1817

Suborder Cladobranchia

Family Aeolidiidae Gray, 1827

Berghia verrucicornis (A. Costa, 1867) (about 15 mm in length) (Figure 6A–C).

Figure 6. *Berghia verrucicornis* (**A**) left antero-lateral view ; (**B**) dorsal view; (**C**) two specimens during mating. *Spurilla neapolitana;* (**D**,**E**) two specimens found on algae; (**F–H**) two individuals found beneath rocks.

Abundance: 4 specimens.

Temperature range: from 18 to 21 °C.

Seasonality: April and May. Almost all *B. verrucicornis* specimens were found in April, a month in which two individuals were observed mating.

Substrates/habitats: the animals were found on the underside of rocks with calcareous red algae, tufts of filamentous red algae, the sponge *Cliona viridis* (Schmidt, 1862), serpulids, and detritus.

Spurilla neapolitana (Delle Chiaje, 1841) (about 15–20 mm in length) (Figure 6D–H).

Abundance: 9 specimens.

Temperature range: from 15 to 27.5 °C.

Seasonality: the individuals were documented in February, March, April, July, September, October, and December.

Substrates/habitats: *S. neapolitana* specimens were found on turfs of filamentous brown algae with *Jania* sp. and *Dictyota dichotoma* (Hudson) J. V. Lamouroux; on *Halopteris scoparia* (Linnaeus) Sauvageau; on turfs of *Ceramium* sp.; on the underside of bare rocks or covered with calcareous red algae, serpulids, detritus, and gravel. The individuals documented under rocks were almost always within rocks' small concavities.

Remarks: during this study, it was noted that all the *S. neapolitana* specimens found under rocks had larger and stubbier bodies (Figure 6F–H) than those previously documented on algae in open environments (Figure 6D,E). Moreover, the former individuals always had a higher number of bright dots on the body surface than the latter.

On one occasion, when disturbed by one of the authors, a specimen peeled off several of its cerata, which became strongly adhesive on the author's glove.

Family Facelinidae Bergh, 1889

Cratena peregrina (Gmelin, 1791) (about 8–25 mm in length) (Figure 7A–C).

Abundance: 16 specimens.

Temperature range: from 15 to 17 °C.

Seasonality: most *C. peregrina* specimens were documented in January. However, this species was also found in February, March and December).

Substrates/habitats: this species was found on turfs of *Jania* sp. with detritus; turfs of filamentous red and brown algae; calcareous red algae; *Peyssonnelia* spp.; *Dictyota dichotoma*; *Caulerpa taxifolia* (M. Vahl) C. Agardh; and under bare rocks.

Facelina annulicornis (Chamisso and Eysenhardt, 1821) (about 15–18 mm in length) (Figure 7D,E).

Figure 7. *Cratena peregrina* (**A–C**) three specimens. *Facelina annulicornis* (**D**) a specimen in the right dorsolateral; (**E**) the wounded individual. *Facelina rubrovittata* (**F**,**G**) two small specimens. (**H**) left view of a specimen; (**I**) left dorsolateral view; (**J**) dorsal view. *Favorinus branchialis* (**K**) a specimen feeding on the egg masses of *Dendrodoris limbata*.

Abundance: 2 specimens.

Temperature range: 16.5 °C.

Seasonality: only in April.

Substrates/habitats: *F. annulicornis* individuals were found on the underside of rocks covered with calcareous red algae, *Peyssonnelia* spp., filamentous algae, and serpulids.

Remarks: the smallest of the two found specimens lacked the posterior part of the body, and some of its cerata were missing or shortened (Figure 7E). Its wounds were probably caused by the clipping action of a small crustacean predator.

Facelina rubrovittata (A. Costa, 1866) (8–15 mm in length) (Figure 7F–J).

Abundance: 29 specimens.

Temperature range: from 15 to 26 °C.

Seasonality: most *F. rubrovittata* specimens were found in January, February, and April. However, this species was also documented in May, September, and December.

Substrates/habitats: this facelinid was found on turfs of filamentous green, red, and brown algae; *Jania* sp.; *Ellisolandia elongata* (J. Ellis and Solander) K. R. Hind and G. W. Saunders; *Peyssonnelia* spp.; red laminar algae; *Halopteris scoparia*; *Dictyota dichotoma*; *Taonia* sp.; *Dictyopteris* sp.; and the sponge *Cliona viridis*.

Favorinus branchialis (Rathke, 1806) (about 4 mm in length) (Figure 7K).

Abundance: 1 specimen.

Temperature range: 16 °C.

Seasonality: this species was only documented in December .

Substrates/habitats: the individual was found on an egg mass of the dendrodorid *Dendrodoris limbata* (Cuvier, 1804) on the underside of a rock.

Family Flabellinidae Bergh, 1889

Calmella cavolini (Vérany, 1846) (about 15–18 mm in length) (Figure 8A–E).

Figure 8. *Calmella cavolini* (**A–E**) Some specimens. *Calmella gaditana* (**F**) dorsal view. *Edmundsella pedata* (**G**) the bluish individual; (**H,I**) normal specimens; (**J**) an egg mass.

Abundance: 8 specimens.

Temperature range: from 16 to 28 °C.

Seasonality: this species was documented in April, July, August, September, November, and December.

Substrates/habitats: *C. cavolini* was found on turfs of filamentous red and brown algae with *Jania* sp., *Cladophora* sp., and detritus; *Caulerpa taxifolia*; rocks covered by algal turfs; and *Taonia* sp.; on the underside of rocks covered by algal turfs, sponges, and detritus.

Calmella gaditana (Cervera, García-Gómez and García, 1987) (about 15–18 mm in length) (Figure 8F).

Abundance: 5 specimens.

Temperature range: from 15 to 21 °C.

Seasonality: this species was found in February, March and May. During this latter, the mating activity was documented.

Substrates/habitats: *C. gaditana* specimens were found only on turfs of filamentous red, brown, and green algae.

Remarks: externally, this species is virtually identical to *Calmella cavolini*. The only difference between these two flabellinids is the presence of bright dots on the distal part of the cerata of *C. gaditana*.

Edmundsella pedata (Montagu, 1816) (about 10–18 mm in length) (Figure 8G–J).

Abundance: 55 specimens.

Temperature range: the animals were present from 14.5 to 16 °C. The egg masses were found in the range of 14.5–15 °C.

Seasonality: *E. pedata* specimens were found from December to March. The egg masses (Figure 8J) were documented in January and February.

Substrates/habitats: this species was found on turfs of *Jania* sp. and filamentous algae; *Peyssonnelia* spp. (also the egg masses); *Halopteris scoparia*; *Dictyota dichotoma*; *Taonia* sp. (also the egg masses); *Cladophora* sp.; *Caulerpa taxifolia*; on and beneath rocks with calcareous red algae, serpulids, *Cliona viridis* and detritus.

Remarks: almost all the *E. pedata* specimens found during this study presented the typical pink-violet body and orange-reddish digestive gland's ramifications of the species (Figure 8H,I). However, one individual, found in December 2023, had a bluish body coloration and very dark-colored digestive gland ramifications (Figure 8G). This difference in body coloration could probably be attributed to genetic causes or different food sources.

Family Dotidae Gray, 1853

Doto paulinae Trinchese, 1881 (about 4–5 mm in length) (Figure 9A–C).

Figure 9. *Doto paulinae* (**A**) a specimen with its egg mass on *Aglaophenia* sp.; (**B**) a small specimen; (**C**) an egg mass. *Trinchesia foliata* (**D**) left lateral view; (**E**) dorsal view. *Trinchesia morrowae* (**F**) dorsal view. *Trinchesia* sp. (**G**) dorsal view. *Aeolid* sp. (**H**) right dorsolateral view. *Candiella manicata* (**I**) right lateral view; (**J**) the same in anterior-left view; (**K**) a specimen on its substrate.

Abundance: 2 specimens.

Temperature range: the individuals were found both at 15 °C. The egg masses (Figure 9C) were documented from 15 to 16 °C.

Seasonality: *Doto* sp. specimens were found in January and February. The egg masses were found in January, February, and December.

Substrates/habitats: this species and its egg masses were exclusively found on the hydrozoan of the genus *Aglaophenia* Lamouroux, 1812 (Figure 9A).

Family Trinchesiidae F. Nordsieck, 1972

Trinchesia foliata (Forbes and Goodsir, 1839) (about 4–5 mm in length) (Figure 9D,E).

Abundance: 2 specimens.

Temperature range: 15 °C.

Seasonality: this species was only found in February.

Substrates/habitats: *T. foliata* was found on the algae *Halopteris scoparia* and *Dictyota dichotoma*.

Trinchesia morrowae Korshunova, Picton, Furfaro, et al., 2019 (about 5–8 mm in length) (Figure 9F).

Abundance: 1 specimen.

Temperature range: 18 °C.

Seasonality: this species was only documented in April.

Substrates/habitats: *T. morrowae* was found on an entanglement of *Cladophora* sp.

Trinchesia sp. (about 5 mm in length) (Figure 9G).

Abundance: 1 specimen.

Temperature range: 15 °C.

Seasonality: the individual was only documented in January.

Substrates/habitats: *Trinchesia* sp. was found on the turf of filamentous red algae.

Remarks: the found specimen had a transparent-gray body coloration with scattered bright cream-azure blotches. On the dorsal area of the head, there was a rhomboidal-shaped blotch, which was white on its anterior part (located between the front edge of the head and the space between rhinophores) and faded grey with scattered bright dots on the posterior part (placed from the space behind the rhinophores and the first group of cerata). Each side of the white area of the head was bordered by an orange stripe. In addition, the rhinophores were transparent grey on their proximal half and stained blue–white in the distal half. The contact area between these two parts was colored orange. Despite a slight external similarity of this specimen with those identified here as *Trinchesia foliata*, the differences in the coloration of the body and rhinophores justify their separation.

Aeolid sp. (about 5 mm in length) (Figure 9H).

Abundance: 1 specimen.

Temperature range: 18 °C.

Seasonality: Aeolid sp. was documented in April.

Substrates/habitats: the individual was found on the turf of filamentous brown algae.

Remarks: the specimen characterized by a grey and white body probably belongs to the family Trinchesiidae.

Family Tritoniidae Lamarck, 1809

Candiella manicata (Deshayes, 1853) (about 8–10 mm in length) (Figure 9I–K).

Abundance: 1 specimen.

Temperature range: 26 °C.

Seasonality: the individual was documented in September.

Substrates/habitats: *C. manicata* was found on the turf of filamentous red algae covered with detritus.

Remarks: the animal was very cryptic on its substrate (Figure 9K).

Suborder Doridina

Family Chromodorididae Bergh, 1891

Felimare gasconi (Ortea, 1996) (about 5–15 mm in length) (Figure 10A–J).

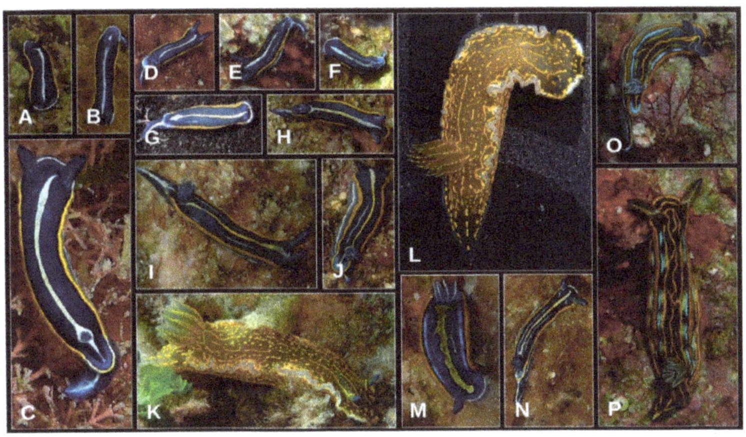

Figure 10. *Felimare gasconi* (**A–C**) specimens with the juvenile chromatic pattern; (**D–F**) individuals with the intermediate chromatic pattern; (**G–J**) specimens with the final chromatic pattern. *Felimare picta* (**K**) right dorsolateral view; (**L**) dorsal view. *Felimare villafranca* (**M**,**N**) specimens with anterior and posterior mantle edges white colored; (**O**,**P**) specimens with darker body color and complicated chromatic pattern.

Abundance: 30 specimens.

Temperature range: from 14.5 to 28 °C.

Seasonality: *F. gasconi* was documented in February, March, April, and from August to December. Most of the individuals were found in February and from August to December.

Substrates/habitats: the favorite substrate of this chromodoridid was the lower face of the rocks colonized by small *Dysidea* spp., *Peyssonnelia* spp., serpulids, and detritus. However, *F. gasconi* was also found on turfs of filamentous red algae; *Jania* sp. intermingled with detritus; *Peyssonnelia* spp.; rocks (bare or with algal turfs); gravel, and detritus.

Remarks: during this study, it was possible to observe the development of the chromatic pattern as *F. gasconi* grew. On the dorsal part of the notum, the smallest individuals showed a white stripe running from the space between the rhinophores to the posterior part of the notum. In this section, the white stripe encircled the gill tuft. Moreover, these juvenile specimens had the edge of the notum differently colored: the frontal mantle border (the one delimiting the notal space from the anterior part of the notum to the level of the rhinophores) was white, the middle edge of the mantle (from the level of the rhinophores to the level of the gill tuft) was yellow, and the posterior edge (from the level of the gill tuft to the posterior notal edge) was white (Figure 10A–C). Subsequently, along one side (or even both) of the central dorsal white stripe, a small, thin white line started to develop in several ways depending on the individual (Figure 10D–F). At the same time, the anterior and posterior white-colored mantle edges turn partially or totally yellow in coloration. The final step was the yellowing of the central white dorsal stripe (Figure 10G–J).

Felimare picta (R. A. Philippi, 1836) (about 70–80 mm in length) (Figure 10K,L).

Abundance: 5 specimens.

Temperature range: from 15 to 27.5 °C.

Seasonality: *F. picta* was documented in January, April, August, and October.

Substrates/habitats: this species was found on turfs of filamentous algae and *Jania* sp. with detritus, among *Padina* spp. and *Dictyota dichotoma* thalli, within a crevice.

Felimare villafranca (Risso, 1818) (about 5–15 mm in length) (Figure 10M–P).

Abundance: 7 specimens.

Temperature range: from 14.5 to 27.5 °C.

Seasonality: *F. villafranca* was documented in January, February, May, June, August, and November.

Substrates/habitats: this species was mainly found on the underside of rocks with *Dysidea* spp. and calcareous encrusting red algae. Moreover, it was also found on *Peyssonnelia* spp. and turfs of filamentous red and brown algae.

Remarks: despite the scarcity of *F. villafranca* specimens found in this study, some conclusions can be drawn about the development of the chromatic pattern of this species. It was noted that the smallest specimens showed the anterior and posterior parts of the mantle edge white, while the central part of the latter was orange (Figure 10M,N). In the largest individuals, the entire notum edge turned orange. Moreover, in small individuals, the general body coloration was bluish, and the pattern of the dorsal stripes was not very chaotic. In the largest animals, the general color of the body was almost black, and there was an evident complication in the pattern of stripes (Figure 10O,P).

Family Discodorididae Bergh, 1891

Aporodoris cf. *millegrana* (Alder and Hancock, 1854) (about 6 mm in length) (Figure 11A).

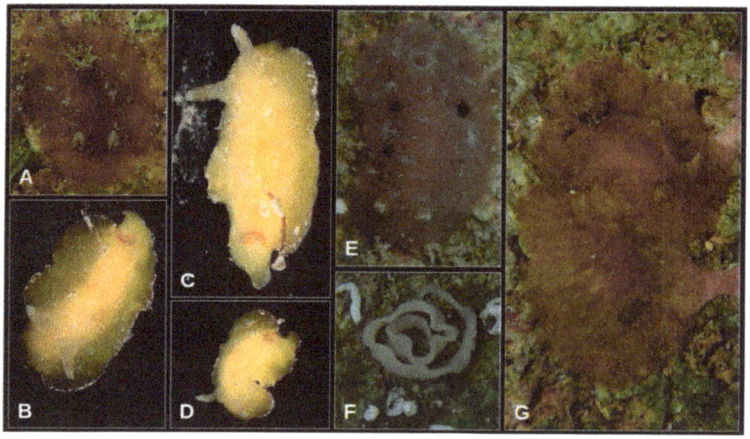

Figure 11. *Aporodoris* cf. *millegrana* (**A**) dorsal view. *Baptodoris cinnabarina* (**B**) dorsal view; (**C**,**D**) detail of the lacking mantle's portions; *Jorunna tomentosa* (**E**) dorsal view; (**F**) egg mass. *Taringa* sp. (**G**) two specimens during mating.

Abundance: 1 specimen.

Temperature range: 27.5 °C.

Seasonality: *A.* cf. *millegrana* was documented only in August.

Substrates/habitats: the individual was found on the underside of a small rock covered by serpulids.

Baptodoris cinnabarina Bergh, 1884 (about 6–8 mm in length) (Figure 11B–D).

Abundance: 1 specimen.

Temperature range: 23 °C.

Seasonality: *B. cinnabarina* was only documented in June.

Substrates/habitats: the animal was found on the underside of a rock.

Remarks: the found specimen lacked some portions of the mantle edge (Figure 11C,D).

Jorunna tomentosa (Cuvier, 1804) (about 15 mm in length) (Figure 11E,F).

Abundance: 1 specimen.

Temperature range: the individual and its egg mass were found at 16.5 °C.

Seasonality: *J. tomentosa* was only documented in April (also one egg mass).

Substrates/habitats: the animal and its egg mass (Figure 11F) were found on the underside of a rock covered by serpulids.

Taringa sp. (about 20 mm in length) (Figure 11G).

Abundance: 2 specimens.

Temperature range: 25 °C.

Seasonality: two *Taringa* sp. individuals were documented in June during mating.

Substrates/habitats: the animals were found on the underside of a rock covered with calcareous red algae.

Tayuva lilacina (A. Gould, 1852) (about 25–35 mm in length) (Figure 12A–D).

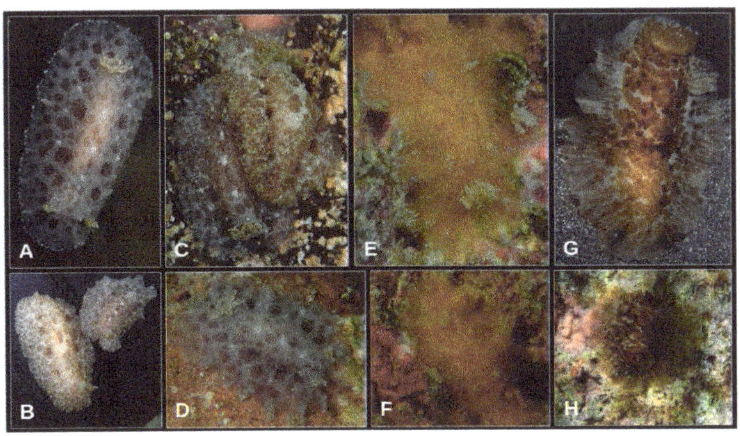

Figure 12. *Tayuva lilacina* (**A**,**B**) specimens in dorsal view; (**C**) two individuals during mating; (**D**) a specimen with wrinkled mantle margin. *Doridina* sp. (**E**,**F**) dorsal view. *Dendrodoris grandiflora* (**G**) dorsal view; (**H**) the same in a semi-resting/contracted body shape.

Abundance: 3 specimens.

Temperature range: from 26 to 27.5 °C.

Seasonality: *T. lilacina* was found in July and September. The mating (Figure 12C) was documented in July.

Substrates/habitats: the animals were found on the underside of rocks with calcareous red algae, serpulids, and tunicates.

Remarks: on one occasion, after exposing a specimen to the light while standing on the underside of a rock, the animal started to wrinkle its mantle margin (Figure 12D).

Compared to the other discodoridid species found during this study, *T. lilacina* responds quickly to disturbances by not contracting its body much and moving away quickly in search of darker areas.

Doridina sp. (about 10 mm in length) (Figure 12E,F).

Abundance: 1 specimen.

Temperature range: 28 °C.

Seasonality: *Doridina* sp. was only documented in August.

Substrates/habitats: this species was found on a yellow–orange sponge on the underside of a rock covered by calcareous red algae and serpulids.

Family Dendrodorididae O'Donoghue, 1924 (1864)

Dendrodoris grandiflora (Rapp, 1827) (about 40 mm in length) (Figure 12G,H).

Abundance: 1 specimen.

Temperature range: 28 °C.

Seasonality: *D. grandiflora* was only documented in July.

Substrates/habitats: this species was found on the underside of a rock covered with calcareous red algae and tufts of filamentous algae.

Dendrodoris limbata (Cuvier, 1804) (about 8–40 mm in length) (Figure 13A–G).

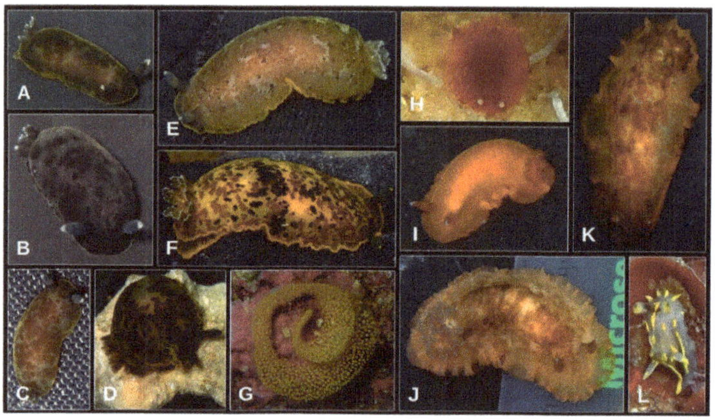

Figure 13. *Dendrodoris limbata* (**A**,**B**) small specimens with the black-brown mottled chromatic pattern; (**C**) an individual with the intermediate chromatic pattern; (**D**–**F**) large specimens with the light-colored chromatic pattern; (**G**) an egg mass. *Dendrodoris temarana* (**H**) a red juvenile; (**I**) an orange intermediate specimen; (**J**,**K**) two adult specimens. *Polycera quadrilineata* (**L**) right dorsolateral view.

Abundance: 13 specimens.

Temperature range: the individuals were found from 15 to 28 °C. Instead, the egg masses (Figure 13G) were found from 16 to 20.5 °C.

Seasonality: this species was documented all year round except in November. The egg masses were only found in November and December.

Substrates/habitats: *D. limbata* was always found on the underside of rocks, which could be totally bare or presented *Cliona viridis*, serpulids, and detritus; calcareous encrusting red algae (also the egg masses); *Peyssonnelia* spp. and remnants of shells.

Remarks: During this study, it was observed that all small *D. limbata* specimens had a general black-brown mottled body coloration. However, the latter was only evident when these animals were placed on a black substrate, which made the mottled appearance very obvious (Figure 13A,B). In contrast, when these small individuals were on their natural substrate, they appeared almost completely black. The larger specimens showed several different chromatic patterns, with the lighter side predominating over the darker one (Figure 13D–F). Consequently, in *D. limbata* specimens, it can be seen that the development of the chromatic pattern moves toward a progressive lightening of the body coloration (dark → light) combined with the separation and narrowing of the dark component (large, clustered group of spots → small, scattered spots). This is easily visible through the finding of individuals with an intermediate chromatic pattern (Figure 13C).

Moreover, egg masses of *D. limbata* were documented to be preyed upon by the facelinid *Favorinus branchialis*.

Dendrodoris temarana Pruvot-Fol, 1953 (about 3–45 mm in length) (Figure 13H–K).

Abundance: 5 specimens.

Temperature range: from 23 to 28 °C.

Seasonality: this species was only documented in June, July, and September.

Substrates/habitats: *D. temarana* was found on the underside of rocks.

Remarks: The specimens of *D. temarana* documented here represent the first occurrence of this species in the central Mediterranean. To date, this species (which has a complex taxonomic history) has been found (under different names) in several areas of the western Mediterranean and eastern Atlantic [24]: in the Western Mediterranean, *D. temarana* has been found along the coasts of France (Thau lagoon), Spain (Catalan coast, El Portil, Cubelles, Ebro Delta, Balearic Islands, and La Línea de la Concepción); in the eastern Atlantic, this dendrodorid has been documented in Spain (Gulf of Cadiz), Portugal (Sado estuary), Morocco (Temara), the Azores, and the Canary Islands.

As reported by Ballesteros [24], this species exhibits considerable morphological variability during both growth and adulthood. Juvenile specimens show a completely red body coloration that changes during development, giving adults different color patterns (red, pink, yellowish, dark brown, and almost black). The color pattern of adults is usually associated with reddish, orange, or brownish spots. However, adults may have uniform black or brown coloration. During this study, two juvenile-colored (completely red) juveniles (Figure 13H), two adult-colored specimens (reddish-orange bodies with dark spots on the back) (Figure 13J,K), and one completely orange specimen (probably a specimen between juvenile and adult) (Figure 13I) were documented. The reddish juveniles were both found in July (one in 2022 and one in 2023), the adults both in June (one in 2022 and one in 2023), and the one with uniformly orange coloration during September 2022.

Family Polyceridae Alder and Hancock, 1845

Polycera quadrilineata (O. F. Müller, 1776) (about 5–6 mm in length) (Figure 13L) **Abundance:** 1 specimen.

Temperature range: 14.5 °C.

Seasonality: *P. quadrilineata* was documented in February.

Substrates/habitats: the individual was found on encrusting bryozoans on a *Peyssonnelia* sp. thallus.

Order Cephalaspidea P. Fischer, 1883

Family Bullidae Gray, 1827

Bulla striata Bruguière, 1792 (only the shell) (about 20–25 mm in length) (Figure 14A–F).

Abundance: 7 specimens.

Temperature range: from 16 to 28 °C.

Seasonality: the shells were found in May, June, August, October, and December.

Substrates/habitats: under and among rocks.

Order Aplysiida

Family Aplysiidae Lamarck, 1809

Aplysia depilans Gmelin, 1791 (about 30 mm in length) (Figure 14G–I).

Abundance: 2 specimens.

Temperature range: the individuals were found from 18 to 20 °C. Instead, the egg masses (Figure 14I) were found at 23 °C.

Seasonality: the animals were documented in April and May. The egg masses were only found in June.

Substrates/habitats: *A. depilans* was only found on the underside of rocks, always together with sea urchins *Arbacia lixula* (Linnaeus, 1758) and *Paracentrotus lividus* (Lamarck, 1816). The egg masses were found on *Peyssonnelia* spp. Thalli.

Figure 14. *Bulla striata* (**A–F**) has several empty shells. *Aplysia depilans* (**G**) right dorsolateral view; (**H**) left dorsolateral view; (**I**) an egg mass. *Elysia gordanae* (**J**) dorsal view. *Elysia timida* (**K**) right dorsolateral view; (**L**) some specimens on their preferred substrate; (**M**) a specimen on the green algae *Acetabularia acetabulum*; (**N**) two individuals during mating; (**O**) the specimen that lacked almost all the body; (**P**) the individual with an additional and smallest parapodial-tail portion behind the tail.

Superorder Sacoglossa
Family Plakobranchidae Gray, 1840
Elysia gordanae T. E. Thompson and Jaklin, 1988 (about 5 mm in length) (Figure 14J).
Abundance: 2 specimens.
Temperature range: from 18 to 21 °C.
Seasonality: *E. gordanae* was documented in April and May.
Substrates/habitats: this species was found on *Caulerpa taxifolia* and *Padina* spp.
Elysia timida (Risso, 1818) (about 5–15 mm in length) (Figure 14K–P).
Abundance: 1781 specimens.
Temperature range: *E. timida* was documented from 14.5 to 28 °C.

Seasonality: this plakobranchid was found all year round. However, the periods with the highest abundance in specimens were the late-end spring, summer, and beginning of winter (Figure 15). The mating (Figure 14N) was all year round except in October.

Substrates/habitats: the favorite substrates of this species were the rocky surfaces covered by very thin (nearly visible) or low turfs of filamentous algae. Often, these algal turfs were mixed with detritus (Figure 14L). Exclusively in April and May, several specimens were seen on the green algae *Acetabularia acetabulum* (Linnaeus) P.C. Silva (Figure 14M). Although this seaweed was present in the study area also in June and July, it was completely disregarded by *E. timida* specimens, probably due to its complete calcification.

Besides the abovementioned substrates, *E. timida* specimens were also found on *Jania* sp; calcareous encrusting red algae; *Peyssonnelia* spp.; *Halopteris scoparia; Dictyota dichotoma; Taonia* sp.; *Padina* spp.; *Flabellia petiolata; Penicillus capitatus* Lamarck; *Caulerpa cylindracea* Sonder; *Posidonia oceanica*'s rhizomes entanglements; bare rocks and coralligenous gravel/sand.

It was also seen that numerous specimens of *E. timida* were stationed (always resting on their substrate) at the air-water interface, often remaining exposed to the air for very short periods of time. This did not bother the animals.

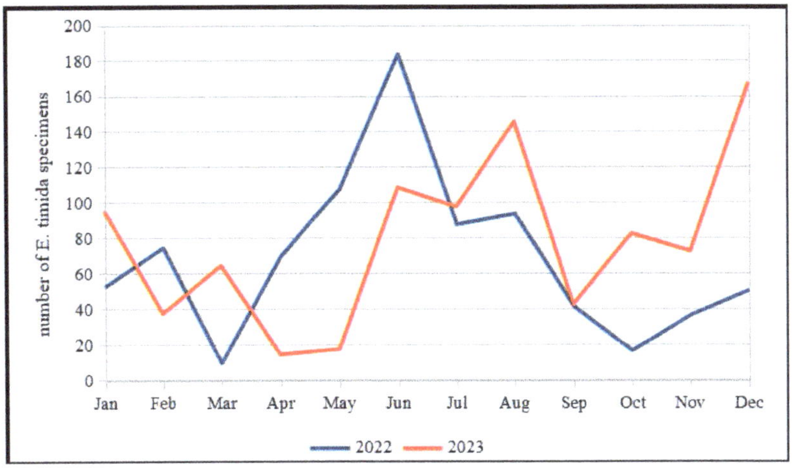

Figure 15. Variation in abundance of *Elysia timida* over the study period.

Remarks: although *Acetabularia acetabulum* appears to be the favorite food of *E. timida* [30,31], only a small portion of the population in the study area was found on thisalgae. On the contrary, most *E. timida* specimens were found on rocky substrates covered by low turfs of filamentous algae. Consequently, as already hypothesized by Lombardo and Marletta [32] for *E. timida* populations of the central-eastern coast of Sicily, this species would base much of its nutrition on other algal material. However, it is important to point out that this species has the ability to maintain and retain functioning chloroplasts of *A. acetabulum* within its digestive gland for at least 45 days by not feeding for long periods of time [31]. Consequently, it is possible that the animals not found in this study on *A. acetabulum* fed on this alga previously while maintaining functioning chloroplasts (kleptoplasts).

During this study, two teratological *E. timida* specimens were found. The first animal was devoid of almost the entire body, presenting only the head and a small portion of the anterior part of the pericardial area and parapodia (Figure 14O). This teratology was almost certainly caused by an attack of a hypothetical predator [33].

The second teratological individual had an additional and smallest parapodial-tail portion behind the tail (Figure 14P). This teratology made the animal twice as long as a normal specimen. Given that this individual had no wounds or scars, it is likely that this strange teratology was genetic in origin.

Finally, it was documented an "attack" suffered from an *E. timida* specimen by a small fish of the genus *Parablennius*, who swallowed the first one whole and then immediately spat it out and ignored it.

Elysia viridis (Montagu, 1804) (about 5–10 mm in length) (Figure 16A,B).

Abundance: 2 specimens.

Temperature range: 15 °C.

Seasonality: *E. viridis* was documented in January and March.

Substrates/habitats: this sacoglossan was found on turfs of filamentous red algae and *Flabellia petiolata*.

Thuridilla hopei (Vérany, 1853) (about 5–10 mm in length) (Figure 16C).

Abundance: 2 specimens.

Temperature range: from 15 °C to 26 °C.

Seasonality: *T. hopei* was documented in January and September.

Substrates/habitats: this species was found on *Halopterisscoparia* and *Jania* sp.

Family Hermaeidae H. Adams et A. Adams, 1854

Cyerce cristallina (Trinchese, 1881) (about 25–30 mm in length) (Figure 16D–H).

Figure 16. *Elysia viridis* (**A**) dorsal view; (**B**) a specimen on *Flabellia petiolata*. *Thuridilla hopei* (**C**) dorsal view. *Cyerce cristallina* (**D**) the two specimens were found together on the underside of a rock; (**E**) the two individuals on an artificial substrate; (**F**) the normal individual; (**G**,**H**) the teratological one.

Abundance: 2 specimens.
Temperature range: 28 °C.
Seasonality: this species was found only in July.
Substrates/habitats: the two specimens were found together on the underside of a rock covered by calcareous encrusting red algae Figure 16D.
Remarks: once exposed to light, the two animals began to rapidly crawl toward the shaded areas. Moreover, one of the animals had three malformations on its body. Specifically, this individual had several cerata of the anterior part of the body that were much smaller than normal, the right oral tentacle bent medially with an additional small tentacle originating from the bent point, and the right eye displaced towards the midline of the body (giving the feeling of being lacking) (Figure 16G,H). All these malformations were probably caused by a severe attack by a putative predator. This was easily detectable by the evident disproportion of the right anterior cerata compared to those of the rest of the body and by the fact that the other malformations were on the same side of the body. The oral tentacle malformation can be explained as a poor regrowth of the previously detached right oral tentacle. However, the movement of the right eye towards the reddish central part of the head is difficult to explain well.

4. Discussion

The sea slug fauna of the Magnisi peninsula comprised 37 species, which were divided into five taxonomic groups: Pleurobranchida (2 species), Nudibranchia (28), Cephalaspidea (1), Aplysiida (1), and Sacoglossa (5). This fauna was dominated by a large population of the sacoglossan *Elysia timida*, which represented the most numerous sea slug population of the area with its 1781 found specimens (829 in 2022 and 952 in 2023). It is notable that none of the other species' populations reached this level of abundance. The other most abundant species of the area were *Edmundsella pedata* (55 specimens), *Felimare gasconi* (30), *Facelina rubrovittata* (29), *Berthellina* cf. *edwardsii* (26), *Cratena peregrina* (16), *Dendrodoris limbata* (13), *Spurilla neapolitana* (9), *Calmella cavolini* (8), *Felimare villafranca* (7), and *Bulla striata* (only the shell) (7). All remaining species were observed on only a few occasions (from one to five times) during the study period (Figure 17).

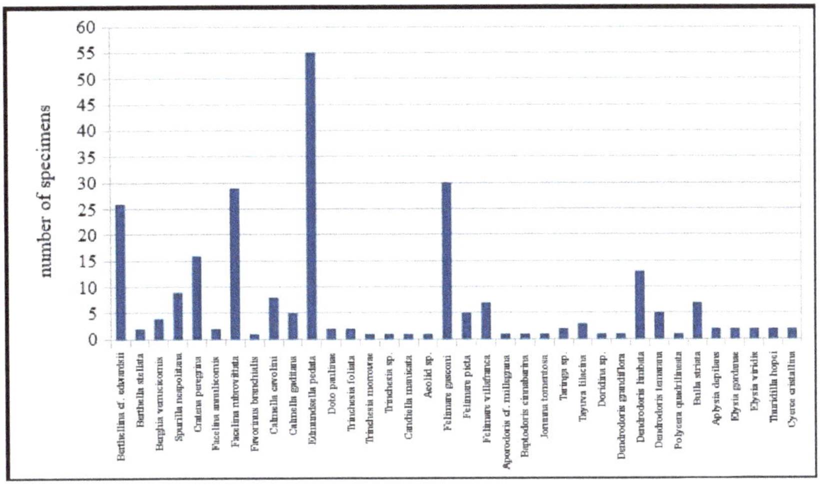

Figure 17. Abundance of the found sea slug species without *Elysia timida*.

This fauna was dominated by a diverse community of sea slug species, indicative of brackish-water/euryhaline habitats and shallow waters (whose favorite habitat was the underside of rocks), which overall co-occurred with some more or less common/widely distributed species. Among the species found, those belonging to the first category were in order of abundance: *Elusia timida*, *Dendrodoris limbata*, *Felimare villafranca*, *Bulla striata*, *Polycera quadrilineata*, and *Elysia viridis*. In particular, *Elusia timida* is known to be a very common sacoglossan of brackish lagoons [34]; the dendrodorid *Dendrodoris limbata* is not only a very common nudibranch in lagoons [34,35], but it also exhibits invasive tendencies in such a type of habitat [36]; *Felimare villafranca*, along with other chromodoridid species, appears to be a common inhabitant of lagoonal habitats [34,35,37] as well as *Bulla striata* [34], *Polycera quadrilineata*, and *Elysia viridis* [35]. In contrast, *Berthella stellata* [38,39], *Calmella gaditana* [40], *Aporodoris* cf. *millegrana* [40], *Baptodoris cinnabarina* [37], *Taringa* sp. (as genus *Taringa* in general) [40], *Doridina* sp. [38], *Dendrodoris grandiflora* [40], *Dendrodoris temarana* [24], and *Cyerce cristallina* [41,42] are shallow water species that typically inhabit habitat beneath rocks. It is notable that several species documented during this study fall under both of these two habitat categories, namely brackish-water/euryhaline and shallow-water species that live under rocks. Examples of this include *Berghia verrucicornis* [35,37], *Spurilla neapolitana* [37,43], *Favorinus branchialis* [34,43] *Calmella cavolini* [34,35,37], *Jorunna tomentosa* [35,37], *Tayuva lilacina* [36,40,44] and *Aplysia depilans* [35,43]. In terms of the category of those species that can be considered more or less common/widely distributed species, the following species were found during this study: *Cratena peregrina* [37], *Facelina rubrovittata* [45], *Edmundsella pedata* [37], *Trinchesia foliata* [45], and *Elysia viridis* [21].

This faunistic composition would reflect the distinctive combination of environmental conditions that characterize the examined area. The coastal stretch under investigation is located in the northernmost part of the gulf, which is formed between the Magnisi peninsula (to the north) and Punta Cannone (to the south). The examined area is situated in the most secluded part of the gulf, which is predominantly characterized by very shallow and extensive sandy bottoms with a depth of less than one meter.

The combination of the concealed location of this site and the sandy bottoms, which almost completely surround the shallow stony studied area, could probably result in a reduction in the seawater flow from and to the open area of the gulf. Furthermore, this coastal stretch is subject to freshwater inflows, which may be attributed to industrial discharges [46]. Consequently, although the examined coastal stretch is not a real brackish-water/lagoon/estuarine area, its environmental conditions are similar to those of brackish environments due to the abovementioned reasons. This is evidenced by the presence in the study area of the green algae *Penicillus capitatus*, which is considered a common algal species of estuarine-marine/brackish habitats [47]. It can thus be argued that the specific environmental conditions present in this area, characterized by a shallow rocky seabed with a considerable amount of scattered pebbles, boulders, and rocks, provide an adequate rationale for the occurrence of both brackish-water/euryhaline species, along with those typical of shallow waters.

The presence of the common/widely distributed species, along with occasional ones, may indicate an enduring connection with the external area, exemplified by the coastal stretch examined at the foot of the southeast small cliff, where the surrounding sandy bottoms deepen by a few meters. This may nullify the stagnation effect observed in the neighboring coastal stretch located northward.

This study revealed that the most representative sea slug group in the area was the order Nudibranchia (28 species). The temporal variations in the number of species within this group can provide valuable insights into the seasonal dynamics of the marine waters of a given coastal area [48]. Similarly, some inferences can be drawn regarding this coastal area.

Given the very shallow nature and the general ecological conditions of the study area, the documented temporal differences in the number of nudibranch species in the study area may be the result and consequence of two environmental parameters: temperature and wave disturbance.

The occurrence of nudibranchs is affected by temperature in both direct and indirect ways. Directly, temperature affects the specific thermal sensitivity of each species (e.g., cold water/northern species vs. warm water/southern species [49,50]) [51]. Indirectly, temperature changes affect the presence of nudibranchs' prey, seawater stratification, phytoplankton blooms, and the arrival of allochthonous nudibranchs' larvae, among other factors [48].

The physical disturbance caused by waves on rocky substrates plays a significant role in determining the level of biodiversity in a given area [52]. In shallow, stony marine environments, wave action results in the displacement of rocks. The frequency of disturbance to stones, rocks, and pebbles determines the time available for sessile benthic suspension feeders (e.g., bryozoans, cnidarians, tunicates, and sponges [53], the favorite preys of nudibranchs [48]) to colonize these specific surfaces [52]. It is notable that the underside of rocks and spaces between them are among the most highly favored habitats for such mollusks, e.g., [23,54]. Therefore, the displacement, overturning, and impacts caused by wave motion play a pivotal role in determining the numbers of species and specimens of these gastropods in very shallow coastal areas.

The wave disturbance observed in this study area, when generated by sea storms, has the potential to be very impactful. In addition to the violent shifting and overturning of rocks, these events can result in the movement and suspension of a considerable amount of sediment.

The present study revealed that the highest number of nudibranch species was observed in both 2022 and 2023 during the winter season, when the lowest temperature values were recorded (Figure 4). This winter species abundance is in accordance with two recent findings concerning the seasonality of Mediterranean nudibranchs, which highlighted that the cooling of marine waters is always associated with an increase in the number of species [48,50]. With the exception of this similarity, the other seasons displayed a different trend over the two years. This discrepancy is likely attributable to the differential wave disturbance observed in 2022 and 2023. In 2022, a single sea storm event occurred in the area during March. Therefore, the species trend observed in 2022 was likely influenced by the seasonal changes in temperature, whose gradual increase determined the constant decline in species number across seasons. In contrast, in 2023, the coastal area under study was subjected to three sea storms, one in winter [February (very strong)] and two in spring (April and May). Consequently, during this year, the typical seasonal trend in the number of nudibranch species was likely offset by the prolonged and intense wave disturbance. Additionally, the impact of this disturbance may have altered the usual temperature trend, delaying the expected trend for a few months (Figure 2). Consequently, the spring of 2023, which was characterized by a near-continuous wave disturbance, exhibited a lower number of species compared to that of 2022. Conversely, the summer of 2023, and subsequently also the autumn of 2023, were distinguished by an exceptionally high number of species.

The temporal variation in nudibranch species number may, therefore, serve as an indicator of the seasonal changes that occur along the water column, including the temporal phytoplanktonic variations and seawater stratification, as recently proposed by Lombardo and Marletta [48]. However, it may also reflect the spatio-temporal variations in temperature and wave disturbance in very shallow environments.

The potential for making comparisons between the sea slug fauna of the Magnisi peninsula and faunas of other Mediterranean coastal areas is constrained by the lack of prior studies (to our knowledge) employing the methods outlined herein. Nevertheless, between 1984 and 1986, a noteworthy study into the sea slug fauna of the Marsala Stagnone, the largest Sicilian lagoon, was conducted using similar methods [34]. In this study, the authors found a total of 23 sea slug species, subdivided into five taxonomic groups: Cephalaspidea (6 species), Sacoglossa (1), Aplysiida (1), Pleurobranchida (2), and Nudibranchia (13). It should be noted that the present study differs from that of Cattaneo-Vietti and Chemello [34] in that the present study examined a single site (approximately 200 m in length), whereas the previous one [34] investigated ten sites distributed across approximately 4 km of the lagoon. From this comparison, it can be highlighted that despite the relatively limited coastal area of just 200 m investigated here, there is almost twice the number of sea slug species found in the examined 4 km of coastline of the Marsala Stagnone. Therefore, given its concealed and shallow location, coupled with the local environmental status, which is affected by industrial pollution, this coastal stretch of eastern Sicily could be considered a small biodiversity hotspot for the sea slug fauna.

5. Conclusions

In conclusion, after 53 years from the latter report [17], the data reported in this study demonstrated that the Magnisi peninsula exhibits markedly high marine mollusk biodiversity compared to similar shallow areas. Notwithstanding the considerable damage inflicted in the past on the marine environment by local industries, the presence of a terrestrial Natural Oriented Reserve (NOR) and Natura 2000 site could serve as a foundation for the conservation and protection of this area, which hosts significant marine habitats and species. As evidence, this study presented findings of species considered rare, such as *Aporodoris* cf. *millegrana* [55]. *Tayuva lilacina* [44], *Cyerce cristallina* [41], and unidentified species may represent new species, such as *Trinchesia* sp., *Aeolid* sp., and *Doridina* sp. Moreover, the area examined here represents the only known occurrence where *Dendrodoris temerana* was found outside its documented range of distribution (see above) with a stable population.

It will become increasingly important in the future to focus attention on the naturalistic significance of this small stretch of coastline and emphasize the enhancement of the natural marine environment of this small gulf, which represents a hotspot of diversity for marine sea slugs.

Author Contributions: Conceptualization, A.L. and G.M.; methodology, A.L.; validation, G.M.; data curation, A.L.; writing—original draft preparation, A.L.; writing—review and editing, G.M. All authors have read and agreed to the published version of the manuscript.

Funding: This research received no external funding.

Institutional Review Board Statement: Not applicable.

Informed Consent Statement: Not applicable.

Data Availability Statement: The original contributions presented in the study are included in the article, further inquiries can be directed to the corresponding author.

Acknowledgments: The authors wish to thank three anonymous reviewers and the journal editor for their constructive comments and suggestions.

Conflicts of Interest: The authors declare no conflicts of interest.

References

1. Mignosa, G. *Priolo Gargallo—Da borgo Feudale a Centro Industriale*, 2nd ed.; Tipografia "V. Tarantello": Priolo Gargallo, Italy, 2006; pp. 1–248.
2. Vaggioli, M.A. THAPSOS (penisola). *Bibliogr. Topogr. Della Colon. Greca Ital. E Nelle Isole Tirreniche* **2011**, *20*, 517–570.
3. Trimarchi, R. Territorio e de-industrializzazione in Sicilia. Un contributo alla rilettura del paesaggio industriale nelle aree dismesse. *Ann. Della Fac. Di Sci. Della Form.* **2004**, *3*, 391–410.
4. Treccani. Available online: https://www.treccani.it/enciclopedia/thapsos_(Enciclopedia-Italiana)/ (accessed on 2 June 2024).
5. Benadusi, M. Oil in Sicily: Petrocapitalist imaginaries in the shadow of old smokestacks. *Econ. Anthropol.* **2018**, *5*, 45–58. [CrossRef]
6. Wikipedia. Available online: https://it.wikipedia.org/wiki/Polo_petrolchimico_siracusano (accessed on 3 June 2024).
7. Costantini, C. Ripristino dei Fondali Degradati e Recupero del Paesaggio Sommerso Nell'area Mediterranea. Ph.D. Thesis, Università degli studi di Palermo, Palermo, Italy, 2015.
8. R.N.O Saline di Priolo. Available online: https://www.salinedipriolo.it/ (accessed on 3 June 2024).
9. Lifeseposso. Available online: https://lifeseposso.eu/?page_id=9168 (accessed on 3 June 2024).
10. Philippi, R.A. *Enumeratio Molluscorum Siciliae cum Viventium tum in Tellure Tertiaria Fossilium, quae in Itinere suo Observavit*; Sumptibus Simonis Schroppi et Sociorum: Berlin, Germany, 1836; pp. 1–305.
11. Philippi, R.A. *Enumeratio Molluscorum Siciliae cum Viventium tum in Tellure Tertiaria Fossilium, quae in Itinere suo Observavit*; Sumptibus Eduardi Anton: Halle, Germany, 1844; Volume 2, pp. 1–323.
12. Aradas, A.; Benoit, L. *Conchiologia Vivente Marina della Sicilia e delle Isole che la Circondano*; Stabilimento Tipografico Di C. Galatola: Catania, Italy, 1870; pp. 1–324.
13. Monterosato, T.A.D. *Notizie Intorno alle Conchiglie Mediterranee*; Ufficio Tipografico di Michele Amenta: Palermo, Italy, 1872; pp. 1–61.
14. Monterosato, T.A.D. Nuova rivista delle conchiglie mediterranee. *Atti Accad. Sci. Lett. Arti Palermo* **1875**, *5*, 1–50.
15. Monterosato, T.A.D. Enumerazione e sinonimia delle conchiglie mediterranee. *G. Delle Sci. Nat. Ed. Econ.* **1878**, *13*, 61–115.
16. Monterosato, T.A.D. *Nomenclatura Generica e Specifica di Alcune Conchiglie Mediterranee*; Stabilimento Tipografico Virzì: Palermo, Italy, 1884; pp. 1–152.
17. De Martino, C. Un itinerario malacologico siciliano: La penisola di Magnisi. *Conchiglie* **1971**, *7*, 59–60.
18. Rudman, W.B.; Willan, R.C. Opisthobranchia. In *Mollusca: The Southern Synthesis. Fauna of Australia*; Beesley, P.L., Ross, G.J.B., Wells, A., Eds.; CSIRO Publishing: Collingwood, Australia, 1998; Volume 5B, pp. 915–942.
19. Prkić, J.; Petani, A.; Iglić, D.; Lanča, L. *Opisthobranchs of the Adriatic Sea: Photographic Atlas and List of Croatian Species*; Ronilački klub Sveti Roko: Bibinje, Croatia, 2018; pp. 1–462.
20. Sea Slug Forum. Available online: http://www.seaslugforum.net/welcome.htm (accessed on 25 June 2024).
21. Trainito, E.; Doneddu, M. *Nudibranchi del Mediterraneo*, 2nd ed.; Il Castello: Cornaredo, Italy, 2014; pp. 1–432.
22. Gosliner, T.M.; Valdés, Á.; Behrens, D.W. *Nudibranch & Sea Slug Identification–Indo-Pacific*, 2nd ed.; New World Publications, Inc.: Jacksonville, FL, USA, 2018; pp. 1–451.
23. Salvador, X. *Guía Práctica para Fotografiar Nudibranquios del Litoral Español*; Marcombo: Barcelona, Spain, 2020; pp. 1–360.
24. OPK Opistobranquis. Available online: https://opistobranquis.info/en/ (accessed on 1 July 2024).
25. Valdés, Á.; Ortea, J.; Ávila, C.; Ballesteros, M. Review of the genus *Dendrodoris* Ehrenberg, 1831 (Gastropoda: Nudibranchia) in the Atlantic Ocean. *J. Molluscan Stud.* **1996**, *62*, 1–31. [CrossRef]
26. Galià-Camps, C.; Cervera, J.L.; Valdés, Á.; Ballesteros, M. Attack on crypsis: Molecular and morphological study of *Dendrodoris* Ehrenberg, 1831 (Mollusca: Gastropoda: Nudibranchia) from the Mediterranean Sea and Northern Atlantic Ocean reinstates *Dendrodoris temarana* Pruvot-Fol. 1953. *Zootaxa* **2022**, *5133*, 383–406. [CrossRef]
27. Trainito, E.; Baldacconi, R. *Atlante di Flora e Fauna del Mediterraneo*, 5th ed.; Il Castello: Cornaredo, Italy, 2014; pp. 1–432.
28. Rodríguez-Prieto, C.; Ballesteros, E.; Boisset, F.; Afonso-Carrillo, J. *Alghe e Fanerogame del Mediterraneo*; Il Castello: Cornaredo, Italy, 2015; pp. 1–656.
29. WoRMS. Available online: https://www.marinespecies.org/ (accessed on 28 June 2024).
30. Marín, A.; Ros, J. Los sacoglosos (Mollusca, Opisthobranchia) del sudeste iberico. Catalogo de las especies y presencia de cloroplastos algales en las mismas. *Iberus* **1989**, *8*, 25–49.
31. Giménez-Casalduero, F.; Muniain, C.; González-Wangüemert, M.; Garrote-Moreno, A. *Elysia timida* (Risso, 1818) three decades of research. *Anim. Biodivers. Conserv.* **2011**, *34*, 217–227. [CrossRef]
32. Lombardo, A.; Marletta, G. The sacoglossans (Gastropoda *Heterobranchia*) of the central-eastern coast of Sicily (Ionian Sea). *Biodivers. J.* **2021**, *12*, 705–718. [CrossRef]
33. Lombardo, A.; Marletta, G. Su alcuni esemplari teratologici di nudibranchi e sacoglossi (Mollusca: Gastropoda) rinvenuti lungo la costa centro-orientale della Sicilia. *Alleryana* **2021**, *39*, 1–4.

34. Cattaneo Vietti, R.; Chemello, R. The opisthobranch fauna of a Mediterranean lagoon (Stagnone di Marsala, Western Sicily). *Malacologia* **1991**, *32*, 291–299.
35. Rufray, X.; Girard, P.; Le Bris, S.; Menut, T.; Santarelli, C.; Salvador, X. Liste commentée des limaces de mer de la lagune de Thau, *Hérault. Cah. Fond. Biotope* **2021**, *36*, 1–189.
36. Vitale, D.; Giacobbe, A.; Spinelli, A.; De Matteo, S.; Cervera, J.L. "Opistobranch" (mollusks) inventory of the Faro Lake: A Sicilian biodiversity hot spot. *Ital. J. Zool.* **2016**, *83*, 524–530. [CrossRef]
37. Cattaneo-Vietti, R.; Chemello, R.; Giannuzzi-Savelli, R. *Atlas of Mediterranean Nudibranchs*; Editrice La Conchiglia: Roma, Italy, 1990; pp. 1–264.
38. Lombardo, A.; Marletta, G. New findings of Nudipleura (Mollusca: Gastropoda) along the central-eastern coast of Sicily (Ionian Sea). *Thalassia Sal.* **2021**, *43*, 71–82.
39. Lombardo, A.; Marletta, G. Diversity of the Marine Heterobranchia Fauna at the Island of Pantelleria, Sicily Channel, Mediterranean Sea: First Contribution. *Acta Zool. Bulg.* **2023**, *75*, 37–48.
40. Ballesteros, M.; Pontes, M.; Madrenas, E. *Els Nudibranquis del Mar Català*; Brau: Girona, Spain, 2019; pp. 1–191.
41. Perrone, A. *Cyerce cristallina* (Trinchese, 1881) (Opisthobranchia: Sacoglossa) nel Golfo di Taranto. *Boll. Malacologico* **1983**, *19*, 145–150.
42. Vayssière, A. Monographie de la famille des pleurobranchidés. I. *Ann. Sci. Nat. Zool. Biol. Anim.* **1898**, *8*, 209–402.
43. Barletta, G. *Gasteropodi Nudi (Pleurobranchomorpha, Sacoglossa, Aplysiomorpha e Nudibranchia)*; Monotipia erredi: Genoa, Italy, 1980; pp. 1–132.
44. Lombardo, A.; Marletta, G. No longer so common: Findings of *Calliopaea bellula* d'Orbigny 1837 (Gastropoda: Sacoglossa) and *Tayuva lilacina* (A. Gould 1852) (Gastropoda: Nudibranchia) along the central-eastern coast of Sicily (Ionian Sea). *Atti Soc. It. Sci. nat. Museo Civ. Stor. Nat. Milano* **2023**, *10*, 79–82. [CrossRef]
45. Lombardo, A. The nudibranchs (Gastropoda *Heterobranchia*) of the central-eastern coast of Sicily I: Suborder Cladobranchia. *Biodivers. J.* **2021**, *12*, 913–935. [CrossRef]
46. Cormaci, M.; Furnari, G.; Giaccone, G.; Colonna, P.; Mannino, A.M. Metodo sinecologico per la valutazione degli apporti inquinanti nella rada di Augusta (Siracusa). *Boll. Acc. Gioenia Sci. Nat.* **1985**, *18*, 829–850.
47. Collado-Vides, L.; Mazzei, V.; Thyberg, T.; Lirman, D. Spatio-temporal patterns and nutrient status of macroalgae in a heavily managed region of Biscayne Bay, Florida, USA. *Bot. Mar.* **2011**, *54*, 377–390. [CrossRef]
48. Lombardo, A.; Marletta, G. Seasonal trend of nudibranchs (Gastropoda, *Heterobranchia*) along the central-eastern coast of Sicily (Mediterranean Sea). *J. Black Sea Mediterr. Environ.* **2023**, *29*, 1–24.
49. Goddard, J.H.R.; Treneman, N.; Pence, W.E.; Mason, D.E.; Dobry, P.M.; Green, B.; Hoover, C. Nudibranch range shifts associated with the 2014 warm anomaly in the Northeast Pacific. *Bull. South. Calif. Acad. Sci.* **2016**, *115*, 15–40. [CrossRef]
50. Betti, F.; Bava, S.; Cattaneo-Vietti, R. Composition and seasonality of a heterobranch assemblage in a sublittoral, unconsolidated, wave-disturbed community in the Mediterranean Sea. *J. Molluscan Stud.* **2017**, *83*, 325–332. [CrossRef]
51. Clark, K.B. Nudibranch life cycles in the Northwest Atlantic and their relationship to the ecology of fouling communities. *Helgoländer Wiss. Meeresunters.* **1975**, *27*, 28–69. [CrossRef]
52. Osman, R.W. The establishment and development of a marine epifaunal community. *Ecol. Monogr.* **1977**, *47*, 37–63. [CrossRef]
53. Coma, R.; Ribes, M.; Gili, J.M.; Zabala, M. (2000) Seasonality in coastal benthic ecosystems. *Trends Ecol. Evol.* **2000**, *15*, 448–453. [CrossRef]
54. Picton, B.; Morrow, C. *Nudibranchs of Britain, Ireland and Northwest Europe*; Princeton University Press: Princeton, NJ, USA, 2023; pp. 1–360.
55. Lombardo, A.; Marletta, G. First record of *Aporodoris millegrana* (Alder et Hancock, 1854) (Gastropoda Heterobranchia Nudibranchia) in the Ionian Sea, central Mediterranean Sea. *Biodivers. J.* **2020**, *11*, 875–878. [CrossRef]

oceans

MDPI

Article

Analyzing Trends in Saharan Dust Concentration and Its Relation to *Sargassum* Blooms in the Eastern Caribbean

José J. Hernández Ayala [1,*] and Rafael Méndez-Tejeda [2]

[1] Graduate School of Planning, University of Puerto Rico Rio Piedras Campus, San Juan 00921, Puerto Rico
[2] Atmospheric Sciences Laboratory, University of Puerto Rico at Carolina, Carolina 00984, Puerto Rico; rafael.mendez@upr.edu
* Correspondence: jose.hernandez180@upr.edu

Abstract: This study investigates the temporal trends and correlations between Saharan dust mass concentration densities (DMCD) and *Sargassum* concentrations (SCT) in the tropical North Atlantic. Average DMCD data for June, July, and August from 1980 to 2022, alongside SCT data for the same months from 2012 to 2022, were analyzed using Mann–Kendall tests for trends and lagged regression models to assess whether higher Saharan dust levels correlate with *Sargassum* outbreaks in the region. A comprehensive analysis reveals a significant upward trend in Saharan dust quantities over the study period, with the summer months of June, July, and August exhibiting consistent increases. Notably, 2018 and 2020 recorded the highest mean DMCD levels, with June showing the most significant increasing trend, peaking in 2019. These findings are consistent with previous studies indicating a continuous elevation in Saharan dust concentrations in the tropical atmosphere of the North Atlantic. Simultaneously, *Sargassum* concentrations also show a notable increasing trend, particularly in 2018, which experienced both peak SCT and elevated DMCD levels. Mann–Kendall tests confirm statistically significant upward trends in both Saharan dust and *Sargassum* concentrations. Simple linear regression and lagged regression analyses reveal positive correlations between DMCD and SCT, highlighting a temporal component with stronger associations observed in July and the overall June–July–August (JJA) period. These results underscore the potential contribution of elevated Saharan dust concentrations to the recent surge in *Sargassum* outbreaks in the tropical North Atlantic. Furthermore, the results from forward stepwise regression (FSR) models indicate that DMCD and chlorophyll (CHLO) are the most critical predictors of SCT for the summer months, while sea surface temperature (SST) was not a significant predictor. These findings emphasize the importance of monitoring Saharan dust and chlorophyll trends in the Eastern Caribbean, as both factors are essential for improving *Sargassum* modeling and prediction in the region. This study provides valuable insights into the climatic factors influencing marine ecosystems and highlights the need for integrated environmental monitoring to manage the impacts on coastal economies.

Keywords: Saharan dust; *Sargassum*; sea surface temperatures; chlorophyll; Caribbean

1. Introduction

Saharan dust, a phenomenon characterized by the transport of fine mineral particles from the arid regions of the Sahara Desert in North Africa across vast distances, exerts substantial influences on ecosystems and human activities in the tropical North Atlantic and the greater Caribbean region [1]. This natural phenomenon, also known as the Saharan Air Layer (SAL), has gained increasing attention due to its far-reaching impacts on air quality, weather patterns, and various sectors such as agriculture, health, and tourism [2–6]. The Saharan dust episodes are often associated with elevated concentrations of particulate matter (Figure 1) and the presence of various pollutants, which can lead to respiratory issues and reduced visibility [7]. Additionally, these dust particles can interact with cloud formation and solar radiation, potentially influencing local weather and climate [8].

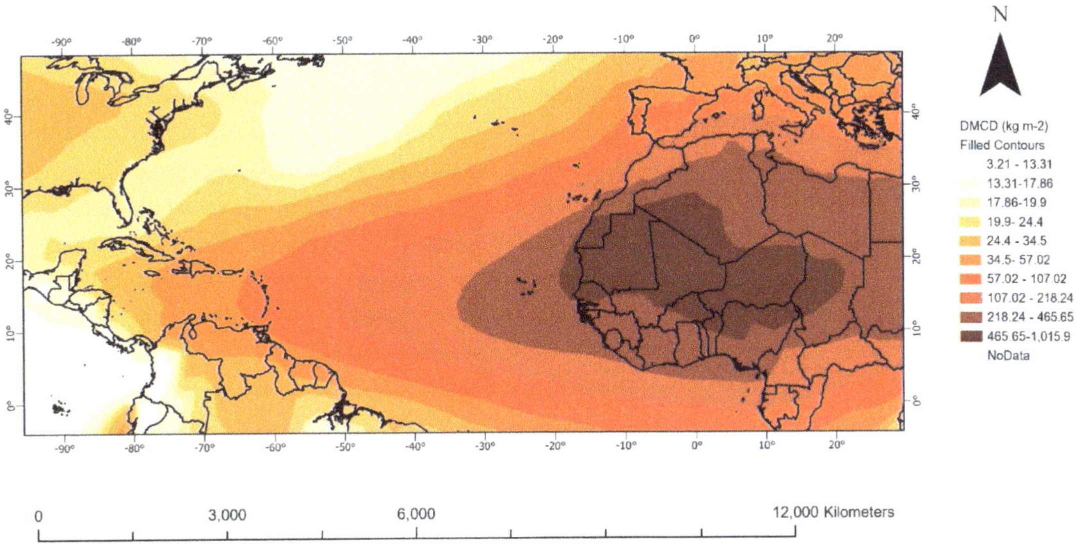

Figure 1. Mean Saharan dust column mass density for 28 June 2018.

As global temperatures rise due to climate change, the Sahara Desert experiences altered wind patterns and increased aridity, intensifying dust emissions [9]. This higher dust load can affect the radiative balance of the Earth by absorbing or reflecting sunlight, potentially contributing to regional and even global climate shifts [10,11]. Furthermore, the nutrient content within the transported dust can fertilize marine ecosystems, including the pelagic *Sargassum* in the tropical Atlantic [1,12,13]. While the exact mechanisms are still being studied, it is suggested that iron and other dust minerals might enhance *Sargassum* growth.

Saharan dust can have positive effects on *Sargassum*. The deposition of Saharan dust on the ocean surface can provide nutrients, including iron and phosphorus, which are essential for the growth of *Sargassum* [14,15]. These dust-derived nutrients act as fertilizers, stimulating primary productivity and potentially enhancing *Sargassum* growth rates. Iron, in particular, is a micronutrient that can be limiting in some oceanic regions, and the input of iron-rich Saharan dust can alleviate this limitation for *Sargassum* and other marine organisms [16].

In recent years, unprecedented SAL events have coincided with *Sargassum* blooms in the tropical waters of the North Atlantic, but little is known about whether Saharan dust deposition contributes to higher concentrations of the algae. Research on the relationship between SAL and pelagic *Sargassum* in the tropical North Atlantic has shed light on their interactions and potential influences, with some studies showing that these dust particles can affect the growth and distribution of the algae (12), even triggering the appearance of large-scale blooms in the Caribbean Sea [17,18].

Some studies suggest that atmospheric dust, which contains essential nutrients such as iron, phosphorus, and nitrogen, can enhance the growth of marine macroalgae, including *Sargassum* [19]. Dust-borne phosphorus also plays a crucial role in marine productivity [20]. There is empirical evidence supporting the hypothesis that nutrient-rich dust can stimulate algal blooms, identifying a correlation between dust transport events and increased *Sargassum* growth in the Caribbean Sea [21]. Such findings suggest that the deposition of nutrient-rich dust may be a significant factor in the proliferation of *Sargassum*.

Other factors that are important drivers of *Sargassum* blooms are sea surface temperature and high nutrient concentrations. Higher sea surface temperatures can promote the proliferation of *Sargassum* as warmer waters provide optimal conditions for its growth and reproduction [22]. On the other hand, higher chlorophyll concentrations in the water column can indicate regions of enhanced primary productivity and nutrients, which may be associated with favorable conditions for the growth and proliferation of *Sargassum* populations [23]. In combination with Saharan dust, sea surface temperatures and elevated nutrient concentrations could explain why some periods reflect higher *Sargassum* blooms.

For that reason, the objective of this study is to further examine the relation between SAL events and *Sargassum* blooms to determine if higher dust deposition leads to larger algae growth in the area. This study examines temporal trends in Saharan dust events in the Eastern Caribbean to determine if high-concentration SAL occurrences have increased over time and if there is any connection with anomalous *Sargassum* blooms in the region. After examining trends in SAL events in the tropical waters of the North Atlantic, the focus shifts to examining whether those events coincide with *Sargassum* outbreaks in the region.

2. Data & Methods

Average dust mass column density (DMCD) data—which refers to the total mass per unit area of a specific substance, such as aerosols or gasses, extending vertically from the Earth's surface to the top of the atmosphere—were obtained from MERRA-2 (Modern-Era Retrospective Analysis for Research and Applications, version 2), a state-of-the-art atmospheric reanalysis model developed by NASA. This variable is calculated by integrating the atmospheric concentration of the aerosols over the entire vertical column above a given point on the Earth's surface. The integration involves accounting for factors such as the vertical distribution of the dust and aerosols, the density of the air, and the altitude variations within the column. The monthly average MCD data were extracted for the Eastern Caribbean region of the tropical North Atlantic (−80.5, 10, −60, 23.5) for the entire period 1980–2022 (Figure 2A). The analysis was limited to that section of the study area to compare DMCD concentrations with *Sargassum* coverage areas (Figure 2B). The DMCD data have been used by previous researchers examining trends and patterns in Saharan dust concentrations [24–26].

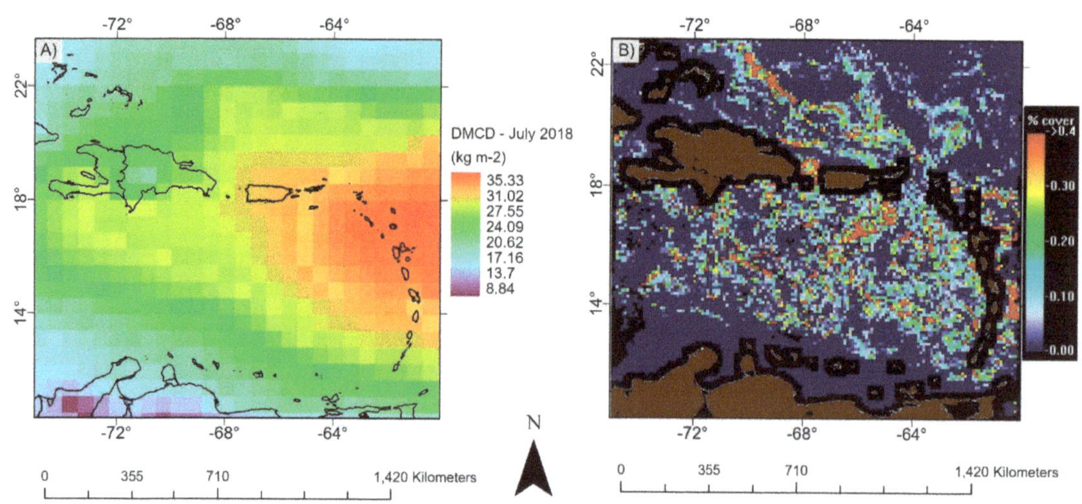

Figure 2. Mean dust mass column density (DMCD) for July of 2018 (**A**) and floating algae density, in terms of percentage of area covered for July 18, 2018 (**B**).

Monthly *Sargassum* concentrations (SCT) in tons for a section of the Eastern Caribbean (−80.5, 10, −60, 23.5) were obtained from the Optical Oceanography Lab at the University of South Florida [27]. The monthly average *Sargassum* in tons was extracted for June, July, and August (JJA), which have been documented to be the period with higher concentrations of the algae [1,12]. The temporal resolution of the *Sargassum* data is from 2010 through 2022. Mann–Kendall (MK) tests for trends were performed on the DMCD and SCT to determine if the dust and *Sargassum* concentrations during the JJA period or individual months were increasing over time. The MK test is a widely used statistical test in the atmospheric sciences that allows for robust trend detection, and several studies have implemented the method to analyze trends in dust aerosols [28–30].

Mean sea surface temperature (SST) and chlorophyll (CHLO) data were also extracted for the study area to determine if those factors, in combination with DMCD, could also explain SCT variability. Mean SST data for June, July, and August were obtained from the NOAA Extended Reconstruction SSTs Version 5. Average chlorophyll concentrations for the same months and region were extracted from the Moderate-Resolution Imaging Spectroradiometer (MODIS).

The DMCD and SCT data were analyzed using simple linear regression models to determine if higher dust amounts coincided with higher *Sargassum* concentrations in the study area. The regression models between Saharan dust and *Sargassum* concentrations were performed on the individual months and the JJA period. The simple linear regression models were also performed with a lag, this was done to consider the possibility that previous Saharan dust deposition in the ocean would later enhance *Sargassum* growth in a later period. This type of regression is particularly useful when you suspect that the current value of a variable depends on its past values; for example, in this case, we hypothesize that higher Saharan dust deposition leads to higher *Sargassum* concentrations.

A one-month lag between dust arrival and *Sargassum* concentration is justified as nutrients from dust, like iron and phosphorus, take several weeks to be absorbed and utilized. This aligns with observations that phytoplankton blooms occur weeks after nutrient deposition. *Sargassum* mats can persist in regions, allowing time to utilize these nutrients [18,31].

Forward stepwise regression (FSR) procedures were used to examine how important DMCD was to SCT variability when compared to other factors like SST and CHLO. FSR is a statistical method that sequentially adds predictor variables to a regression model based on their contribution to improving the model's fit, aiming to identify the most relevant factors for explaining the variability in the response variable. In this case, SCT is the response variable that we are interested in predicting using the predictors DMCD, SST, and CHLO.

3. Results and Discussion

3.1. *Trends in Saharan Dust and Sargassum Concentrations*

When trends in monthly and seasonal average DMCD concentrations are examined, we find that Saharan layer dust quantities in the atmosphere of the tropical North Atlantic have been increasing over time (Figure 3A). When the summer months of JJA are analyzed together, the results show a steadily increasing trend of mean DMCD for the tropical North Atlantic for the 1980–2022 period (Figure 3A). The years with the highest mean DMCD concentrations were 2018 and 2020. Of the three summer months (JJA) with higher mean DMCD concentrations, June was found to have the most significant increasing trend in Saharan dust from 1980 through 2022 (Figure 3A). For the June series, 2019 was found to be the year with the highest mean DMCD. The individual months of July and August show increasing trends in average DMCD concentrations (Figure 3A), yet those were not as significant as the positive trend in Saharan dust quantities that June reflected. Of all of the years analyzed, July 2018 was found to be the period with the highest average DMCD. These results confirm what some studies have been suggesting, that Saharan layer dust concentrations have been increasing over time in the tropical atmosphere of the North Atlantic [9].

The time series plots for *Sargassum* concentrations also suggest that the algae have been experiencing an increasing trend in total quantity and area in the region (Figure 3B). The results for the JJA period show that *Sargassum* concentrations have been increasing from 2010 through 2022, with 2018 being the year with the highest agglomerations detected (Figure 3B). It is important to note that 2018 was also the year with the highest DMCD concentrations in the region, so both the highest *Sargassum* agglomerations and Saharan dust quantities coincided during that period. When the individual months are examined, we find that all of them exhibited significant increasing trends in *Sargassum* concentration (Figure 3B). June and July of 2018 and August of 2021 were the periods with the highest mean concentrations of *Sargassum* in the region (Figure 3B). These results coincide with the findings of other studies that have detected a recent increase in *Sargassum* concentrations and total area covered [18], which might be associated with higher dust deposition from the Sahara [12].

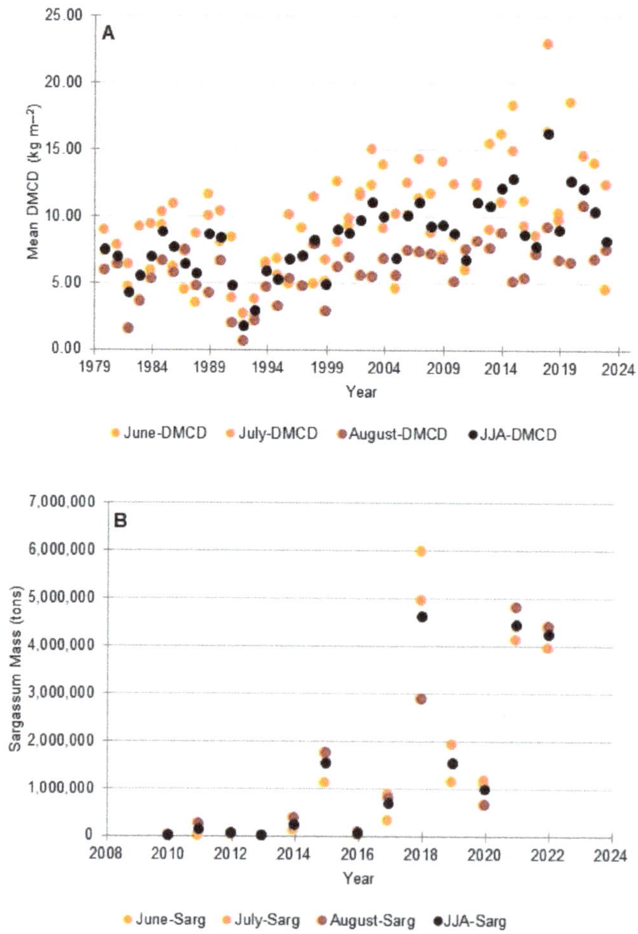

Figure 3. Average monthly dust mass concentrations in kg m^{-2} for the 1980–2022 period (**A**) and average June, July, and August *Sargassum* concentrations in tons from 2010–2022 (**B**).

When both DMCD and SCT time series are analyzed using Mann–Kendall tests for trends, we find that Saharan dust and *Sargassum* concentrations have been increasing over time in the tropical atmosphere of the North Atlantic (Table 1). The month of June was found to have the most statistically significant increase in DMCD and SCT concentrations. When the three months are examined together, the results show that both Saharan dust and *Sargassum* concentrations reflect statistically significant trends. Even though July and August show weaker increasing trends in mean DMCD when compared to June and the JJA period, they still exhibit statistically significant trends (Table 1). The Mann–Kendall results for SCT show that all of the individual months and the JJA period exhibited statistically significant increasing trends in *Sargassum* concentrations, with June showing the strongest trend.

Table 1. Results of Mann–Kendall tests for trends for Saharan dust (DMCD) and *Sargassum* concentrations in tons (SCT) for different periods.

Period (1980–2022)	Tau-Coeff Dust	*p*-Value Dust
June	0.451	0.001
July	0.258	0.015
August	0.287	0.006
JJA	0.433	0.001
Period (2012–2022)	**Tau-Coeff Sarg**	***p*-Value Sarg**
June	0.636	0.001
July	0.576	0.011
August	0.58	0.011
JJA	0.57	0.011

3.2. Statistical Modeling Results

The results of the simple linear regression models between DMCD and SCT show an overall positive correlation between the two variables (Figure 4); as Saharan dust concentrations increase, so do *Sargassum* concentrations. When the analysis is done for the JJA (Figure 4A), the results show a very strong positive correlation (R2 0.429) between DMCD and SCT. The results for the individual months show the month of July (Figure 4C) with the strongest correlation, while weaker correlations were found for June and August (Figure 4B,D). These findings suggest that the relationship between Saharan dust and *Sargassum* concentrations has a particular temporal component, with weaker correlations in June and August, and stronger correlations in July and the JJA period overall.

The results of the lagged regression models also show positive correlations between DMCD and SCT (Figure 4E,D). When the lagged regression was done using DMCD for June and SCT for July, we found a positive yet weak correlation between Saharan dust and *Sargassum* (Figure 4E). However, when the analysis was performed using DMCD for July and SCT for August, the results yielded a stronger positive correlation between the two factors (Figure 4D). These findings suggest that higher dust concentrations in June do not necessarily materialize into higher *Sargassum* quantities in July. Yet, greater amounts of dust in July show a stronger statistical relationship with elevated *Sargassum* concentrations. These results are similar to those found in previous studies [12,14], showing that higher Saharan dust concentrations likely enhanced some of the recent *Sargassum* outbreaks in the region.

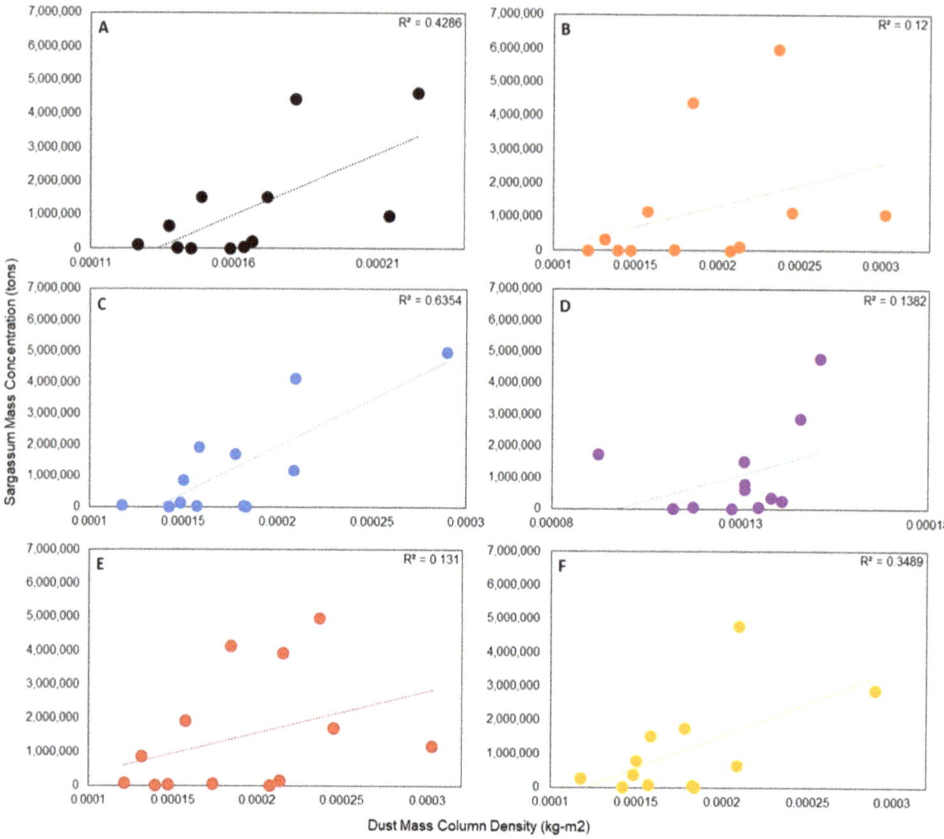

Figure 4. Simple linear regression models between mean monthly DMCD and total SCT data for the JJA period (**A**), June (**B**), July (**C**), and August (**D**). Lagged regression models between total *Sargassum* concentrations in tons (y-variable) and mean Saharan dust concentrations for June (dust) and July (*Sargassum*) (**E**) and July (dust) and August (*Sargassum*) (**F**).

The results of the forward stepwise regression (FSR) models suggest that DMCD is one of the most important predictors of SCT in the Eastern Caribbean region that was analyzed in this study (Table 2). The FSR results for June selected average CHLO in May as the most important factor behind higher mean SCT, suggesting that elevated chlorophyll concentrations often indicate increased phytoplankton growth. While phytoplankton and Sargassum are both primary producers and compete for nutrients, the presence of phytoplankton, indicated by elevated chlorophyll levels, suggests nutrient-rich conditions that can support Sargassum growth [18]. FSR results for July and August show DMCD as the most important predictor of SCT, suggesting that higher Saharan dust deposition during those months is the key driver of higher *Sargassum* concentrations (Table 2). The FSR results for the JJA period identified both CHLO and DMCD as the most important factors that were examined, explaining higher SCT in the region. CHLO alone accounted for a third of the variability in SCT, while the addition of DMCD into the model increased the adjusted R-square to 0.563. These results suggest that Saharan dust and chlorophyll are the most important predictors of SCT in the Eastern Caribbean region. It is important to note that SST was not selected as an important factor in the FSR models, suggesting that ocean temperatures in this region might not be as critical to higher *Sargassum* concentrations as Saharan dust and chlorophyll.

Table 2. Forward stepwise regression (FSR) models of mean *Sargassum* concentration and several factors, including the means of dust column mass density, sea surface temperatures, and chlorophyll.

Forward Stepwise Regression Models						
Avg. Sargassum Concentra-Tions—June						
Step	Entered	Adj. R-Square	R-Square	C(p)	AIC	RMSE
1	Avg. CHLO (May)	0.3968	0.3420	4.1159	413.3	1,678,885.8
Avg. Sargassum Concentrations—July						
Step	Entered	Adj. R-Square	R-Square	C(p)	AIC	RMSE
1	Avg. DMCD (July)	0.4385	0.3875	1.2860	408.6	1,397,085.3
Avg. Sargassum Concentrations—August						
Step	Entered	Adj. R-Square	R-Square	C(p)	AIC	RMSE
1	Avg. DMCD (Aug)	0.2148	0.1434	0.8582	411.2	1,548,442.6
Avg. Sargassum Concentrations—Jja						
Step	Entered	Adj. R-Square	R-Square	C(p)	AIC	RMSE
1	Avg. CHLO (JJA)	0.3536	0.2948	7.2811	410.7	1,514,045.2
2	Avg. DMCD (JJA)	0.6364	0.5637	2.1574	405.2	1,190,910.5

4. Conclusions

In this study, the average Saharan dust mass concentration densities (DMCD) and mean *Sargassum* concentrations (SCT) in the tropical North Atlantic were analyzed to identify any coinciding increasing trends over the examined period. Mann–Kendall tests for trends and lagged regression models were used to examine the relationship between DMCD and SCT to determine if higher Saharan dust concentrations were associated with *Sargassum* outbreaks in the region. The analysis aimed to understand the potential impact of Saharan dust on the proliferation of *Sargassum*.

The analysis revealed a significant upward trend in Saharan dust quantities from 1980 to 2022, particularly during the summer months of June, July, and August (JJA). Notably, the years 2018 and 2020 recorded the highest levels of DMCD, with June exhibiting the most significant increasing trend, peaking in 2019. These findings are consistent with existing studies that indicate a continuous rise in Saharan dust concentrations in the tropical atmosphere of the North Atlantic, suggesting an ongoing elevation of dust levels in the region.

Simultaneously, *Sargassum* concentrations showed a parallel increasing trend, with 2018 marked by both peak SCT agglomerations and heightened DMCD concentrations. The Mann–Kendall tests confirmed statistically significant increasing trends in both Saharan dust and *Sargassum* concentrations. Additionally, simple linear regression models and lagged regression analyses established positive correlations between DMCD and SCT, revealing a temporal component with stronger correlations observed in July and the overall JJA period. These results support the hypothesis that Saharan dust contributes to Sargassum outbreaks in the tropical North Atlantic.

While this study provides valuable insights into the relationship between Saharan dust and *Sargassum* blooms, several limitations should be considered. The use of monthly average data may obscure short-term variations and specific events that could offer a more nuanced understanding of the relationship between dust deposition and *Sargassum* growth. The study's focus on a specific section of the tropical North Atlantic may not capture the full spatial variability of both dust deposition and Sargassum blooms across the entire region.

Moreover, while correlations between DMCD and SCT were identified, the specific nutrient dynamics and biological mechanisms by which Saharan dust influences *Sargassum* growth remain complex and are not fully explored. Other factors, such as ocean currents, local nutrient sources, and anthropogenic impacts, were not comprehensively analyzed and could also play significant roles.

Future research should address these limitations by incorporating higher temporal resolution data, expanding the spatial scope of analysis, and investigating the detailed nutrient dynamics and biological mechanisms involved. Additionally, exploring the interplay between Saharan dust, local nutrient sources, and other environmental factors will provide a more comprehensive understanding of the drivers behind *Sargassum* blooms.

Author Contributions: Conceptualization, J.J.H.A. and R.M.-T.; methodology, J.J.H.A.; software, J.J.H.A.; validation, J.J.H.A. and R.M.-T.; formal analysis, J.J.H.A.; investigation, J.J.H.A. and R.M.-T.; resources, R.M.-T.; data curation, J.J.H.A.; writing—original draft preparation, J.J.H.A. and R.M.-T.; writing—review and editing, J.J.H.A. and R.M.-T.; visualization, J.J.H.A.; supervision, R.M.-T.; project administration, R.M.-T.; funding acquisition, R.M.-T. All authors have read and agreed to the published version of the manuscript.

Funding: This research received no external funding.

Institutional Review Board Statement: Not applicable.

Data Availability Statement: Average monthly dust mass column density data was obtained from the MERRA-2 (Modern-Era Retrospective Analysis for Research and Applications, version 2) https://gmao.gsfc.nasa.gov/reanalysis/MERRA-2/data_access/, (accessed on 10 August 2024). Monthly *Sargassum* concentrations (SCT) in tons for a section of the Eastern Caribbean (−80.5, 10, −60, 23.5) were obtained from the Optical Oceanography Lab at the University of South Florida (https://optics.marine.usf.edu/projects/saws.html, accessed on 30 August 2024). Mean monthly SST data were obtained from the NOAA Extended Reconstruction SSTs Version 5 (https://psl.noaa.gov/data/gridded/data.noaa.ersst.v5.html, accessed on 20 October 2023) Average chlorophyll concentrations for the same months and region were extracted from the Moderate-Resolution Imaging Spectroradiometer (https://oceancolor.gsfc.nasa.gov/resources/atbd/chlor_a/, accessed on 30 August 2024).

Conflicts of Interest: The authors declare no conflict of interest.

References

1. Méndez-Tejeda, R.; Rosado-Jimenez, G.A. Influence of climatic factors on Sargassum arrivals to the coasts of the Dominican Republic. *J. Oceanogr. Mar. Sci.* **2019**, *10*, 22–32. [CrossRef]
2. Prospero, J.M.; Ginoux, P.; Torres, O.; Nicholson, S.E.; Gill, T.E. Environmental characterization of global sources of atmospheric soil dust identified with the Nimbus 7 Total Ozone Mapping Spectrometer (TOMS) absorbing aerosol product. *Rev. Geophys.* **2002**, *40*, 1002. [CrossRef]
3. Bergametti, G.; Liousse, C.; Dulac, F. Composition of aerosol collected in the free troposphere during the ACE 2 project. *Tellus B. Chem. Phys. Meteorol.* **2007**, *59*, 368–377.
4. Monteil, M.A. Saharan dust clouds and human health in the English-speaking Caribbean: What we know and don't know. *Environ. Geochem. Health* **2008**, *30*, 339–343. [CrossRef]
5. Akpinar-Elci, M.; Martin, F.E.; Behr, J.G.; Diaz, R. Saharan dust, climate variability, and asthma in Grenada, the Caribbean. *Int. J. Biometeorol.* **2015**, *59*, 1667–1671. [CrossRef]
6. Xu-Yang, Y.; Dessert, C.; Losno, R. Atmospheric deposition over the Caribbean region: Sea salt and Saharan dust are sources of essential elements on the island of Guadeloupe. *J. Geophys. Res. Atmos.* **2022**, *127*, E2022jd037175. [CrossRef]
7. Tagliabue, A.; Bowie, A.R.; Boyd, P.W.; Buck, K.N.; Johnson, K.S.; Saito, M.A. The integral role of iron in ocean biogeochemistry. *Nature* **2017**, *543*, 51–59. [CrossRef]
8. Gyan, K.; Henry, W.; Lacaille, S.; Laloo, A.; Lamsee-Ebanks, C.; McKay, S.; Monteil, M.A. African dust clouds are associated with increased pediatric asthma accident and emergency admissions on the Caribbean island of Trinidad. *Int. J. Biometeorol.* **2005**, *49*, 371–376. [CrossRef]
9. Sakhamuri, S.; Cummings, S. Increasing trans-Atlantic intrusion of Sahara dust: A cause of concern? *Lancet Planet. Health* **2019**, *3*, e242–e243. [CrossRef]
10. Tanré, D.; Haywood, J.; Pelon, J.; Léon, J.F.; Chatenet, B.; Formenti, P.; Myhre, G. Measurement and modeling of the Saharan dust radiative impact: Overview of the Saharan Dust Experiment (SHADE). *J. Geophys. Res. Atmos.* **2003**, *108*. [CrossRef]
11. Slingo, A.; Ackerman, T.P.; Allan, R.P.; Kassianov, E.I.; McFarlane, S.A.; Robinson, G.J.; Dewitte, S. Observations of the impact of a major Saharan dust storm on the atmospheric radiation balance. *Geophys. Res. Lett.* **2006**, *33*. [CrossRef]

12. Lapointe, B.E.; Brewton, R.A.; Herren, L.W.; Wang, M.; Hu, C.; McGillicuddy Jr, D.J.; Morton, P.L. Nutrient content and stoichiometry of pelagic *Sargassum* reflects increasing nitrogen availability in the Atlantic Basin. *Nat. Commun.* **2021**, *12*, 3060. [CrossRef]
13. Skliris, N.; Marsh, R.; Appeaning Addo, K.; Oxenford, H. Physical drivers of pelagic *Sargassum* bloom interannual variability in the Central West Atlantic over 2010–2020. *Ocean. Dyn.* **2022**, *72*, 383–404. [CrossRef]
14. Baker, A.R.; Jickells, T.D.; Witt, M. Atmospheric deposition of soluble iron to the global oceans: Implications for iron limitation of phytoplankton. *Annu. Rev. Mar. Sci.* **2020**, *12*, 523–550.
15. Meskhidze, N.; Xu, J.; Gantt, B. Iron in atmospheric mineral dust: Biogeochemical cycling and impact on ocean productivity. *Glob. Biogeochem. Cycles* **2021**, *35*, e2020GB006864.
16. Liriano-Gómez, R.E.; Gómez, A.C.; Bolívar, M.G.; González-Hernández, Y.; Jauregui-Haza, U.J. Characterization of *Sargassum* accumulated on Dominican beaches in 2021: Analysis of heavy, alkaline, and alkaline-earth metals, proteins, and fats. *Mar. Pollut. Bull.* **2023**, *193*, 115120. [CrossRef]
17. Wang, M.; Hu, C.; Barnes, B.B.; Mitchum, G.T. The expanding *Sargassum* horizons: Mechanisms, detection, and management. *Remote Sens. Environ.* **2019**, *231*, 111215. [CrossRef]
18. Johnson, M.T.; Hu, C.; Schollaert Uz, S.; Zhao, J.; Muller-Karger, F.E. The Ecology of Harmful Algal Blooms in the Caribbean and Western Atlantic. *Front. Mar. Sci.* **2020**, *7*, 342.
19. Measures, C.I.; Vink, S. On the use of dissolved aluminum in surface waters to estimate dust deposition to the ocean. *Glob. Biogeochem. Cycles* **2000**, *14*, 317–327. [CrossRef]
20. Mahowald, N.M.; Jickells, T.D.; Baker, A.R.; Artaxo, P.; Benitez-Nelson, C.R.; Bergametti, G.; Bond, T.C.; Chen, Y.; Cohen, D.D.; Herut, B.; et al. Global distribution of atmospheric phosphorus sources, concentrations and deposition rates, and anthropogenic impacts. *Glob. Biogeochem. Cycles* **2008**, *22*, GB4026. [CrossRef]
21. Gower, J.; Young, E.; King, S. Satellite Images Suggest a New Sargassum Source Region in 2011. *Remote Sens. Lett.* **2013**, *4*, 764–773. [CrossRef]
22. Djakouré, S.; Araujo, M.; Hounsou-Gbo, A.; Noriega, C.; Bourlès, B. On the potential causes of the recent Pelagic *Sargassum* blooms events in the tropical NorthAtlantic Ocean. *Biogeosciences Discuss.* **2017**, 1–20.
23. Oviatt, C.A.; Huizenga, K.; Rogers, C.S.; Miller, W.J. What nutrient sources support anomalous growth and the recent *Sargassum* mass stranding on Caribbean beaches? A review. *Mar. Pollut. Bull.* **2019**, *145*, 517–525. [CrossRef]
24. Asutosh, A.; Vinoj, V.; Murukesh, N.; Ramisetty, R.; Mittal, N. Investigation of June 2020 giant Saharan dust storm using remote sensing observations and model reanalysis. *Sci. Rep.* **2022**, *12*, 6114. [CrossRef] [PubMed]
25. Bencherif, H.; Bounhir, A.; Bègue, N.; Millet, T.; Benkhaldoun, Z.; Lamy, K.; Gadouali, F. Aerosol Distributions and Sahara Dust Transport in Southern Morocco from Ground-Based and Satellite Observations. *Remote Sens.* **2022**, *14*, 2454. [CrossRef]
26. Hu, C.; Zhang, S.; Barnes, B.B.; Xie, Y.; Wang, M.; Cannizzaro, J.P.; English, D.C. Mapping and quantifying pelagic *Sargassum* in the Atlantic Ocean using multi-band medium-resolution satellite data and deep learning. *Remote Sens. Environ.* **2023**, *289*, 113515. [CrossRef]
27. Kang, L.; Huang, J.; Chen, S.; Wang, X. Long-term trends of dust events over the Tibetan Plateau during 1961–2010. *Atmos. Environ.* **2016**, *125*, 188–198. [CrossRef]
28. Gavrouzou, M.; Hatzianastassiou, N.; Gkikas, A.; Korras-Carraca, M.B.; Mihalopoulos, N. A global climatology of dust aerosols based on satellite data: Spatial, seasonal and inter-annual patterns over the period 2005–2019. *Remote Sens.* **2021**, *13*, 359. [CrossRef]
29. Jafari, R.; Amiri, M.; Asgari, F.; Tarkesh, M. Dust source susceptibility mapping based on remote sensing and machine learning techniques. *Ecol. Inform.* **2022**, *72*, 101872. [CrossRef]
30. Jickells, T.D.; An, Z.S.; Andersen, K.K.; Baker, A.R.; Bergametti, G.; Brooks, N.; Cao, J.; Boyd, P.W.; Duce, R.A.; Hunter, K.A.; et al. Global Iron Connections Between Desert Dust, Ocean Biogeochemistry, and Climate. *Science* **2005**, *308*, 67–71. [CrossRef]
31. Lapointe, B.E. A Comparison of Nutrient-Limited Productivity in Sargassum natans from Different Water Masses in the Western North Atlantic Ocean. *Limnol. Oceanogr.* **1995**, *40*, 625–633. [CrossRef]

oceans

MDPI

Article

Trajectory Planning of a Mother Ship Considering Seakeeping Indices to Enhance Launch and Recovery Operations of Autonomous Drones

Salvatore Rosario Bassolillo, Egidio D'Amato, Salvatore Iacono, Silvia Pennino * and Antonio Scamardella

Department of Science and Technology, University of Naples "Parthenope", Centro Direzionale Isola C4, 80143 Naples, Italy; salvatorerosario.bassolillo@collaboratori.uniparthenope.it (S.R.B.); egidio.damato@uniparthenope.it (E.D.); salvatore.iacono001@studenti.uniparthenope.it (S.I.); antonio.scamardella@uniparthenope.it (A.S.)

* Correspondence: silvia.pennino@uniparthenope.it; Tel.: +39-081-547-6686

Abstract: This research focuses on integrating seakeeping indices into the trajectory planning of a mother ship in order to minimize risks during UAV (unmanned aerial vehicle) takeoff and landing in challenging sea conditions. By considering vessel dynamics and environmental factors, the proposed trajectory planning algorithm computes optimal paths that prioritize the stability and safety of the ship, mitigating the impact of adverse weather on UAV operations. Specifically, the new adaptive weather routing model presented is based on a genetic algorithm. The model uses the previously evaluated response amplitude operators (RAOs) for the reference ship at different velocities and encounter angles, along with weather forecast data provided by the global wave model (GWAM). Preliminary evaluations confirm the effectiveness of the presented model in significantly improving the reliability of autonomous UAV operations from a mother ship across all encountered sea state conditions, particularly when compared with a graph-based solution. The current results clearly demonstrate that it is possible to achieve appreciable improvements in ship seakeeping performance, thereby making UAV-related operations safer.

Keywords: autonomous navigation; adaptive weather routing; seakeeping performance; route optimization; Dijkstra algorithm; genetic algorithm; UAV operations; drone recovery operations; mother ship; ship's trajectory planning

1. Introduction

The field of autonomous navigation, situated at the convergence of robotics, artificial intelligence, and transport engineering, offers significant potential to transform transportation in numerous sectors. Leveraging state-of-the-art technologies about sensors, algorithms for machine learning, and complex control systems, guidance, control, and navigation (GNC) systems allow vehicles to function and make choices on their own, eliminating the need for direct human control. Ranging from autonomous cars to drones and ships, these technologies stand on the brink of altering the conventional methods of movement and goods transportation.

In the context of UAV operations on a mother ship, integration and coordination between the unmanned aerial vehicles (UAVs) and the vessel become paramount. The mother ship serves as a mobile base of operations, providing a stable platform for launching and recovering UAVs in various maritime environments. However, the challenges inherent in conducting these operations at sea necessitate careful consideration of factors such as vessel motion, deck dynamics, and environmental conditions.

During launch and recovery procedures, the UAVs must navigate the dynamic environment of the mother ship deck with precision and agility. This requires sophisticated control systems that can adapt to the vessel motion and compensate for any disturbances induced by waves, wind, or other external forces. Furthermore, the UAVs must be equipped with robust sensing capabilities to accurately assess the relative motion between the aircraft and the ship, enabling them to make real-time adjustments to their flight path and trajectory.

For UAV recovery operations, the mother ship typically employs a variety of systems and mechanisms to safely retrieve the aircraft upon completion of their mission. This may include the use of recovery nets, landing pads, or even autonomous landing systems that guide the UAVs to a designated touchdown area on the ship deck. In addition to ensuring the safety of the UAVs themselves, these systems must also take into account the safety of personnel and equipment on board the vessel, particularly in rough sea conditions where precision and stability are critical.

By integrating seakeeping criteria into the trajectory planning process, this research aims to reduce the dangers associated with the landing and departure operations of UAVs, optimizing the route followed by the mother ship in order to maximize stability and reduce exposure to motions during navigation in adverse marine conditions.

Through a combination of advanced algorithms, the goal is to enable autonomous UAV operations from a mother ship platform that are safe, efficient, and capable of supporting a wide range of maritime missions.

The suggested method relies on a new adaptive weather routing system made to find the best route, improving how well ships perform in various sea conditions. This model focuses on particular factors that keep ships steady, ensuring that it picks routes that make the most of the ship's abilities in the ever-changing sea conditions. Within this framework, this research explores the application of a genetic algorithm-based planning methodology.

The algorithm is applied to a designated mother ship, serving as the benchmark vessel, to investigate the impact of selected seakeeping variables and vessel speed on the optimal route selection for launch and recovery operations. Consequently, the primary objective of this study is to develop an adaptive weather routing model with the following aims: (i) improve the vessel seakeeping performance across varying sea states; (ii) increase safety in autonomous launch and recovery procedures, particularly in adverse weather conditions; and (iii) conduct a comprehensive sensitivity analysis to verify the influence of individual seakeeping parameters on the optimal route determination. A campaign of numerical results is considered to prove the effectiveness of the proposed procedure, with a comparative analysis with a Djikstra-based algorithm being used to evaluate the effectiveness of each model in facing the intricate challenges posed by adverse marine conditions.

Although the existing literature on weather routing mainly concentrates on route changes to optimize fuel consumption, the impact of weather routing on fuel savings can be considered only marginal, as discussed in previous papers [1,2]. For this reason, this paper focuses on the seakeeping performance impact in weather routing, taking into account some specific parameters useful for improving the takeoff and landing operations of UAVs. The proposed procedure is a genetic-based optimization algorithm to overcome graph-based solutions and consider seakeeping performance as nonlinear constraints.

The key contributions of the paper are: (i) development of an adaptive weather routing system introducing a new model that enhances ship performance in various sea state conditions by optimizing the route of the mother ship, focusing on reducing ship motions; (ii) application of a genetic algorithm-based planning methodology to optimize the trajectory of the ship, taking into account seakeeping performance limits as constraints; (iii) the limits are defined in order to improve UAV takeoff and landing operations; and (iv) a comprehensive sensitivity analysis is conducted to verify the influence of individual seakeeping index on optimal route determination.

2. Literature Review

Understanding the intricacies of weather routing is foundational to appreciating the complexities and solutions in the operations of unmanned aerial vehicles (UAVs) on a mother ship, where the seamless integration and coordination between UAVs and the vessel are paramount. Within this framework, ship weather routing (SWR) emerges as an important operational-level decision tool, entailing the determination of the optimal route for a vessel navigating from an originating port to a designated final port, taking into account variable weather conditions such as winds, waves, and sea currents.

In recent years, there has been a proposal to adopt the term environmental routing as a more apt descriptor as it encompasses environmental factors in the pursuit of the most efficient or economical route. For example, in [3], authors highlight that the terms environmental routing and weather routing are often used interchangeably, but the latter can be considered as a subset of the former. Weather conditions, such as winds, waves, and currents, constitute elements of the environment where ships operate, significantly impacting their performance. It is important to clarify that in this context, environmental routing pertains to the prevailing weather conditions and environmental factors (such as sea depth and designated areas necessitating specific fuel usage) in which ships navigate, distinct from the environmental performance of a voyage itself. The selection of a route optimizing vessel performance, such as minimizing ship emissions, can be addressed as part of the green ship routing problem [4].

In the context of weather routing, risk can be considered in several ways, which is contingent upon the concerns of the ship operator seeking to mitigate adverse weather-related impacts. For instance, in [5], authors identify wind as a primary safety hazard during voyages and aim to reduce the risk of prolonged exposure to high-wind areas. However, their utilization of multicriteria optimization may inadvertently prolong total sailing time, which could serve as an alternative objective in ship weather routing. Ref. [6] assesses safety risk based on the probability of encountering waves exceeding 4 m in height. In [7], navigation risk is evaluated according to the sailing conditions outlined in the International Maritime Organization (IMO) guidelines for navigators. The determination of weather conditions, such as wave height, length, and wind, that constitute a safety risk is contingent upon several factors specific to the voyage, including ship type, condition, and geographical region. These environmental considerations can be integrated into ship weather routing models by minimizing fuel consumption as it is feasible to estimate fuel consumption in areas affected by adverse weather conditions. A comprehensive analysis of risk assessment models in narrow maritime waterways is presented in [8], focusing on accident frequency and severity, albeit without exploring the influence of weather. Alternatively, risk definitions may concentrate on the financial implications of a voyage, such as assessing the risk of failing to reach the next port within a specified timeframe, potentially resulting in contractual penalties [9], or increased operating costs due to heightened fuel consumption in adverse weather conditions [10].

Weather routing not only addresses the risk of maritime incidents but also plays a crucial role in boosting the reliability of maritime services and reducing the operating costs of voyages. For example, although it is not the unique factor in preventing container losses at sea, it significantly aids in this effort. Economically, navigating through adverse weather conditions can lead to heightened fuel consumption and the possibility of voyage delays, potentially incurring additional demurrage charges [11]. Environmentally, lower fuel consumption results in decreased emissions of CO_2 and SO_x. Nonetheless, it is important to recognize potential environmental tradeoffs; for example, even with reduced fuel consumption, emissions of particulate matter may increase due to engines operating at lower loads [12]. In [13], various strategies aimed at optimizing vessel operations to reduce carbon emissions are reviewed, identifying weather routing as a key operational alternative.

Nevertheless, route planning presents considerable challenges due to the intricate and diverse nature of environments. This issue is frequently articulated as an optimization problem, wherein the most efficient trajectory traverses a series of predetermined waypoints, taking into account obstacles, potential traffic, constraints arising from the ship behavior in rough seas, and the characteristics of the encountered sea conditions. A foundational approach to solving such complex, multi-stage decision-making problems was introduced in [14], where the author provided a mathematical framework for optimizing decisions across different fields, introducing optimization techniques for problems involving sequential decisions.

The extensive body of literature on 2D path planning problems benefits from interdisciplinary exchanges with the robotics and automotive sectors, sharing algorithms and resolution strategies. These methodologies can be categorized into several groups: variational methods, optimal control strategies, geometric methodologies, graph optimization techniques, artificial potential field frameworks, and nature-inspired optimization algorithms.

Variational methods for path planning are intuitively appealing but become challenging in complex scenarios, particularly when coupled with flight dynamics constraints [15,16]. Optimal control strategies, designing closed-loop algorithms that converge to locally suboptimal trajectories, may struggle to find a global optimum and require extensive computational resources when dealing with nonlinear optimization algorithms [17–20].

Geometric approaches to path planning, which define paths as sequences of segments, arcs, or template curves, include solutions such as Dubins' car problem [21] or clothoid curves [22]. These methods usually need to be integrated with other algorithms to manage obstacles and no-fly zones [23,24].

Graph optimization, employing edges designed through geometric methods, requires discretizing environments into grids and often utilizes visibility graphs, Voronoi diagrams, rapidly exploring random trees (RRTs), tangent graphs, sparse tangential networks (SPARTANs), and roadmaps [25–37]. Algorithms such as Dijkstra's, A*, and D* algorithms are commonly employed to find the shortest paths [38–40].

Artificial potential field techniques, while effective for real-time guidance, become problematic with increased obstacles or "potential sources", necessitating strategies to circumvent emerging singularities [41,42].

Lastly, nature-inspired optimization methods enable the design of complex models based on flight dynamics, optimizing sequences of feasible maneuvers to reach target points. These methods, while sophisticated, are computationally demanding and typically suited for offline optimization [43–47].

Concurrently, the literature on ship weather routing presents various approaches, broadly categorized into exact methods, heuristic approaches, and multi-objective evolutionary methods. Early works typically employed exact methods, while recent studies have leaned towards heuristics. This overview outlines these approaches and highlights key objective functions, such as economic concerns, climate considerations, and regional fairness.

One of the earliest techniques is the isochrone method, devised by James in 1957 [48], which prioritizes minimizing travel time and facilitates manual route planning. Subsequent advancements, like those by Hagiwara and Spaans in 1987 [49], incorporated fuel consumption considerations and computerized the method. The calculus of variations method, introduced by Haltiner et al. [50], provides an analytical approach to weather routing, determining both the ship path and engine power. In Chen et al. [51], five improved methods are proposed and compared to investigate the most effective enhancement strategy for achieving robustness and practicality in real-time applications. Papadakis [52] develops general methodologies for the minimal time routing problem of a vessel moving in stationary or time-dependent environments.

Commonly employed methods include multi-objective genetic algorithms, which probabilistically tackle a discretized nonlinear optimization problem, utilizing the ship course and velocity profile represented by parametric curves. Most existing weather route optimization methods generate the optimal route to minimize fuel oil consumption; however, in the work of Pennino et al. [1,2], a new adaptive weather routing model was presented based on Dijkstra's shortest path algorithm, which aims to select the optimal route that maximizes the ship performance during navigation. Szłapczynska and Smierzchalski [5] introduced a multi-criteria evolutionary weather routing algorithm, employing an iterative population development process to yield a Pareto-optimal set of solutions. Wei and Zhou [53] devised a three-dimensional (3D) dynamic programming approach based on De Wit's original work [54], focusing on optimizing ship speed and heading angle during route planning. Zaccone et al. [55] developed a 3D dynamic-programming-based method for ship voyage optimization, aiming to select the optimal path and speed profile considering expected weather conditions along the route. Another technique for weather routing optimization is pathfinding. Padhy et al. [56] applied the Dijkstra algorithm to address the ship routing problem, while Veneti [57] proposed an enhanced solution based on the time-dependent bi-objective shortest path algorithm. This improved method optimizes conflicting objectives, such as fuel consumption and expected risk, along the route. Huu [58] also proposed a weather routing method to obtain the optimal fuel route based on the Dijkstra algorithm and neural network.

Other methods, including dynamic programming approaches by Motte and Calvert [59], and those utilizing shortest path algorithms by Montes [60], Panigrahi et al. [61], and Mannarini et al. [62], have also been explored. van Dooren et al. [63] reformulated an optimal control problem using convex functions to convert it into a convex optimization problem, which they initially solved assuming that environmental conditions were constant over time, and then addressed the problem under time-varying conditions. In [64], the authors enhance the original FOPSO (fractional order particle swarm optimization) algorithm by introducing adaptive coefficients in the fractional order velocity update formula to avoid local optima and premature convergence.

Evolutionary approaches like genetic algorithms (GAs) offer efficient solutions to pathfinding problems by accommodating a wide range of constraints and objectives. For instance, Walther et al. [65] proposed a GA optimizing fuel costs while supporting variable ship speed. Wang et al. [66] introduced a real-coded GA, albeit with limitations around waypoint flexibility. Similarly, simulated annealing algorithms, such as the one introduced by Yuankui et al. [67], have been utilized to minimize vessel travel time.

Multi-objective evolutionary methods enable the simultaneous optimization of multiple objectives. Tsou [68] and Azariadis [69] combined evolutionary algorithms with modified A* algorithms to optimize routes considering safety and economic factors. Li and Zhang [70] proposed a GA focusing on minimizing turning variation and fuel consumption, with fixed vessel orientations at the start and end positions. Other studies, like those by Szlapczynska [71,72] and Vettor and Guedes Soares [73], applied multi-objective evolutionary algorithms to minimize passage time, fuel consumption, and route risk. Zhao et al. [74] introduce a hybrid multicriteria ship route planning method that combines enhanced particle swarm optimization with genetic algorithms to optimize meteorological risk, fuel consumption, and navigation time.

Several problems exist that are similar to the vessel weather routing problem. One of them is the path planning of unmanned aerial vehicles (UAVs). Arantes et al. [75], for example, used a genetic algorithm (GA) to plan the paths of drones and mention no-fly zones and obstacles that were similar to the land masses in the weather routing problem. They included the position uncertainty of the aircraft due to turbulence, which does not occur in the weather routing problem. The multi-population genetic algorithm is combined with a visibility graph maintaining all feasible paths for the drone, which would not be possible for the weather routing problem because of the much larger solution space. Hasircioglu et al. [76] also used a GA to plan paths offline for UAVs. The GA uses three different mutation operators that update, insert, and delete control points visited by the UAV. Hence, there are no operators that are specific to the case of UAV path planning. Ragusa et al. [77] also investigate a GA for "micro aerial vehicles". However, for the problem of weather routing, intersections are infeasible, which is different than in the approach of Ragusa et al. In their approach, intersections are allowed, and the algorithm focuses on minimizing the degree of intersection with obstacles. Furthermore, the routes used by Ragusa et al. have fewer waypoints than in the case of ship routing because the routes are much shorter and need fewer waypoints.

Recent research activities have examined cooperative path-following control algorithms for an unmanned surface vessel and unmanned aerial vehicles (USV-UAVs) that simplify design complexity and transmission requirements. In [78], the research focuses on reducing communication overhead and increasing fault tolerance in sensor systems by presenting an adaptive control strategy that adjusts based on sensor faults and varying data transmission needs. In [79], a cooperative path control strategy is explored that simplifies design complexity and minimizes command transmission between USVs and UAVs.

Certainly, up to now, the problem of planning the ship voyage by adapting the route to weather conditions has not been addressed by linking it to the trajectory planning of UAVs. The goal of this work is to leverage weather routing techniques to enable the safest possible takeoff and landing from the mother ship.

3. Modeling of Ship Behavior in Seaways

The performance of a ship during operations can markedly differ when transitioning from calm waters to navigating through waves up to the conditions of a storm. Influenced by environmental elements, predominantly waves, wind, and occasionally currents, a ship undergoes increased and dynamic loading. This typically leads to an increase in resistance, which consequently reduces the vessel forward speed and decreases the efficiency of the propulsion system. The lack of efficiency is largely due to altered water flow around the propeller, which can lead to the propeller emergence and ventilation, along with oscillatory movements of the ship hull across all six degrees of freedom. These movements can potentially create unsafe conditions for both the crew and the cargo, necessitating either a course alteration or a deliberate reduction in speed.

During the navigation of a mother ship, various weather and sea conditions can significantly affect the safety of UAV landing and takeoff operations. Pitch, roll, and heave motions, vertical plane movements, and water on the deck can create adverse conditions that make UAV landing and takeoff challenging and hazardous. To address this challenge, an adaptive weather routing model has been specifically designed to assist the mother ship in identifying safe routes during various weather and sea conditions. This model takes into account weather forecasts and the specific characteristics of the ship to determine optimal routes that minimize the risk of UAV operations. By using this approach, the mother ship can navigate more safely and reliably, avoiding adverse weather conditions that could compromise the stability of the ship and the safety of UAV operations. Additionally, the adaptive weather routing model allows the mother ship to dynamically adapt to changing weather and sea conditions along the route, ensuring that landing and takeoff operations are conducted safely and efficiently at any time and in any location.

Seakeeping Performances Index

The present evaluation assesses seakeeping performance using five key reference criteria: (i) pitch motion amplitude, (ii) heave motion amplitude, (iii) roll motion amplitude, (iv) probability of green water on deck, and (v) vertical plane movement. This determination follows a specified equation.

$$SPI = max\left\{0; \left(1-\frac{rms_p}{rms_{p,lim}}\right)\cdot\left(1-\frac{rms_h}{rms_{h,lim}}\right)\cdot\left(1-\frac{rms_r}{rms_{r,lim}}\right)\cdot\left(1-\frac{P_{wd}}{P_{wd,lim}}\right)\cdot\left(1-\frac{rms_V}{rms_{V,lim}}\right)\right\} \quad (1)$$

where rms_p is the RMS, root mean square, of the pitch motion amplitude; rms_h is the root mean square of the heave motion amplitude; rms_r is the root mean square of the roll motion amplitude; rms_V is the RMS of the vertical acceleration; and P_{wd} is the probability of green water on deck. The denominators represent the relevant limit values. This relation is structured to ensure the optimal seaworthiness performance for the ship by minimizing the complement of the criteria once their adherence is confirmed. The limit value of the pitch amplitude is based on NATO STANAG Standardization Agreement 4154 criteria, while the RMS of roll amplitude and green water on deck probability comply with the NORDFORSK 1987 criteria, as detailed in Table 1. The limit values for the RMS of the heave motion amplitude and the vertical plane motion, not being covered in the previously reported regulatory references, have been hypothesized by assuming 10% of the draft as the limit value for the heave. Obviously, all limit values can be adjusted according to the needs.

Table 1. General operability limiting criteria for ships.

Criteria	Limit Value
RMS of pitch amplitude	1.5 degrees
RMS of heave amplitude	0.15 m
RMS of roll amplitude	4 degrees
Green water on deck probability	0.05
RMS of vertical plane movement	0.2 m

A seakeeping analysis for a displacing ship in rough seas is commonly performed in the frequency domain. The response spectrum for the six ship motions is obtained from the following equations.

$$S_{\eta_j}(\omega_e) = \left|H_j(\omega, \omega_e, U, \mu)\right|^2 S_\zeta(\omega_e) \qquad for\ j = 1,\ 2,\ 3 \tag{2}$$

$$S_{\eta_j}(\omega_e) = \left|H_j(\omega, \omega_e, U, \mu)\right|^2 S'_\zeta(\omega_e) \qquad for\ j = 4,\ 5,\ 6 \tag{3}$$

where H_j are the frequency transfer functions and the modulus of these transfer functions is defined as the response amplitude (RAO). The absolute wave frequency is indicated with ω, ω_e is the encounter wave frequency, U represents the vessel speed, and μ is the encounter angle between the ship heading and the prevailing sea state direction. $S_\zeta(\omega_e)$ denotes the wave spectrum, while $S'_\zeta(\omega_e)$ is defined as:

$$S'_\zeta(\omega_e) = k^2 S_\zeta(\omega_e) \tag{4}$$

where k is the wave number. In this paper, the JONSWAP [80] spectrum is applied, thanks to its reliability to resemble wind-generated sea state conditions, and it is defined as follows:

$$S_\zeta^J(\omega) = 0.658 S_\zeta^B(\omega)\gamma^{exp[\frac{-(\omega-\omega_m)^2}{2\sigma^2 \omega_m{}^2}]} \tag{5}$$

where S_ζ^B is the Bretschneider [81] spectral density function, whose expression is a function of the significant height; ω_m is the angular spectral peak frequency; and γ is the peak enhancement factor that expresses, less than a constant, the ratio between the maxima of the JONSWAP to the corresponding Bretschneider spectrum.

The root mean square (RMS) of the pitch motion amplitude is calculated using the following equation, which relies on the ship response amplitude operator.

$$rms_p = \sqrt{\int_0^\infty S_\zeta(\omega_e) \left|H_{pitch}(\omega_e)\right|^2 \left(\frac{\omega^2}{g}\right)^2 d\omega_e} \tag{6}$$

where H_{pitch}, also commonly referred to as H_5, is the pitch motion transfer function. S_ζ is the wave spectrum, and ω_e is the encounter wave frequency that satisfies the Doppler shift equation depending on the absolute wave frequency ω and the factor $\psi = U\ cos\mu/g$; the vessel speed is denoted by U, and μ, as indicated previously, denotes the encounter angle between the ship heading and the prevailing sea state direction. ω^2/g is the wave number satisfying the deep water condition. Similarly, the root mean square (RMS) of the heave motion amplitude is calculated using the following equation.

$$rms_h = \sqrt{\int_0^\infty S_\zeta(\omega_e)\ |H_{heave}(\omega_e)|^2 d\omega_e} \tag{7}$$

where H_{heave}, also commonly referred to as H_3, is the heave motion transfer function. The root mean square (RMS) of the roll motion amplitude is defined by the following equation.

$$rms_r = \sqrt{\int_0^{\infty} S_\zeta(\omega_e)\,|H_{roll}(\omega_e)|^2 \left(\frac{\omega^2}{g}\right)^2 d\omega_e} \tag{8}$$

where H_{roll}, also commonly referred to as H_4, is the roll motion transfer function. Regarding the probability of green water on the forward deck structures, its determination follows a similar approach.

$$P_{wd} = e^{-\frac{f_b^2}{2C_s^2 m_{0,r}}} \tag{9}$$

where C_s is the swell-up coefficient equal to one, $m_{0,r}$ is the zero spectral moment of the ship relative motion with respect to the sea surface, and f_b is the freeboard at the ship forward perpendicular. In the end, the root mean square of the relative vertical acceleration is obtained by combining the heave, pitch and roll motions at the ship forward perpendicular.

$$rms_V = \sqrt{\int_0^{\infty} \left|H_{heave}(\omega_e) + \overline{y}H_{roll}(\omega_e) - \overline{x}H_{pitch}(\omega_e) - e^{-ik(\overline{x}cos\mu+\overline{y}sin\mu)}\right|^2 \omega_e^4 S_\zeta(\omega_e)d\omega_e} \tag{10}$$

where $\overline{x}$ is the longitudinal distance (fwd+) of the ship forward perpendicular from its center of mass; $\overline{y}$ is the transverse distance (stbd+), equal to zero since the considered point lies on the diametral plane; and $k = \omega^2/g$ is the wave number satisfying the deep water condition.

To conduct a comprehensive analysis of the ship seaworthiness, the Motion module of Maxsurf was utilized. This module calculates and analyzes the ship response to user-defined sea conditions. Input data included initial velocity and various heading angles. The Motion module played a crucial role in defining response amplitude operators (RAOs) for roll, pitch, and heave motions, essential for evaluating seakeeping criteria along the optimized route. Additionally, added resistance values were computed by Motion, aiding in determining speed reduction due to varying sea states. In particular, by assuming an engine brake horsepower and total propulsion efficiency, the effective power was derived and, consequently, the ship resistance as well. By adding the values of added resistance provided by the Motion module to the resistance, it was possible to determine the speed reduction in various sea conditions, which is crucial for assessing the impact of sea conditions on the ship performance.

The indices were selected based on the type of ship and the specific service which it is intended for. The objective is to enhance safety in the launch and landing operations of autonomous drones. The developed code provides the ability to identify the motions and, consequently, the indices to analyze and consider according to specific needs. Therefore, the indices can be modified and adapted based on the specific needs of the ship and the priorities dictated by the service it must perform.

4. GA-Based Weather Routing

In this study, an optimization methodology employing the genetic algorithm was utilized. Genetic algorithms are an optimization technique inspired by the principles of natural selection and biological evolution. These algorithms simulate the process of natural evolution by generating a population of candidate solutions to a problem and evolving these solutions over successive generations. Each candidate solution, often referred to as an individual, is evaluated using a fitness function that measures its quality or performance concerning the optimization objective. The key principles of genetic algorithms include selection, crossover, and mutation. Selection involves choosing the fittest individuals from the current population to participate in the creation of the next generation. Crossover, or recombination, combines pairs of selected individuals to produce offspring that inherit characteristics from both parents. Mutation introduces random changes to individual solutions, promoting diversity and preventing premature convergence to local optima. In this paper, we refer to an implementation in Matlab of NSGA-II [82].

In the context of optimizing the ship trajectory, the genetic algorithm was implemented to find the most efficient route considering various sea states and ship responses.

Consider $A = (x_A, y_A)$ and $B = (x_B, y_B)$ being the departure and target points, respectively. Let $S = X \times Y = [x_A, x_B] \times [y_A, y_B]$ be the search space and $\mathcal{P} = \{P_0, P_1, \ldots, P_N, P_{N+1}\}$ be the set of $N + 2$ waypoints, which are the passage points of the trajectory such that $P_0 = A$, $P_{N+1} = B$ and $P_i = (x_i, y_i) \in S\ \forall i$.

Defining waypoints solely by their coordinates without considering the order and relative positions among them can result in inefficient and impractical routes. This approach can cause the trajectory to intersect itself or loop back to previously traversed points, leading to unnecessary detours and an overall suboptimal path.

In order to achieve a more effective solution for route planning, it is advantageous to employ a local reference system for defining the coordinates of each waypoint. Each point can be described in terms of its direction and distance relative to the preceding waypoint rather than using absolute coordinates. This method simplifies the definition of constraints on the distances between waypoints and the changes in direction, thereby preventing the generation of inefficient trajectories with intersections or backward movements. Using a local reference system ensures a more structured and feasible approach for defining the path from the starting point to the destination.

Consider $T = M \times \Phi = [0, 1] \times [0, 1]$ as the local search space in which the design variables $m_i \in M$, $\phi_i \in \Phi$ are defined on. The position of the i-th waypoint $P_i = (x_i, y_i)$ can be computed as follows:

$$P_i = P_{i-1} + m_i \begin{Bmatrix} \cos \alpha_i \\ \sin \alpha_i \end{Bmatrix} (P_{N+1} - P_{i-1})$$

where $\alpha_i = \alpha_m + \phi_i(\alpha_M - \alpha_m)$, with α_m and α_M being computed in order to consider a vision cone whose axis is the line joining the points P_{N+1} and P_{i-1} (see Figure 1).

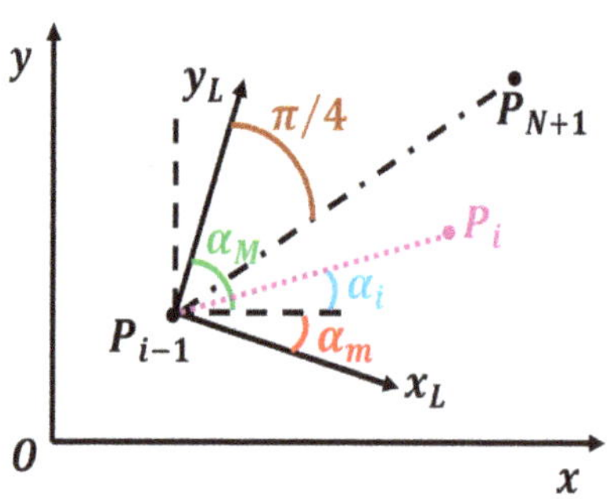

Figure 1. Local reference system for waypoint definition.

The objective function J is defined as the sum of all distances between consecutive waypoints:

$$J = \sum_{i=0}^{N} d(P_i, P_{i+1})$$

where $d(\cdot, \cdot)$ indicates the arc distance between two points.

The optimization procedure seeks to minimize the objective function J while ensuring that the resulting route adheres to constraints related to the ship performance and safety in varying sea conditions by calculating the seakeeping performance index (SPI) for various sea states along the route. Routes with SPI values exceeding a critical threshold are penalized, encouraging the selection of safer and more efficient paths. This iterative process continues until an optimal or satisfactory route is found, balancing the tradeoffs between route efficiency and seakeeping performance.

The proposed adaptive weather routing procedure operates through a series of well-defined steps:

(1) Input data reception—Initially, the procedure receives crucial input data, including weather information, vessel details (such as dimensions, response amplitude operators, and added resistance values), initial speed, and specific coordinates for the departure and arrival points of the trip.

(2) Trajectory optimization process—For each considered trajectory in the optimization problem, the following steps are performed:

   (a) Waypoint definition—The coordinates of waypoints are computed.
   (b) Speed reduction estimation—Using the meteorological data, the procedure estimates the reduction in speed that the ship will experience in various sea states. This step is essential for understanding how weather conditions affect the vessel's speed.
   (c) JONSWAP sea spectrum calculation—The JONSWAP sea spectrum is computed based on the received meteorological data. This spectrum characterizes the energy distribution of sea waves, which is crucial for further analysis.
   (d) Vessel response evaluation—With the JONSWAP spectrum, the vessel dynamic response to different combinations of heading angles and speeds is computed. The response in terms of roll, heave, and pitch motions can computed by simulation to predict the behavior of the vessel in open sea under various conditions.
   (e) SPI function determination—The values about each criterion along the trajectory can be computed and used to determine the overall value of the SPI function.
   (f) Fitness evaluation—The objective function is evaluated and the fitness is computed taking into account the constraint on the SPI.

(3) Optimal route provision—After the optimization process is complete, the procedure provides the optimal route. This route is the safest and most efficient for the ship, taking into account expected weather and sea conditions and the vessel's specific characteristics.

A flowchart of the proposed adaptive weather routing procedure is summarized in Figure 2.

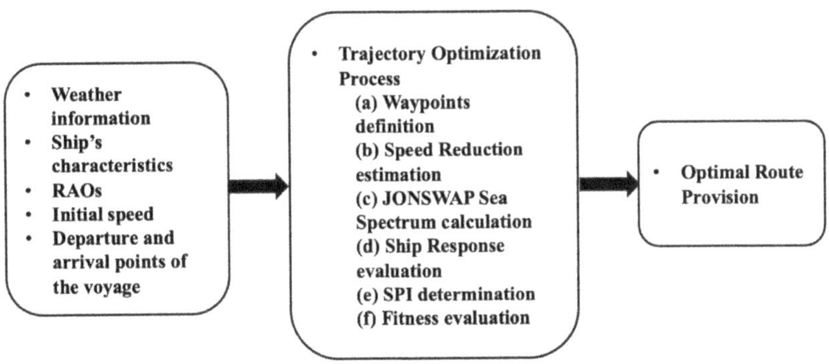

Figure 2. Flow chart: code structure.

5. Input Data

The reference ship is a vessel that serves as a UAV mother ship, and its main characteristics are detailed in Table 2.

Table 2. Main details of the UAV mother ship.

Dimension	Value
Length between perpendiculars	30 m
Breadth	7.7 m
Design draught	1.5 m
Displacement	114.2 t
Pitch moment of inertia	7343.56 tm^2
Roll moment of inertia	1025.75 tm^2
Waterplane area	181.99 m^2
Longitudinal metacentric radius	98.674 m
Transversal metacentric radius	6.006 m
Block coefficient	0.32
Shaft Power	2000 kW
Efficiency	0.6

The response operator details of the above ship was ascertained by defining the representative model using Maxsurf 2023 Update 7 software. Maxsurf is an integrated naval architecture used to design, analyze, and build different types of vessels. The detailed analysis of the vessel's seaworthiness was conducted using the Maxsurf Motion module, which computes and analyzes the ship response in user-defined sea conditions. It is well known for simulating the dynamic response of the ship to various forces and effects. For the relevant vessel, the Motion module played a key role in defining the RAOs for the rolling, pitching, and heaving motions that enabled the assessment of the seakeeping criteria over segments of the optimized route.

The selected route is in the North Atlantic Ocean, one of the most important shipping zones in the world, being the corridor where all the infrastructure of Western Europe's key ports has been regularly and significantly upgraded to keep up with current and anticipated increases in shipping. Weather conditions in this area can vary through the seasons and affect the final outcomes. The coordinates of the departure point are 14.4° N 46.3° W, and the target is at 30.8° N 24.6° W. Figure 3 presents the selected route to be followed, showing the great circle distance between points. The great circle route is 1549 nm long, and the rhumb-line route is 1555 nm long. Although the great circle is shorter, it requires the captain to correct the heading frequently. On the other hand, rhumb-line, although a little longer, does not require changes in the absence of wind and sea current, making it more practical.

The daily operational meteorological forecast data are a good estimation of the reference scenario needed to test the functionality of the adaptive weather routing algorithm. These data are crucial for assessing the various scenarios that the algorithm must handle. With this objective, several software packages are available, with different resolutions, domains (from regional to global), and qualities. These applications are validated by the agreements that shipping companies have with meteorological institutes to ensure the data are both current and accurate.

Currently, the community adheres to the standardization established by the World Meteorological Organization, which mandates that all information be presented in the self-explanatory GRIB (GRIdded Binary) format. This standardization facilitates the reading and utilization of input data, significantly easing the integration process into weather routing systems. In the current analysis, data are obtained from the third-generation global wave model (GWAM), abbreviated as GWAM, initially developed in the mid-1980s by an international group of wave modelers [83]. This model is extensively recognized and utilized by research institutions worldwide due to its ability to explicitly solve the wave transport equation without assuming the wave spectrum's shape. The GWAM can be applied to any regional or global grid, tailored to specific topographic data sets, with adjustable grid resolutions in space and time. The wave propagation analysis can be performed on latitudinal–longitudinal or Cartesian grids, making it highly adaptable to various geographical contexts. The outputs of the GWAM model, which are essential to the route optimization process, include the significant wave height, the mean wave period, and the prevailing sea state direction.

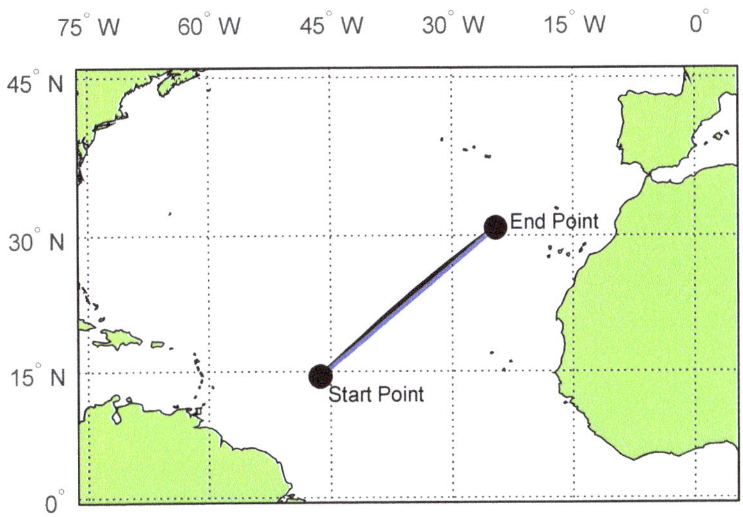

Figure 3. Reference route.

In this study, the GRIB files were downloaded with a time step of 3 h and a duration of 7 days for each sequence from 17 March to 24 March 2024, with 0.25° × 0.25° grid spacing. An example of the meteorological images that depict changes in environmental conditions during a certain period in the selected patch can be seen in Figure 4, which contains a photo series of GRIB files showing the trend of the significant wave height from 03:00 UTC on 17 March 2024 up to 00:00 UTC on 18 March 2024. In particular, the significant wave height is defined as:

$$H_s = \sqrt{{H_{s,wind}}^2 + {H_{s,swell}}^2}$$

where the wind wave component is denoted by $H_{s,wind}$ while the swell component is indicated as $H_{s,swell}$. GRIB files can be easily read using open-source software like XyGrib v1.2.6 or, as in this case, imported into Matlab 2023a to be used as the input for subsequent simulations. The parameters of interest extracted include: the significant height, the prevailing direction, and the mean period of the wind wave surface; and also the significant height, the prevailing direction, and the mean period of the swell wave surface. These data represent the environmental factors used as the input in the optimization procedure, giving us the possibility to use either a single-peak spectrum, such as the JONSWAP spectrum, or a two-peak spectrum, such as the Torsethaugen spectrum [84]. By modifying the encounter sea state direction and accounting for speed variations caused by changes in added resistance, the encounter frequency is varied, which subsequently impacts the ship's response to rough seas.

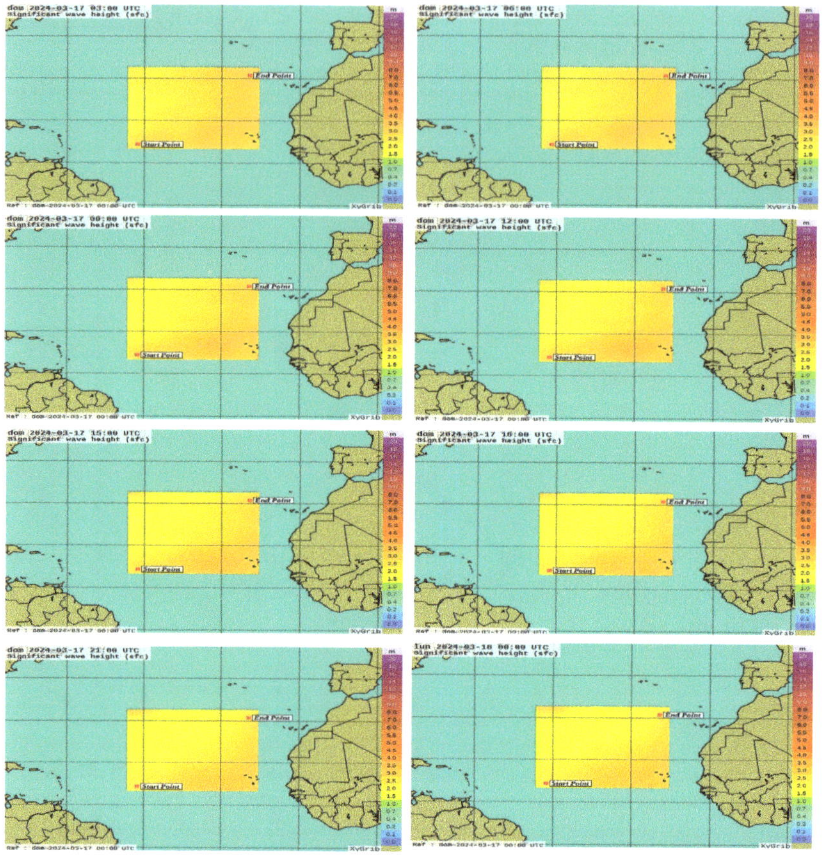

Figure 4. Time trend of the significant wave height from 03:00 UTC on 17 March 2024 up to 00:00 UTC on 18 March 2024.

6. Case Study

After input data definition, the route optimization algorithm was initiated, testing three different starting speed values of 5, 10, and 15 knots, corresponding to Froude numbers of 0.15, 0.3, and 0.45, respectively.

To prove the effectiveness of the proposed solution, it was compared with a Dijkstra-based route planner [1,2].

The genetic algorithm, unlike Dijkstra's algorithm, does not rely on a fixed grid of points but looks for waypoints defining the course between start and end points. The location of the waypoints is optimized in relative polar coordinates as defined above. The results shown in this section were obtained using the parameters in Table 3. In particular, the optimization process involves 100 epochs with a population size of 100 individuals, resulting in a total of 10,000 objective function evaluations. Being the solution that depended on a probabilistic evolution, the process was repeated 10 times to be sure that an optimal solution was found. No stopping criterion was considered to maintain a constant number of objective function evaluations.

To compare the proposed optimization process with an alternative solution available in the literature, a technique using the Dijkstra algorithm was investigated, known for its speed in yielding results. The Dijkstra algorithm identifies the shortest path between two points with known coordinates by maximizing a non-negative cost function, in this case the seakeeping performance index (SPI). Before assessing the ship performance, it is necessary to set up a network of nodes between the start and endpoint to define a region that constrains the ship's route. This network, described in detail in [1,2], is a special form of a Dijkstra graph consisting of nodes and arcs, and it is defined following the steps reported in Lee et al. [85]. Specifically, the central point of the great circle is determined along with a set of center points along a rhumb line perpendicular to the great circle. Subsequently, a network of nodes is generated, considering a series of perpendicular rhumb lines in the first half of the great circle and a similarly occurring series in the second half. Each node is characterized by static information, such as geographic coordinates, and a dynamic grid with meteorological data, which are periodically updated. Once the network of nodes is defined, arcs connecting adjacent grid points are created, and a cost function based on the SPI values at the ends of the arcs is assigned to determine the optimal route. The algorithm is iteratively updated by changing the starting point based on elapsed time and the path already travelled by the ship.

Table 3. Main genetic algorithm parameters.

Genetic Algorithm Parameter	Value
Number of epochs/generations	100
Population size	100
Mutation rate	0.05
Crossover rate	0.95

Figure 5a,b provide the optimal paths with reference to $F_N = 0.15$. In both cases, the great circle route is highlighted in black, the optimal route is represented in blue, and the black circles represent the departure and arrival points.

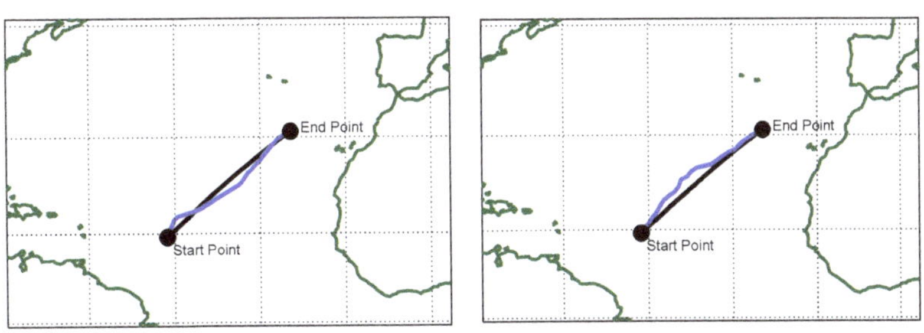

(**a**) (**b**)

Figure 5. Minimum distance (great circle) and optimal route detection, $F_N = 0.15$: (**a**) genetic algorithm; (**b**) Dijkstra algorithm.

To evaluate the performance of the optimization procedure, several performance indices can be defined, taking into account the seakeeping criteria considered in Section 3 as follows:

$$\phi_p = \frac{rms_{p,GC} - rms_{p,GA}}{rms_{p,GC}} \qquad \phi_p = \frac{rms_{p,GC} - rms_{p,D}}{rms_{p,GC}}$$

$$\phi_h = \frac{rms_{h,GC} - rms_{h,GA}}{rms_{h,GC}} \qquad \phi_h = \frac{rms_{h,GC} - rms_{h,D}}{rms_{h,GC}}$$

$$\phi_r = \frac{rms_{r,GC} - rms_{r,GA}}{rms_{r,GC}} \qquad \phi_r = \frac{rms_{r,GC} - rms_{r,D}}{rms_{r,GC}}$$

$$\phi_V = \frac{rms_{V,GC} - rms_{V,GA}}{rms_{V,GC}} \qquad \phi_V = \frac{rms_{V,GC} - rms_{V,D}}{rms_{V,GC}}$$

$$\phi_{wd} = \frac{P_{wd,GC} - P_{wd,GA}}{P_{wd,GC}} \qquad \phi_{wd} = \frac{P_{wd,GC} - P_{wd,D}}{P_{wd,GC}}$$

where the subscript GC indicates the values related to the great circle trajectory, with GA being the genetic procedure result and D being the Dijkstra-based procedure resulting path.

Table 4 provides the values of the indices, revealing a significant performance improvement given by the proposed genetic procedure. Opting for the optimized trajectory over the great circle results in an overall improvement exceeding 57%, notably enhancing vertical plane movement and green water on deck probability management, thereby positively impacting navigation safety. On the other hand, the optimized route results in an increased travelled distance of about 3.8% with respect to the great circle route, equal to 59 nautical miles. However, the comparison with the Dijkstra-based procedure allows for an appreciation of the good performance of the proposed optimizer, with a loss in terms of trajectory length (and consequently fuel consumption) of only 0.9%.

Table 4. Results of optimal route detection, $F_N = 0.15$.

Seakeeping Criteria	Genetic Algorithm (%)	Dijkstra Algorithm (%)
RMS pitch	21.1	10.3
RMS heave	25.5	22.5
RMS roll	39.3	32.0
Green water on deck probability	57.5	19.7
RMS vertical plane movement	89.9	44.8
SPI	57.8	50.2
Increased distance	3.8	2.9

Figure 6a,b and Table 5 provide the results obtained with $F_N = 0.30$. They confirm the obtained improvements, even if in this case the increase in the path length is 3.7%. The genetic-based route shows a 7.3% increase in the travelled distance compared with the great circle route, equal to 114 nautical miles.

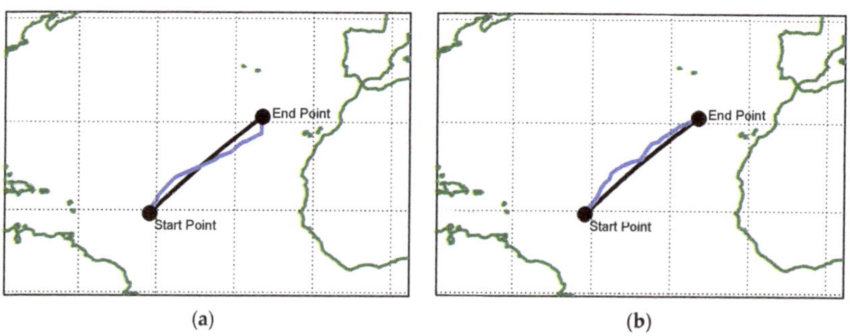

Figure 6. Minimum distance (great circle) and optimal route detection, $F_N = 0.30$: (**a**) genetic algorithm; (**b**) Dijkstra algorithm.

Table 5. Results of optimal route detection, $F_N = 0.30$.

Seakeeping Criteria	Genetic Algorithm (%)	Dijkstra Algorithm (%)
RMS pitch	15.6	0.9
RMS heave	31.5	18.0
RMS roll	18.1	16.6
Green water on deck Probability	35.3	4.5
RMS vertical plane movement	34.4	6.1
SPI	55.7	47.5
Increased distance	7.3	3.6

Finally, Figure 7a,b and Table 6 provide results obtained with $F_N = 0.45$. They confirm the improvements obtained using the proposed procedure. Furthermore, the analysis of the optimized routes reveals a 5.3% increase in distance traveled compared with the great circle route, equal to 82 nautical miles, while the Dijkstra-based results indicate only a 43 nautical mile increase over the great circle route, equivalent to 2.8%.

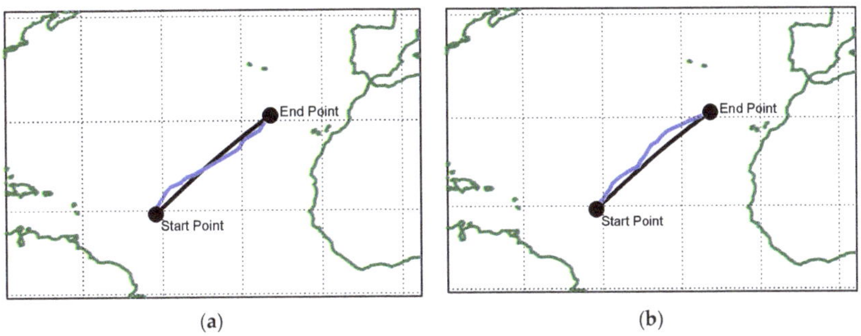

Figure 7. Minimum distance (great circle) and optimal route detection, $F_N = 0.45$: (**a**) genetic algorithm; (**b**) Dijkstra algorithm.

Table 6. Results of optimal route detection, $F_N = 0.45$.

Seakeeping Criteria	Genetic Algorithm (%)	Dijkstra Algorithm (%)
RMS pitch	15.8	7.7
RMS heave	32.2	11.5
RMS roll	43.3	0.4
Green water on deck Probability	26.0	0.1
RMS vertical plane movement	31.4	17.4
SPI	43.1	22.5
Increased distance	5.3	2.8

The developed code was also applied by reversing the travel direction, therefore inverting the departure and arrival coordinates. Obviously, the results differ because the marine conditions encountered when traveling in the opposite direction vary. The outcomes are presented for a Froude number of 0.15 in Figure 8a,b and Table 7, demonstrating that the optimized trajectory compared with the great circle results in an overall improvement of more than 84%, particularly in the management of roll and pitch motions. However, the optimized route leads to an increased travel distance of about 3.7%, equivalent to 56.6 nautical miles. Meanwhile, the Dijkstra algorithm results in a smaller improvement in seakeeping indices and also an increased travel distance of 4.7%, approximately equal to 73 nautical miles.

Table 7. Results of optimal route detection, $F_N = 0.15$.

Seakeeping Criteria	Genetic Algorithm (%)	Dijkstra Algorithm (%)
RMS pitch	58.4	1.3
RMS heave	0.7	70.1
RMS roll	63.2	55.2
Green water on deck Probability	8.4	4.5
RMS vertical plane movement	0.9	12.8
SPI	84.4	75.5
Increased distance	3.7	4.7

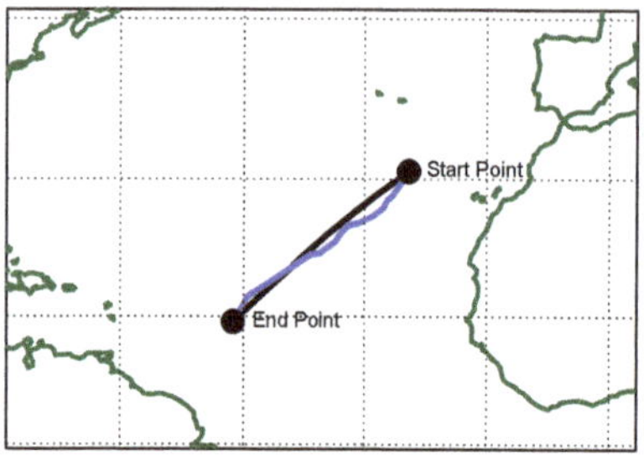

(**a**)

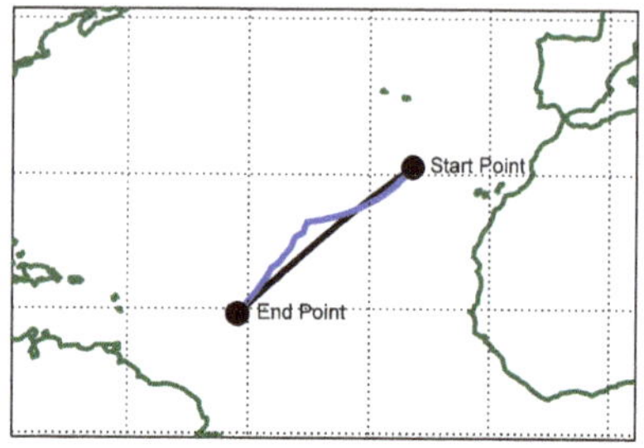

(**b**)

Figure 8. Minimum distance (great circle) and optimal route detection, F_N = 0.15: (**a**) genetic algorithm; (**b**) Dijkstra algorithm.

The results for F_N = 0.30, presented in Figure 9a,b and Table 8, indicate an improvement in this case as well. The SPI index improved by 45.4% compared with 19.7% for the Dijkstra algorithm, with a 7.7% increase in travel distance, equivalent to 120.5 nautical miles.

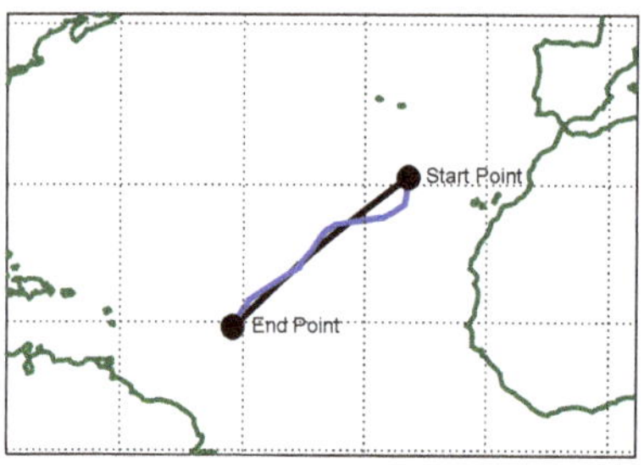

(**a**)

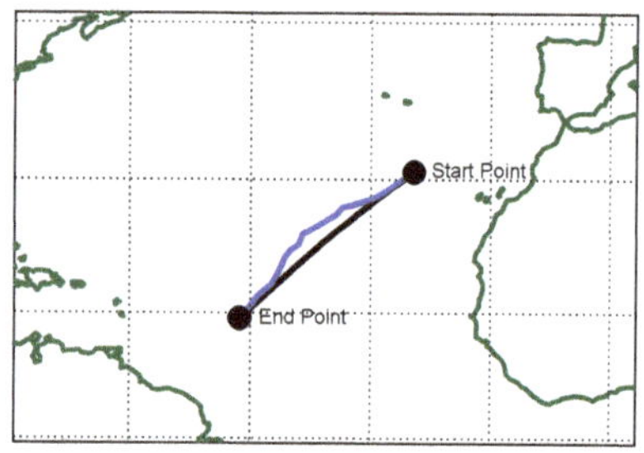

(**b**)

Figure 9. Minimum distance (great circle) and optimal route detection, F_N = 0.30: (**a**) genetic algorithm; (**b**) Dijkstra algorithm.

Table 8. Results of optimal route detection, F_N = 0.30.

Seakeeping Criteria	Genetic Algorithm (%)	Dijkstra Algorithm (%)
RMS pitch	7.0	21.5
RMS heave	4.3	19.8
RMS roll	11.6	37.2
Green water on deck Probability	43.7	17.5
RMS vertical plane movement	0.2	19.2
SPI	45.4	19.7
Increased distance	7.7	3.3

Finally, the case for F_N = 0.45 is presented in Figure 10 and Table 9. In this instance, the overall index improves by approximately 27%, with a notable reduction in roll motion. The increase in direction is only 1.1%, corresponding to 16.8 nautical miles, and the increase in distance is less than that obtained using the Dijkstra algorithm.

The analysis of the results obtained through the application of the genetic and Dijkstra algorithms, in the context of weather routing, highlights that both algorithms are valuable tools for improving ship performance in adverse marine conditions. While Dijkstra's algorithm quickly identifies optimal routes due to its predefined grid, the genetic algorithm, unrestricted by a grid, takes about 40 min to produce a solution, but it excels in optimizing seakeeping criteria. This comparison underlines a tradeoff between time efficiency and solution quality, suggesting that the choice between the algorithms should be based on specific project needs, use, and priorities.

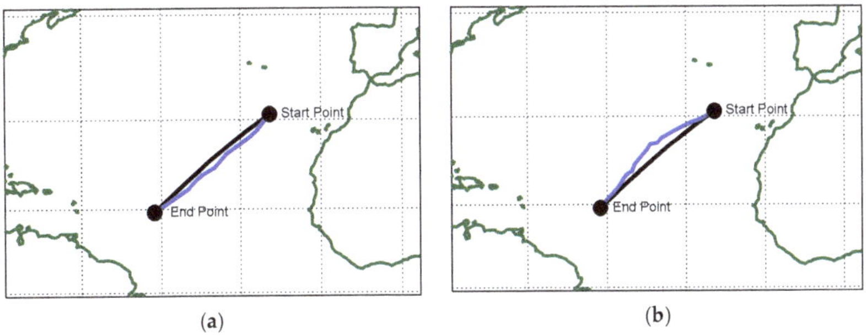

Figure 10. Minimum distance (great circle) and optimal route detection, $F_N = 0.45$: (**a**) genetic algorithm; (**b**) Dijkstra algorithm.

Table 9. Results of optimal route detection, $F_N = 0.45$.

Seakeeping Criteria	Genetic Algorithm (%)	Dijkstra Algorithm (%)
RMS pitch	3.5	3.9
RMS heave	16.3	18.9
RMS roll	60.9	0.8
Green water on deck Probability	9.1	4.5
RMS vertical plane movement	13.2	13.0
SPI	26.8	24.9
Increased distance	1.1	2.7

Route optimization could be used at the design stage, considering a usual route and the most likely meteorological and marine conditions in the area, so that hydrodynamic performance can be used for optimizing the hull shape [86]. More commonly, a course optimization system can be used both for planning the ship's path and during navigation. In this case, the initial condition response spectra can be obtained from measurements of ship motions in the time domain which, as seen in [87,88], can be easily obtained from low-cost measurement systems such as common smartphones. Indeed, these devices are generally equipped with several built-in sensors, providing raw data with a high sampling rate, and a GNSS receiver, which provides the device's position and velocity in a global context.

7. Conclusions

This paper focuses on the integration of seakeeping indices into the trajectory planning of a mother ship dedicated to the launch and recovery operations of an autonomous drone. It aims to adapt seakeeping indices to minimize risks during UAV takeoff and landing in challenging sea conditions by considering vessel dynamics and environmental factors. The proposed trajectory planning algorithm computes optimal paths prioritizing ship stability and safety, thereby mitigating the adverse effects of weather on UAV operations.

The proposed genetic-based procedure, without the restriction of a predefined grid, increases the computational burden to produce a solution, but it is able to find a notable advantage in optimizing seakeeping criteria.

Based on the results obtained, the current outcomes are encouraging for continuing further research activities to make landing and takeoff operations safer. The developed algorithm can be integrated on board by modifying and varying the seakeeping indices to be considered, depending on the service the ship must perform and the utility and importance of using a specific index according to the specific ship mission. This approach can provide valuable support to navigation for both crewed and autonomous ships, thus achieving safer and more efficient navigation.

Future developments will focus on enhancing the model by integrating additional seakeeping indices, tailored to the specific needs of the ship and its mission. These indices will serve as constraints to refine trajectory planning under various conditions. The implemented model will be validated through onboard experimental tests to verify its effectiveness and assess its impact on the ship's seakeeping and consequently on the safety of UAV landing and takeoff operations; more generally, the model's contribution to safer and more efficient maritime navigation for any type of ship will be assessed. Additionally, the goal of the tests is to define a simulation model of ship-drone dynamics in rough seas, entirely developed in the time domain, in such a way as to define a robust cooperative path-following control algorithm.

Author Contributions: Conceptualization, E.D., S.P. and A.S.; methodology, S.R.B., E.D. and S.P.; software, S.R.B., E.D., S.I. and S.P.; formal analysis, S.R.B. and S.I.; writing—original draft preparation, S.R.B., S.I. and S.P.; writing—review and editing, E.D. and A.S.; supervision, E.D. and A.S. All authors have read and agreed to the published version of the manuscript.

Funding: This work was supported by the research project—ID:P2022XER7W "CHEMSYS: Cooperative Heterogeneous Multi-drone SYStem for disaster prevention and first response" granted by the Italian Ministry of University and Research (MUR) within the PRIN 2022 PNRR program, funded by the European Union through the PNRR program.

Data Availability Statement: The data presented in this study are available on request from the authors.

Conflicts of Interest: The authors declare no conflicts of interest.

References

1. Pennino, S.; Gaglione, S.; Innac, A.; Piscopo, V.; Scamardella, A. Development of a new ship adaptive weather routing model based on seakeeping analysis and optimization. *J. Mar. Sci. Eng.* **2020**, *8*, 270. [CrossRef]
2. Pennino, S.; Scamardella, A. Weather Routing Model for Ship Motions Reduction and Fuel Saving. In Proceedings of the NAV 2022, International Conference on Ships and Maritime Research, Genoa and La Spezia, Italy, 15–17 June 2022.
3. Christiansen, M.; Fagerholt, K.; Nygreen, B.; Ronen, D. Maritime transportation. In *Handbooks in Operations Research and Management Science*; Elsevier: Amsterdam, The Netherlands, 2017; Volume 14, pp. 189–284.
4. Kontovas, C.A. The green ship routing and scheduling problem (GSRSP): A conceptual approach. *Transp. Res. Part D Transp. Environ.* **2014**, *31*, 61–69. [CrossRef]
5. Szłapczynska, J.; Smierzchalski, R. Multicriteria optimisation in weather routing. *Mar. Navig. Saf. Sea Transp.* **2009**, *3*, 393–400.
6. Delitala, A.M.; Gallino, S.; Villa, L.A.; Lagouvardos, K.; Drago, A. Weather routing in long-distance Mediterranean routes. *Theor. Appl. Climatol.* **2010**, *102*, 125–137. [CrossRef]
7. Fabbri, T.; Vicen-Bueno, R.; Hunter, A. Multi-criteria weather routing optimization based on ship navigation resistance, risk and travel time. In Proceedings of the 2018 International Conference on Computational Science and Computational Intelligence (CSCI), Las Vegas, NV, USA, 12–14 December 2018; IEEE: Piscataway, NJ, USA, 2018.
8. Li, Y. An overview of the DSRC/WAVE technology. In *Quality, Reliability, Security and Robustness in Heterogeneous Networks: 7th International Conference on Heterogeneous Networking for Quality, Reliability, Security and Robustness, QShine 2010, and Dedicated Short Range Communications Workshop, DSRC 2010, Houston, TX, USA, 17–19 November 2010*; Revised Selected Papers 7; Springer: Berlin/Heidelberg, Germany, 2012.
9. Fisher, M.L.; Rosenwein, M.B. An interactive optimization system for bulk-cargo ship scheduling. *Nav. Res. Logist. (NRL)* **1989**, *36*, 27–42. [CrossRef]
10. Lindstad, H.; Asbjørnslett, B.E.; Jullumstrø, E. Assessment of profit, cost and emissions by varying speed as a function of sea conditions and freight market. *Transp. Res. Part D Transp. Environ.* **2013**, *19*, 5–12. [CrossRef]
11. Gershanik, V.I. Weather routing optimisation–challenges and rewards. *J. Mar. Eng. Technol.* **2011**, *10*, 29–40. [CrossRef]
12. Zis, T.; North, R.J.; Angeloudis, P.; Ochieng, W.Y.; Bell, M.G.H. Evaluation of cold ironing and speed reduction policies to reduce ship emissions near and at ports. *Marit. Econ. Logist.* **2014**, *16*, 371–398. [CrossRef]
13. Armstrong, V.N. Vessel optimisation for low carbon shipping. *Ocean Eng.* **2013**, *73*, 195–207. [CrossRef]
14. Bellman, R. *Dynamic Programming*; Princeton University Press: Princeton, NJ, USA, 1957; 340p.
15. Dasgupta, B.; Gupta, A.; Singla, E. A variational approach to path planning for hyper-redundant manipulators. *Robot. Auton. Syst.* **2009**, *57*, 194–201. [CrossRef]

16. Shukla, A.; Singla, E.; Wahi, P.; Dasgupta, B. A direct variational method for planning monotonically optimal paths for redundant manipulators in constrained workspaces. *Robot. Auton. Syst.* **2013**, *61*, 209–220. [CrossRef]
17. la Cour-Harbo, A.; Bisgaard, M. State-control trajectory generation for helicopter slung load system using optimal control. In Proceedings of the AIAA Guidance, Navigation, and Control Conference, Chicago, IL, USA, 10–13 August 2009.
18. Harada, M.; Nagata, H.; Simond, J.; Bollino, K. Optimal trajectory generation and tracking control of a single coaxial rotor UAV. In Proceedings of the AIAA Guidance, Navigation, and Control (GNC) Conference, Boston, MA, USA, 19–22 August 2013.
19. Xu, N.; Cai, G.; Kang, W.; Chen, B.M. Minimum-time trajectory planning for helicopter uavs using computational dynamic optimization. In Proceedings of the 2012 IEEE International Conference on Systems, Man, and Cybernetics (SMC), Seoul, Republic of Korea, 14–17 October 2012; IEEE: Piscataway, NJ, USA, 2012.
20. Dai, S.-L.; He, S.; Ma, Y.; Yuan, C. Distributed cooperative learning control of uncertain multiagent systems with prescribed performance and preserved connectivity. *IEEE Trans. Neural Netw. Learn. Syst.* **2020**, *32*, 3217–3229. [CrossRef] [PubMed]
21. Dubins, L.E. On curves of minimal length with a constraint on average curvature, and with prescribed initial and terminal positions and tangents. *Am. J. Math.* **1957**, *79*, 497–516. [CrossRef]
22. Babel, L. Coordinated target assignment and UAV path planning with timing constraints. *J. Intell. Robot. Syst.* **2018**, *94*, 857–869. [CrossRef]
23. Liu, P.; Yu, H.; Cang, S. Geometric analysis-based trajectory planning and control for underactuated capsule systems with viscoelastic property. *Trans. Inst. Meas. Control.* **2017**, *40*, 2416–2427. [CrossRef]
24. Duan, H.; Zhao, J.; Deng, Y.; Shi, Y.; Ding, X. Dynamic discrete pigeon-inspired optimization for multi-UAV cooperative search-attack mission planning. *IEEE Trans. Aerosp. Electron. Syst.* **2020**, *57*, 706–720. [CrossRef]
25. Scherer, S.; Singh, S.; Chamberlain, L.; Elgersma, M. Flying fast and low among obstacles: Methodology and experiments. *Int. J. Robot. Res.* **2008**, *27*, 549–574. [CrossRef]
26. Scholer, F.; la Cour-Harbo, A.; Bisgaard, M. Configuration space and visibility graph generation from geometric workspaces for uavs. In Proceedings of the AIAA Guidance, Navigation, and Control Conference, Portland, OR, USA, 8–11 August 2011.
27. Scholer, F.; la Cour-Harbo, A.; Bisgaard, M. Generating approximative minimum length paths in 3D for UAVs. In Proceedings of the 2012 IEEE Intelligent Vehicles Symposium, Alcala de Henares, Spain, 3–7 June 2012; IEEE: Piscataway, NJ, USA, 2012.
28. Maini, P.; Sujit, P.B. Path planning for a UAV with kinematic constraints in the presence of polygonal obstacles. In Proceedings of the 2016 International Conference on Unmanned Aircraft Systems (ICUAS), Arlington, VA, USA, 7–10 June 2016; IEEE: Piscataway, NJ, USA, 2016.
29. Bortoff, S.A. Path planning for UAVs. In Proceedings of the 2000 American Control Conference, Chicago, IL, USA, 28–30 June 2000; ACC (IEEE Cat. No. 00CH36334); IEEE: Piscataway, NJ, USA, 2000; Volume 1.
30. Pehlivanoglu, Y.V. A new vibrational genetic algorithm enhanced with a Voronoi diagram for path planning of autonomous UAV. *Aerosp. Sci. Technol.* **2012**, *16*, 47–55. [CrossRef]
31. Lin, Y.; Saripalli, S. Path planning using 3D dubins curve for unmanned aerial vehicles. In Proceedings of the 2014 International Conference on Unmanned Aircraft Systems (ICUAS), Orlando, FL, USA, 27–30 May 2014; IEEE: Piscataway, NJ, USA, 2014.
32. Yang, J. *Real-Time Trajectory Planning for Ground and Aerial Vehicles in a Dynamic Environment*; University of Central Florida: Orlando, FL, USA, 2008.
33. Véras, L.G.; Medeiros, F.L.; Guimarães, L.N. Rapidly exploring Random Tree* with a sampling method based on Sukharev grids and convex vertices of safety hulls of obstacles. *Int. J. Adv. Robot. Syst.* **2019**, *16*, 1729881419825941. [CrossRef]
34. Liu, Y.-H.; Arimoto, S. Proposal of tangent graph and extended tangent graph for path planning of mobile robots. In Proceedings of the 1991 IEEE International Conference on Robotics and Automation, Sacramento, CA, USA, 9–11 April 1991; IEEE Computer Society: Piscataway, NJ, USA, 1991.
35. Cover, H.; Choudhury, S.; Scherer, S.; Singh, S. Sparse tangential network (SPARTAN): Motion planning for micro aerial vehicles. In Proceedings of the 2013 IEEE International Conference on Robotics and Automation, Karlsruhe, Germany, 6–10 May 2013; IEEE: Piscataway, NJ, USA, 2013.
36. Babel, L. Curvature-constrained traveling salesman tours for aerial surveillance in scenarios with obstacles. *Eur. J. Oper. Res.* **2017**, *262*, 335–346. [CrossRef]
37. Yan, F.; Liu, Y.-S.; Xiao, J.-Z. Path planning in complex 3D environments using a probabilistic roadmap method. *Int. J. Autom. Comput.* **2013**, *10*, 525–533. [CrossRef]
38. Musliman, I.A.; Rahman, A.A.; Coors, V. Implementing 3D network analysis in 3D-GIS. *Int. Arch. ISPRS* **2008**, *37*, 913–918.
39. de Heer, H.D.; Balcazar, H.G.; Castro, F.; Schulz, L. A path analysis of a randomized promotora de salud cardiovascular disease–prevention trial among at-risk Hispanic adults. *Health Educ. Behav.* **2012**, *39*, 77–86. [CrossRef] [PubMed]
40. Carsten, J.; Ferguson, D.; Stentz, A. 3D Field D: Improved Path Planning and Replanning in Three Dimensions. In Proceedings of the 2006 IEEE/RSJ International Conference on Intelligent Robots and Systems, Beijing, China, 9–13 October 2006; IEEE: Piscataway, NJ, USA, 2006.
41. Eun, Y.; Bang, H. Cooperative control of multiple unmanned aerial vehicles using the potential field theory. *J. Aircr.* **2006**, *43*, 1805–1814. [CrossRef]
42. Kitamura, Y.; Tanaka, T.; Kishino, F.; Yachida, M. 3-D path planning in a dynamic environment using an octree and an artificial potential field. In Proceedings of the 1995 IEEE/RSJ International Conference on Intelligent Robots and Systems, Human Robot Interaction and Cooperative Robots, Pittsburgh, PA, USA, 5–9 August 1995; IEEE: Piscataway, NJ, USA, 1995; Volume 2.

43. Roberge, V.; Tarbouchi, M.; Labonte, G. Fast genetic algorithm path planner for fixed-wing military UAV using GPU. *IEEE Trans. Aerosp. Electron. Syst.* **2018**, *54*, 2105–2117. [CrossRef]
44. Chai, R.; Tsourdos, A.; Savvaris, A.; Chai, S.; Xia, Y. Solving constrained trajectory planning problems using biased particle swarm optimization. *IEEE Trans. Aerosp. Electron. Syst.* **2021**, *57*, 1685–1701. [CrossRef]
45. Dever, C.; Mettler, B.; Feron, E.; Popovic, J.; McConley, M. Nonlinear trajectory generation for autonomous vehicles via parametrized maneuver classes. *J. Guid. Control. Dyn.* **2006**, *29*, 289–302. [CrossRef]
46. Frazzoli, E.; Dahleh, M.A.; Feron, E. Real-time motion planning for agile autonomous vehicles. *J. Guid. Control. Dyn.* **2002**, *25*, 116–129. [CrossRef]
47. Blasi, L.; Barbato, S.; D'Amato, E. A mixed probabilistic–geometric strategy for UAV optimum flight path identification based on bit-coded basic manoeuvres. *Aerosp. Sci. Technol.* **2017**, *71*, 1–11. [CrossRef]
48. James, R.W. *Application of Wave Forecasts to Marine Navigation*; New York University: New York, NY, USA, 1957.
49. Hagiwara, H.; Spaans, J.A. Practical weather routing of sail-assisted motor vessels. *J. Navig.* **1987**, *40*, 96–119. [CrossRef]
50. Haltiner, G.; Hamilton, H.; Arnason, G. Minimal-time ship routing. *J. Appl. Meteorol.* **1962**, *1*, 1–7. [CrossRef]
51. Chen, Y.; Tian, W.; Mao, W. Strategies to improve the isochrone algorithm for ship voyage optimisation. *Ships Offshore Struct.* **2024**, 1–13. [CrossRef]
52. Papadakis, N. Minimal Time Vessel Routing Problem. Ph.D. Thesis, University of Michigan, Ann Arbor, MI, USA, 1988.
53. Wei, S.; Zhou, P. Development of a 3D Dynamic Programming Method for Weather Routing. In *Methods and Algorithms in Navigatiom*, 1st ed.; Taylor & Francis Group: Abingdon, UK, 2011.
54. de Wit, C. Proposal for Low Cost Ocean Weather Routeing. *J. Navig.* **1990**, *43*, 428–439. [CrossRef]
55. Zaccone, R.; Ottaviani, E.; Figari, M.; Altosole, M. Ship voyage optimization for safe and energy-efficient navigation: A dynamic programming approach. *Ocean Eng.* **2018**, *153*, 215–224. [CrossRef]
56. Padhy, C.P.; Sen, D.; Bhaskaran, P.K. Application of wave model for weather routing of ships in the North Indian Ocean. *Nat. Hazards* **2007**, *44*, 373–385. [CrossRef]
57. Veneti, A.; Makrygiorgos, A.; Konstantopoulos, C.; Pantziou, G.; Vetsikas, I.A. Minimizing the fuel consumption and the risk in maritime transportation: A bi-objective weather routing approach. *Comput. Oper. Res.* **2017**, *88*, 220–236. [CrossRef]
58. Huu, K.D. Research on Ship Weather Routing Method Based on Dijkstra Algorithm and Neural Network. In Proceedings of the 2024 International Conference on Industrial Engineering, Applications and Manufacturing, ICIEAM, Sochi, Russia, 20–24 May 2024.
59. Motte, R.H.; Calvert, S. On the selection of discrete grid systems for on-board micro-based weather routeing. *J. Navig.* **1990**, *43*, 104–117. [CrossRef]
60. Montes, A.A. Network Shortest Path Application for Optimum Track Ship Routing. Master's Thesis, Naval Postgraduate School, Monterey, CA, USA, 2005.
61. Panigrahi, J.K.; Padhy, C.P.; Sen, D.; Swain, J.; Larsen, O. Optimal ship tracking on a navigation route between two ports: A hydrodynamics approach. *J. Mar. Sci. Technol.* **2012**, *17*, 59–67. [CrossRef]
62. Mannarini, G.; Pinardi, N.; Coppini, G.; Oddo, P.; Iafrati, A. VISIR-I: Small vessels-least-time nautical routes using wave forecasts. *Geosci. Model Dev.* **2016**, *9*, 1597–1625. [CrossRef]
63. van Dooren, S.; Duhr, P.; Onder, C.H. Convex modelling for ship speed optimisation. *Ocean Eng.* **2023**, *28815*, 11594. [CrossRef]
64. Du, W.; Li, Y.; Zhang, G.; Wang, C.; Zhu, B.; Qiao, J. Ship weather routing optimization based on improved fractional order particle swarm optimization. *Ocean. Eng.* **2022**, *24815*, 110680. [CrossRef]
65. Walther, L.; Shetty, S.; Rizvanolli, A.; Jahn, C. Comparing two optimization approaches for ship weather routing. In *Operations Research Proceedings 2016: Selected Papers of the Annual International Conference of the German Operations Research Society (GOR), Helmut Schmidt University, Hamburg, Germany, 30 August–2 September 2016*; Springer: Berlin/Heidelberg, Germany, 2018; pp. 337–342.
66. Wang, H.-B.; Li, X.-G.; Li, P.-F.; Veremey, E.I.; Sotnikova, M.V. Application of real-coded genetic algorithm in ship weather routing. *J. Navig.* **2018**, *71*, 989–1010. [CrossRef]
67. Yuankui, L.; Yingjun, Z.; Feixiang, Z. Minimal time route for wind-assisted ships. *Mar. Technol. Soc. J.* **2014**, *48*, 115–124. [CrossRef]
68. Tsou, M.-C. Integration of a geographic information system and evolutionary computation for automatic routing in coastal navigation. *J. Navig.* **2010**, *63*, 323–341. [CrossRef]
69. Azariadis, P. On using density maps for the calculation of ship routes. *Evol. Syst.* **2017**, *8*, 135–145. [CrossRef]
70. Li, G.; Zhang, H. A Bézier curve based ship trajectory optimization for close-range maritime operations. In Proceedings of the ASME 2017 36th International Conference on Ocean, Offshore and Arctic Engineering, American Society of Mechanical Engineers, Trondheim, Norway, 25–30 June 2017.
71. Szlapczynska, J. Multiobjective approach to weather routing. *TransNav Int. J. Mar. Navig. Saf. Sea Transp.* **2007**, *1*, 273–278.
72. Szlapczynska, J. Multi-objective weather routing with customised criteria and constraints. *J. Navig.* **2015**, *68*, 338–354. [CrossRef]
73. Vettor, R.; Soares, C.G. Development of a ship weather routing system. *Ocean Eng.* **2016**, *123*, 1–14. [CrossRef]
74. Zhao, W.; Wang, Y.; Zhang, Z.; Wang, H. Multicriteria ship route planning method based on improved particle swarm optimization–genetic algorithm. *J. Mar. Sci. Eng.* **2021**, *9*, 357. [CrossRef]
75. Arantes, M.D.S.; Arantes, J.D.S.; Toledo, C.F.M.; Williams, B.C. A hybrid multi-population genetic algorithm for UAV path planning. In Proceedings of the Genetic and Evolutionary Computation Conference 2016, Denver, CO, USA, 20–24 July 2016; ACM: New York, NY, USA, 2016; pp. 853–860.

76. Hasircioglu, I.; Topcuoglu, H.R.; Ermis, M. 3-D path planning for the navigation of unmanned aerial vehicles by using evolutionary algorithms. In Proceedings of the 10th Annual Conference on Genetic and Evolutionary Computation 2008, Atlanta, GA, USA, 12–16 July 2008; pp. 1499–1506.
77. Ragusa, V.R.; Mathias, H.D.; Kazakova, V.A.; Wu, A.S. Enhanced genetic path planning for autonomous flight. In Proceedings of the Genetic and Evolutionary Computation Conference 2017, Berlin, Germany, 15–19 July 2017; pp. 1208–1215.
78. Li, J.; Zhang, G.; Zhang, X.; Zhang, W. Integrating Dynamic Event-Triggered and Sensor-Tolerant Control: Application to USV-UAVs Cooperative Formation System for Maritime Parallel Search. *IEEE Trans. Intell. Transp. Syst.* **2023**, *25*, 3986–3998. [CrossRef]
79. Li, J.; Zhang, G.; Zhang, W.; Shan, Q.; Zhang, W. Cooperative Path Following Control of USV-UAVs Considering Low Design Complexity and Command Transmission Requirement. *IEEE Trans. Intell. Veh.* **2024**, *9*, 715–724. [CrossRef]
80. Hasselmann, K.; Barnett, T.P.; Bouws, E.; Carlson, H.; Cartwright, D.E.; Enke, K.; Ewing, J.A.; Gienapp, H.; Hasselmann, D.E.; Kruseman, P.; et al. Measurement of Wind Wave Growth and Swell Decay during the Joint North Sea Wave Project (JONSWAP). *Ergaenzungsheft Dtsch. Hydrogr. Z. Reihe A* **1973**, *12*, 1–95.
81. Bretschneider, C.L. The generation and decay of wind waves in deep water. *Trans. Am. Geophys. Union* **1952**, *33*, 381–389. [CrossRef]
82. Deb, K.; Pratap, A.; Agarwal, S.; Meyarivan, T. A fast and elitist multiobjective genetic algorithm: NSGA-II. *IEEE Trans. Evol. Comput.* **2002**, *6*, 182–197. [CrossRef]
83. Komen, G.J.; Cavaleri, L.; Donelan, M.; Hasselmann, K.; Janssen, P.A.E.M.; Hasselmann, S. *Dynamics and Modelling of Ocean Waves*; Cambridge University Press (CUP): Cambridge, UK, 1994.
84. Torsethaugen, K. A two-peak wave spectral model. In Proceedings of the 12th International Conference on Offshore Mechanics and Arctic Engineering, Glasgow, UK, 20–24 June 1993; Volume 2, pp. 175–180.
85. Lee, H.; Kong, G.; Kim, S.; Kim, C.; Lee, J. Optimum Ship Routing and It's Implementation on the Web. In *Computer Vision*; Springer Science and Business Media LLC: Berlin/Heidelberg, Germany, 2002; Volume 2402, pp. 125–136.
86. Grigoropoulos, G.J. Hull Form Optimization for Hydrodynamic Performance. *Mar. Technol. Mar. Technol.* **2004**, *41*, 167–182. [CrossRef]
87. Pennino, S.; Scamardella, A. Motions Assessment Using a Time Domain Approach for a Research Ship in Antarctic Waters. *J. Mar. Sci. Eng.* **2023**, *11*, 558. [CrossRef]
88. Pennino, S.; Angrisano, A.; Della Corte, V.; Ferraioli, G.; Gaglione, S.; Innac, A.; Martellato, E.; Palumbo, P.; Piscopo, V.; Rotundi, A.; et al. Sea state monitoring by ship motion measurements onboard a research ship in the antarctic waters. *J. Mar. Sci. Eng.* **2021**, *9*, 64. [CrossRef]

oceans

MDPI

Article

Identification of Risk Patterns by Type of Ship Through Correspondence Analysis of Port State Control: A Differentiated Approach to Inspection to Enhance Maritime Safety and Pollution Prevention

Jose Manuel Prieto [1], David Almorza [2], Víctor Amor-Esteban [3], Juan J. Muñoz-Perez [4] and Bismarck Jigena-Antelo [1,*]

1 CASEM (Andalusian Higher Marine Studies Center), Department of Nautical Sciences and Shipbuilding, University of Cádiz, 11510 Puerto Real, Cadiz, Spain; josemanuel.prieto@gm.uca.es
2 Department of Statistics and Operational Research, University of Cadiz, 11003 Cadiz, Spain; david.almorza@uca.es
3 Department of Statistics, University of Salamanca, 37008 Salamanca, Spain; vamor@usal.es
4 CASEM (Andalusian Higher Marine Studies Center), Department of Applied Physics, University of Cadiz, 11510 Puerto Real, Cadiz, Spain; juanjose.munoz@uca.es
* Correspondence: bismarck.jigena@gm.uca.es

Abstract: This study analyzes the results of Port State Control (PSC) inspections carried out under the Paris Memorandum of Understanding between 2018 and 2022. Through a correspondence analysis, the most frequent deficiencies were identified according to the type of ship being inspected. The study sample included 186,255 inspections obtained from the THETIS platform. The results revealed significant heterogeneity in deficiency profiles across ship types, highlighting specific patterns associated with each category. Container ships, oil tankers and bulk carriers, for instance, exhibited distinctive deficiency profiles. The study emphasizes the necessity for a tailored approach to PSC inspections, with the objective of optimizing resources through the utilization of risk zone indicators for the inspector. The identification of specific risk indicators would not only facilitate the work of inspectors but also enable the earlier detection of potential problems and more effective intervention. The study provides a solid foundation for future research and decision-making on PSC inspections, with the aim of enhancing maritime safety and pollution prevention.

Keywords: maritime safety; port state control; Paris MoU; deficiencies of ship; correspondence analysis; risk indicators

1. Introduction

Maritime transport has been a key driver of globalization over the last three decades, facilitating the exchange of goods on a global scale. However, this exponential growth has generated significant challenges in terms of safety and sustainability. Ecological disasters caused by maritime accidents, such as the "*Torrey Canyon*", "*Amoco Cadiz*" and "*Exxon Valdez*", are a constant reminder of the serious consequences that a lack of safety in maritime transport can cause [1]. These events have not only caused irreversible environmental damage but have also generated significant economic losses and put the lives of numerous people at risk. It is therefore essential to implement effective measures to prevent such incidents and ensure the protection of the marine environment [2].

In view of the growing concern about maritime accidents and their serious environmental consequences, the International Maritime Organization (IMO) established, through Resolution A.682(17) [3], a regional cooperation framework to strengthen controls of ships in port. This system, known as Port State Control (PSC), aims to ensure that foreign ships comply with international standards of safety and protection of the marine environment. Through rigorous inspections, member states' maritime administrations assess compliance with international conventions, such as Convention for the Safety of Life at Sea (SOLAS), Convention for the Prevention of Pollution from Ships (MARPOL) and Convention on Standards of Training, Certification and Watchkeeping for Seafarers (STCW), among others.

The evolution of the PSC inspection system has led to the creation of several Memoranda of Understanding (MoU) at regional level, such as the Paris MoU [4]. These agreements, based on IMO guidelines, have the main objective of ensuring maritime safety and protecting the environment through the harmonization of inspection practices. The Paris MoU, which covers the countries of the European Union, is an outstanding example of this international cooperation and has contributed significantly to improving safety standards in maritime navigation.

This study aims to analyze the results of inspections carried out in the framework of the Paris MoU, in order to identify the most common types of safety deficiencies found on ships after a PSC inspection. Through a correspondence analysis of the inspection data, the aim is to establish relationships between the deficiencies detected according to the use of the ship (container ship, tanker, passenger ship, etc.), with the purpose of optimizing the inspection processes and prioritizing the areas of greatest risk, which can be an additional indication for MoU inspectors.

In order to better understand the role of PSC inspections in improving maritime safety, the following describes the inspection procedures set out in the Paris MoU and presents a review of the latest scientific literature.

1.1. PSC Inspections Procedure in the Paris MoU

PSC inspections are carried out by Port State Control Officers (PSCOs), duly qualified professionals authorized by the European Maritime Safety Agency (EMSA) [5] and by the Maritime Authority of each country. These officers act under the responsibility of that authority and are trained to carry out inspections in accordance with the requirements of the Paris MoU [6].

In 2011, a New System of Inspections (NIR) was implemented in the Paris MoU by Directive 2009/16 of the European Parliament and of the Council of 23 April 2009 (EU, 2009), called the ERIKA III package [7]. This new system employs the concept of "*Prioritized inspections*", which consists of determining a "*Risk Profile*" for ships, based on the shipping company, flag and the results of previous inspections. This approach allows prioritization of inspections and targeting resources on those ships most likely to fail to meet safety standards. In addition, the new system helps to harmonize inspection practices in all Paris MoU ports, ensuring a more equitable treatment of ships and avoiding unequal situations.

By implementing the NIR, each time a foreign ship calls at a Paris MoU port, a PSCO can carry out an inspection. Initially, it uses the software application called The Hybrid European Targeting and Inspection System (THETIS) [8], managed by EMSA. This application checks all ship security inspections carried out in EU ports, generating a risk profile of the ship and determining inspection frequency and priorities. After examining the documentation on board, the inspector can decide to carry out a more thorough inspection if there are indications of non-compliance with international conventions.

The new system allows inspections to be classified into three categories according to the ship's risk profile: initial, more detailed and expanded. If deficiencies are found, the ship may be subject to corrective action, including detention. Such deficiencies are coded and arranged in the Paris MoU in various categories, depending on their nature and where they have been detected [9]. The results of inspections including such deficiencies and ship characteristics are stored in THETIS, in order to keep track of the condition of the ship over time and determine its risk profile. This will be the database we will use for this study, analyzing the results of the inspections carried out and the deficiencies found after the inspections.

1.2. Review of the Latest Research on Safety Controls

The authors Brooks [10] and Håvold [11] both commented on the privatization of maritime safety oversight and the importance of quantitative risk analysis. Brooks noted a clear shift towards privatization, while Håvold highlighted the crucial role of quantitative risk analysis in developing maritime safety measures.

Cariou et al. [12] tried to figure out which factors should be considered when choosing ships for inspection. They used data from 4080 inspections performed by the Swedish Maritime Administration between 1996 and 2001. They used different statistical models to analyze the data and found that the relationship between a ship's age and the number of deficiencies found is not the same for all types of ships. In this respect, our study may provide some novelty, as we found correlations between the type of ship and the deficiencies found after inspection.

The research by Li and Zheng [13] evaluated the effectiveness of the methods used by regional PSC agreements to select ships for inspection.

Hänninen et al. [14] analyzed PSC inspection data using Bayesian network algorithms to discover interactions between the number of different types of deficiencies found on ships and their involvement in maritime accidents and incidents. The results suggested that knowledge of the type of ship, the type of PSC inspection and the number of structural deficiencies are among those that provide the most information about involvement in accidents. This study provides the basis for this article as it confirms that deficiencies found on ships following a PSC inspection are good indicators of maritime risk.

Ravira and Piniella [6] also concluded that professional training and teamwork influence inspection results. Their study emphasized the importance of these factors in successful inspections.

Graziano et al. [15] examined 25 inspection reports produced by the EMSA, focusing on inspections carried out by European Union Member States. They assessed how well these inspections complied with Directive 2009/16/EC [7] and identified any inconsistencies between the requirements of the Directive and the way inspections were actually carried out.

Wang et al. [16] developed a Bayesian network classifier to identify high-risk foreign ships in ports. This tool aids PSC authorities in detecting ships with lower compliance standards, enabling more efficient allocation of inspection resources. This study is closely linked to our analysis, as our results also help to optimize inspection resources by showing the PSCO the most common deficiencies according to the type of ship they are working on, allowing them to set clear monitoring targets. Chen et al. [17] conducted a comprehensive analysis of detention data from Asia-Pacific port states. Their research focused on providing practical guidelines for port states to enhance the effectiveness of ship safety inspections.

Chung et al. [18] explored how data mining can improve ship safety inspections in Taiwan's ports. By analyzing past inspection data, the research identified patterns and relationships between different safety issues found on ships. The research showed that analyzing specific ship types, like oil tankers, provided more useful information than simply looking at the classification society that certified the ship. These findings can help inspectors focus their efforts on the most critical areas during inspections, making them more efficient and effective. The approach used in this study can be applied to other ports to improve their own inspection practices.

Knapp et al.'s [19] findings, as well as those presented in this study, underscored the need to refine methods for identifying and prioritizing ships for inspection. Our research makes an additional contribution by identifying the specific areas of each vessel that require particular attention depending on its use.

Fu et al. [20] investigated how to improve maritime safety by predicting potential ship deficiencies in advance. To achieve this, the research analyzed past port state control inspections to identify patterns and relationships between different types of safety issues found on ships. The findings of this study can help inspectors prioritize their checks during inspections, making them more efficient and effective in identifying and addressing potential safety hazards.

Shen et al. [21] proposed a new approach to assessing the risk of deficiencies on ships using data from the Tokyo MoU and the MCDM technique. By developing a risk assessment scale, ships with deficiencies were identified with 83.3% accuracy in subsequent inspections. The results obtained demonstrated the usefulness of this system in assisting inspectors in their decision-making and in improving the effectiveness of port state inspections by indicating to the PSCO the priority areas for inspection, which is fully in line with this article.

Yuhong-Wang et al. [22] laid the foundation for understanding the factors influencing PSC inspections by identifying safety deficiencies and technical characteristics of ships as the main determinants. Our research is in line with these findings and goes deeper into the relationship between ship type and technical characteristics, as well as the impact of deficiencies found during PSC inspections. This allows us to validate the relevance of the variables selected for our analysis.

Prieto et al. [23] conducted a detailed analysis of inspections carried out between 2013 and 2018 in the main European ports of the Paris MoU, using multivariate statistical techniques (STATIS). The results provided a valuable indicator for maritime administrations to assess the risk profile of ships and make informed decisions on inspection priorities; in this sense, our study can also provide guidance on inspection priorities according to ship type.

Demirci et al. [24] developed I-SIA, an intelligent system using fuzzy c-means clustering and the Apriori algorithm, to analyze ship inspection data. In a test case, I-SIA identified the most critical inspection areas and items for a specific ship based on its history, providing a customized approach to improve inspection effectiveness. It is in this respect that this article contributes something new.

The work of Lai et al. [25] demonstrated the usefulness of PSC inspection data in identifying substandard ships by analyzing the type of ship and the deficiencies found on these ships. Building on this, our study provides an accurate tool to assess the risk associated with each ship type and each deficiency code. These results can help inspection bodies to optimize their resources and focus their efforts on the areas of greatest risk.

Yan et al. [26] proposed a data-driven bi-objective non-linear programming model, called SRP+ model, to optimize the thresholds in the original SRP framework for the sealing of high-risk ships in the PSC inspections. The results suggested that the threshold of total weighting points to classify a ship as high risk should be slightly increased, the thresholds of ship age should be significantly increased, the threshold of historical number of defects should be increased, while the threshold of historical ship detention time should be decreased.

In their study, Almorza et al. [27] utilized the HJ-Biplot, a multivariate statistical method, to analyze PSC inspections in ten major European ports. By examining ship characteristics, the analysis effectively classified ships, aligning with existing risk assessments. This provided valuable insights for maritime authorities, including improved risk profiling for inspection prioritization. Furthermore, the study categorized ports based on their maritime traffic and identified characteristics associated with substandard ships, enhancing the understanding of PSC within the European port system.

While the PSC system offers benefits in relation to the safety control of ships, it is necessary to optimize the distribution of resources to focus on those most likely to be in non-compliance with maritime regulations. Therefore, the aim of this paper is to improve the efficiency [20] of PSC inspections by identifying the areas that require the most attention, which will allow PSCO to concentrate their efforts on the most critical areas according to the type of vessel inspected and reduce the inspection time and resources required for the inspection. This document presents a detailed description of the database used, as well as the methodologies and techniques used for the analysis, in Section 2. The results obtained are presented in Section 3, while Section 4 provides a summary of the most relevant conclusions.

2. Materials and Methods

The sample of this study consists of PSC inspections carried out in the main European ports from January 2018 to December 2022. The data from these inspections were obtained from the THETIS platform.

The variables used for this study are two: on the one hand, the type of ship according to its use (container ship, tanker, passenger ship, etc.), and the other variable, the type of deficiencies found after passing the PSC inspection. The dataset taken comes from a total of 186,255 PSC inspections, where it is worth noting that in the years 2018–2022 we obtain an average number of PSC inspections, observing a drastic decrease in the main year of the pandemic, with 27,178 inspections in 2020, thus showing one of its main consequences [28].

In Figure 1, we can see a graphical representation of the total number of inspections between 2018 and 2022 carried out according to the PSC MoU Paris, showing a clear decrease in 2020, with a smaller decrease in 2021.

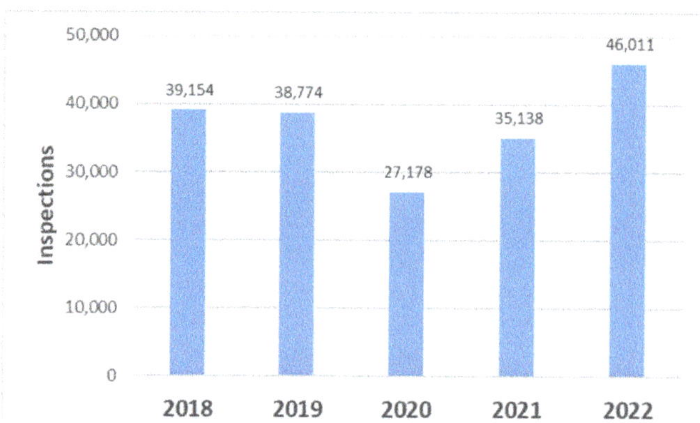

Figure 1. PSC MoU Paris inspections. Total number of inspections between 2018 and 2022.

In this study, the main objective is to identify deficiency patterns in PSC inspections based on the type of vessel, regardless of the year in which the inspections were conducted. Although the data cover the 2018–2022 period, the analysis is not broken down by year since temporal variability does not significantly affect the relationship between vessel types and the deficiencies found (see Figure S1, Supplementary Materials). Furthermore, the sharp decrease in inspections in 2020 due to the COVID-19 pandemic could lead to a biased interpretation of the evolution of deficiencies over time. Therefore, the dataset has been considered as a whole, ensuring a more robust analysis of structural risk patterns in maritime safety.

2.1. Correspondence Analysis

Correspondence analysis [29] is a statistical technique used to explore and visualize relationships between two qualitative variables: one variable whose categories appear in rows and the other variable whose categories are represented in columns.

Let X and Y be two categorical variables with values $\{x_e, x_2, \ldots, x_i, \ldots, e_k\}$ and $\{y_1, y_2, \ldots, y_j, \ldots y_m\}$, where k and m are the categories of X and Y, respectively; the data are organized in a contingency table, where n_{ij} represents the observed frequency of the category combination k of variable X and the category m of variable Y. Table 1 shows the frequency distribution, where the row of $N_{k\cdot}$ corresponds to the marginal totals for the rows, $N_{\cdot m}$ to the marginal totals for the columns and $N_{\cdot\cdot}$ to the total sample size.

Table 1. Frequency distribution for two qualitative variables.

		Variable Y (m Categories)						
e		y_1	*e*	...	y_j	...	y_m	*Total Row*
	x_1	n_{11}	n_{12}	...	n_{1j}	...	n_{1m}	$N_{1\cdot}$
	x_2	n_{21}	n_{22}	...	n_{2j}	...	n_{2m}	$N_{2\cdot}$
Variable X	...	...	...	...	...	...	...	...
(k	x_i	n_{i1}	n_{i2}	...	n_{ij}	...	n_{im}	$N_{i\cdot}$
categories)	...	...	...	...	...	...	...	...
	x_k	n_{k1}	n_{k2}	...	n_{kj}	...	n_{km}	$N_{k\cdot}$
	Total Column	$N_{\cdot 1}$	$N_{\cdot 2}$	...	$N_{\cdot j}$	...	$N_{\cdot m}$	$N_{\cdot\cdot}$

The main purpose of a correspondence analysis is to transform a table with numerical information into a graphical representation that facilitates the interpretation of the information, looking for axes (*hypothetical variables*) that explain the distribution of frequencies. To do this, we do not work directly with the frequencies, we calculate the row and column profiles. A profile is the distribution of relative frequencies of a line in the table, whether row or column, in relation to its marginal total. The $i-th$ profile is $\left(\frac{n_{i1}}{N_{i\cdot}}, \frac{n_{i2}}{N_{i\cdot}}, \ldots, \frac{n_{ij}}{N_{i\cdot}}\right)$ and the $j-th$ profile is $\left(\frac{n_{1j}}{N_{\cdot j}}, \frac{n_{2j}}{N_{\cdot j}}, \ldots, \frac{n_{ij}}{N_{\cdot j}}\right)$. Another important concept is that of mass, each profile (row or column) is assigned a mass, which is the quotient between the marginal total of the line (row or column) and the global total, being the $i-th$ mass $r_i = \frac{N_{i\cdot}}{N_{\cdot\cdot}}$ and the $-th$ $r_j = \frac{N_{\cdot j}}{N_{\cdot\cdot}}$. Not all rows have the same importance, since the profile of each row is referred to a different marginal total. Therefore, when representing the profiles, we should give more importance to those with greater mass (r_i), same for the columns (r_j). The use of profiles facilitates interpretation but can produce a mistaken view of the relationship between variables. To solve this problem, a distance is used that does not ignore the differences between the profiles of each line. The chi-square distance is a weighted Euclidean distance used to measure distances between profiles. It is calculated in such a way that the categories that are more frequent are weighted less and those that are less frequent are weighted more.

So, given two points $P_i = (X_{i1}, X_{i2})$ and $P_{i'} = (X_{i'1}, X_{i'2})$, the Euclidean distance between them is $d^2\left(P_i, P_{i'}\right) = (x_{i1} - x_{i'1})^2 + (x_{i2} - x_{i'2})^2$. Therefore, given two profiles, $i : \left(\frac{n_{i1}}{N_{i\cdot}}, \frac{n_{i2}}{N_{i\cdot}}, \ldots, \frac{n_{ij}}{N_{i\cdot}}\right)$ and $i' : \left(\frac{n_{i'1}}{N_{i'\cdot}}, \frac{n_{i'2}}{N_{i'\cdot}}, \ldots, \frac{n_{i'j}}{N_{i'\cdot}}\right)$, the chi-square distance between them will be $d^2\left(i, i'\right) = \sum_{j=1}^{p} \left(\frac{1}{N_{\cdot j}}\right)\left(\frac{n_{ij}}{N_{i\cdot}} - \frac{n_{i'j}}{N_{i'\cdot}}\right)^2$, $d^2(j, j') = \sum_{i=1}^{n} \left(\frac{1}{N_{i\cdot}}\right)\left(\frac{n_{ij}}{N_{\cdot j}} - \frac{n_{ij'}}{N_{\cdot j'}}\right)^2$. Distances are not measured between rows or columns but relative to an average row or column profile, that is, relative to the average of the coordinates of that line weighted by its mass. This average profile is located at the origin of the coordinates and is called the centre of gravity or centroid. The inertia of a point cloud, which represents the profiles, is a measure of the dispersion of the profiles in a multidimensional space. It is calculated as the sum, for each point, of its mass times its distance to the centroid squared. A low value of inertia means that all products are located close to the centroid and are therefore similar, while high values of inertia indicate the opposite. The total inertia of a table coincides with the corresponding chi-square value.

As a preliminary step to correspondence analysis, we must determine whether the data can be fitted to a model, that is, determine whether there is a relationship between the variables we wish to study. It would make no sense to perform correspondence analysis if there was no relationship between the variables. This is achieved with the chi-square test on the contingency table to test the goodness of fit of the data to the model, which is given by the standardized residuals, what is known as contributions $c_{ij} = \frac{\left(no_{ij} - ne_{ij}\right)^2}{ne_{ij}}$, being no_{ij} the observed frequencies (the data sample) and ne_{ij} the frequencies that would be expected if the variables were independent.

Thus, the method consists of finding the singular value decomposition of the matrix $C = \left(c_{ij}\right)$, to build a coordinate system (usually two-dimensional) associated with the rows and columns of the contingency table, which reflects the relationships existing between said rows and columns. In the representation, the aforementioned chi-square distances between profiles play an important role, which the correspondence analysis attempts to reproduce in its graphic representations.

Once the basic concepts of correspondence analysis have been explained, the analysis proceeds to diagonalize the variance–covariance matrix (C) in order to obtain the vectors and eigenvalues that will define the new axes on which the point cloud will be projected. This method attempts to find two arrays called A and B of cartesian coordinates.

$A = \begin{pmatrix} a'_1 \\ a'_2 \\ \cdots \\ a'_k \end{pmatrix}$ representing the row points with: $a_i = (a_{i1}, \ldots, a_{ih})'$, $B = \begin{pmatrix} b'_1 \\ b'_2 \\ \cdots \\ b'_m \end{pmatrix}$ representing the column points with: $b_j = (b_{j1}, \ldots, b_{jh})'$ (generally $h = 2$ dimensions).

There are several ways to calculate matrices A and B, known as normalizations. A widely used way is known as symmetric or canonical normalization, which seeks to satisfy that the scalar product $\left(a_i \cdot b_j\right)$ is proportional to the standardized residuals c_{ij}.

Correspondence analysis is therefore a technique used to represent the rows and columns of a contingency table as points in a low-dimensional vector space [30], so that the corresponding spaces can be superimposed to obtain a joint representation, associating the proximity between the points in the graph with the relationship between the categories (Figure 2). This method, in its different versions, is commonly applied in research in different fields such as circular economy [31], economics [32], nutrition [33], medicine [34,35] and business [36]; moreover, there are multiple computer implementations for its application [37]. In our case, using the free software R version 4.3.1, we are going to superimpose in the same subspace the type of vessel and the number of deficiencies found for each of them, allowing us to find the relationships between the categories of both variables based on their proximity in this plane (Figure 2).

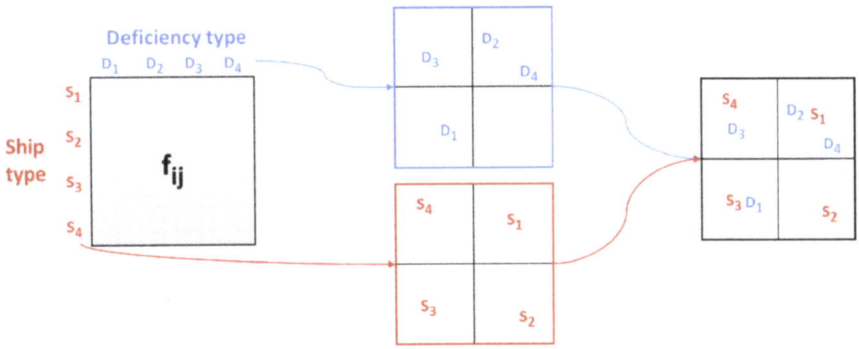

Figure 2. Illustrative example of the graphic results of a simple correspondence analysis.

2.2. Description and Processing of Data

Next, we examined the total data we worked with using frequency distribution tables for each variable from 2018 to 2022, selecting the most representative ones, excluding those with lower frequency.

The Paris MoU agreement establishes a hierarchical classification for both the deficiencies detected in ship inspections and for the types of ships. Deficiencies are categorized using numerical codes: general categories (2–3 digits) and specific categories (4–5 digits). The classification of ships is based on their use, using a specific coding system.

Figure 3 shows the number of inspections by type of vessel, considering only the type of vessels that have more than 1000 inspections/year.

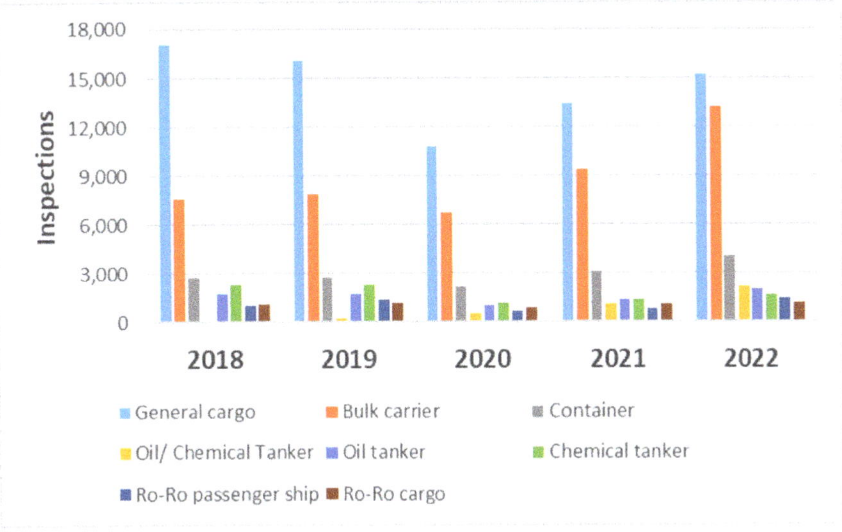

Figure 3. PSC MoU Paris number of inspections by type of ship between 2018 and 2022.

Table S1 (Supplementary Materials) presents the distribution of vessel types according to their use and the study years. The General Cargo/Multipurpose (72,543 inspections) and Bulk Carrier (44,763 inspections) types account for 63% of the total sample. Given the substantial variation in frequencies among vessel types, we selected only those above the 50th percentile, with at least 2000 inspections. An exception was made for the commercial yacht (1103 inspections) due to its relevance in the study, as its inclusion allows for a more comprehensive comparison between vessel types and prevents the loss of information on this specific category. As a result, the final sample consists of 183,154 inspections, representing a 1.7% reduction compared to the total sample of 186,255 inspections. This selection ensures a more robust statistical analysis by minimizing the influence of vessel categories with very low representation. Table S2 (Supplementary Materials) shows the frequency distribution of deficiency types, grouped by categories (DIC 2 Díg column) and subcategories (DIC 3 Díg column).

By categories, the most repeated type of deficiencies is "*18 MLC*" with 16% of the whole sample (29,792 inspections) and within this, the subcategories "*183 Accommodation, recreational facilities, food and catering*" with 10,918 inspections and "*184 Health protection, medical care, social security*" with 17,138 inspections stand out.

The second most repeated category is "*01 Certificates & Documentation*" with 15% of the sample; this is followed by "*07 Fire safety*" with 13%, followed by "*10 Safety of Navigation*" with 11%, "*11 Lifesaving appliances*" with 8%, "*14 Pollution Prevention*" with 7%, highlighting the subcategories "*MARPOL Annex I, IV, V and VI*" and "*04 Emergency Systems*" with 6%. The remaining categories have fewer than 10,000 inspections and have percentages of less than 5%.

Figure 4 shows the percentages by impairment type category of PSC inspections identified in the sample. Therefore, we are going to perform a simple correspondence analysis for the different categories, grouped in two digits, of the type of deficiency, highlighting the most important relationships.

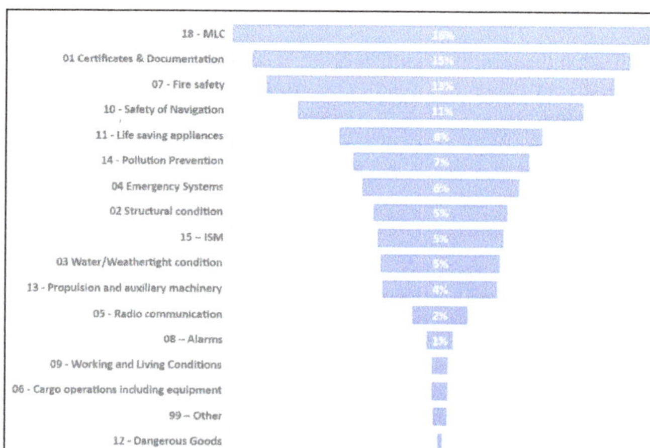

Figure 4. Percentage of identified PSC inspections by deficiency type category.

3. Results and Discussion

In this section, we begin with the analysis of the variables, using correspondence tables, which show the frequencies and their visualization on a factorial plane, with their inertia percentages described for each axis (Dim 1, Dim 2), obtaining good representation qualities in all the figures shown. In order to facilitate the interpretation and understanding of the figures, each image will be accompanied by the codes of the corresponding deficiencies according to the Paris MoU classification.

We start by studying the deficiencies, but grouping them by categories with the first two digits (Table S2). Figure 5 shows such a representation, where we observe notable differences in the type of deficiencies per ship.

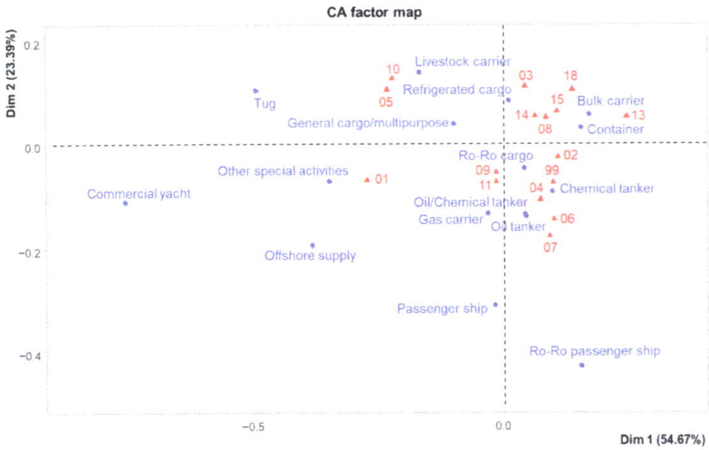

Figure 5. Correspondence analysis of ship type and deficiencies by category (2 digits).

Other special activities and offshore supply are close to *"01 certificates and documentation"*, with a similar profile in the commercial yacht.

The Livestock Carrier vessels are characterized by the deficiencies *"05 radio communication"* and *"10 safety of navigation"*, somewhat less in the Tug and General cargo vessels.

Bulk Carrier, Container and Refrigerated cargo vessels are characterized by deficiencies in categories *"03 Water/Weathertight condition"*, *"08 Alarms"*, *"13 Propulsion and auxiliary machinery"*, *"14 Pollution Prevention"*, *"15 ISM"* and *"18 MLC"*.

The Chemical tanker, Gas carrier, Oil tanker, Oil/Chemical tanker and Ro-Ro cargo vessels stand out in the categories: *"02 Structural condition"*, *"04 Emergency Systems"*, *"09 Working and Living Conditions"*, *"11 Lifesaving appliances"* and *"99 Other"* and to a lesser extent in *"06 Cargo operations including equipment"* and *"07 Fire safety"*.

Finally, the Passenger ship and Ro-Ro Passenger ship, which are separated from the rest of the ships, heading to *"07 Fire safety"*.

This representation of the deficiency profiles for each type of ship has revealed a notable heterogeneity. These initial results suggest that the different deficiencies vary significantly between each type of ship. Therefore, a more detailed analysis at the subcategory level is proposed to identify specific deficiencies and develop customized inspection strategies [24], with the aim of improving the operational efficiency of inspectors. These subcategories (Table S3, Column DIC 4–5 Dig.) are codified in the Paris Memorandum, and we have used their digits to represent them in the figures, making it easier to visualize and interpret the results. The codification corresponding to each of them is included in Table S3 of the Supplementary Material.

We start with *"01 Certificates & Documentation"* with its representation in Figure 6. This category represents 14.6% of the PSC inspections in the sample, being the second highest (below category 18 MLC with 16%). The General cargo/multipurpose vessel stands out from the rest and is located close to many of the deficiencies such as 1101, 1123, 1218, 1305, 1306, 1314 or 1320; the Livestock Carrier with 1315. Others such as Bulk Carrier or some more polluting ones such as Oil tanker with 1220, 1308, 1330. Finally, we could highlight the Container, close to 1136, 1139, 1199, 1310.

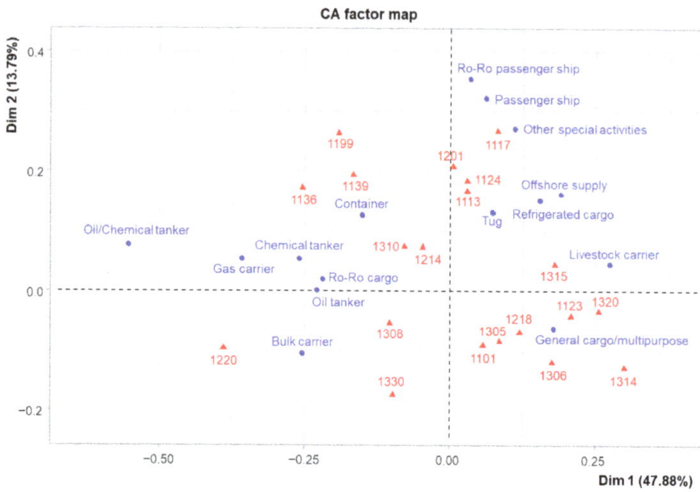

Figure 6. Correspondence analysis of the type of vessel and deficiencies in category 01 Certificates & Documentation.

It is worth noting in this section as an important result that General Cargo/multipurpose ships are the ones with the most deficiencies with respect *to "01 Certificates & Documentation"*, ranking close to many categories and separated from the rest. The analysis revealed that general cargo and multipurpose ships require improvement in their compliance with documentary requirements, especially with regard to the updating of certificates.

The category *"02 Structural condition"* represents 5% of the inspections and its representation is shown in Figure 7. In this case, important relationships can be observed, such as that of the Ro-Ro Passenger ship with deficiency 2122; that of the oil/chemical tanker with 2109; that of the offshore supply with 2101; that of the bulk carrier with 2103; that of the oil and chemical tanker with 2107, 2108, 2199 and 2117 and container with 2105, among others.

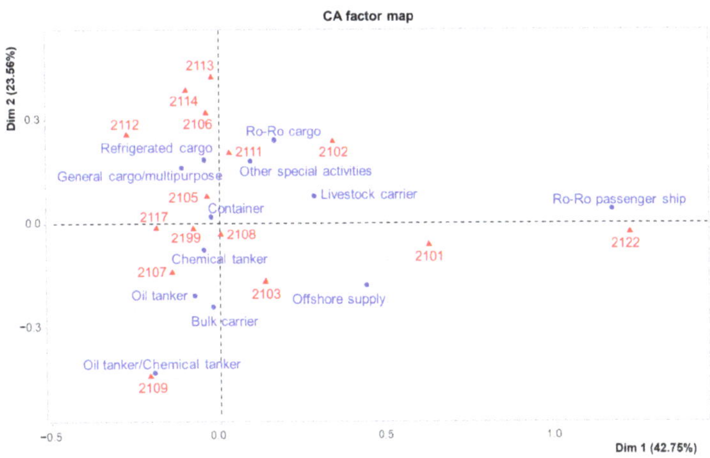

Figure 7. Correspondence analysis of the type of vessel and deficiencies in category 02 Structural condition.

The relationship between Ro-Ro Passenger ships and structural deficiencies is one of the most important results obtained from this analysis [14]. These ships are designed to transport both cargo and passengers, the latter being regulated by the International Convention SOLAS and mandatory on all ships. More specifically, the deficiency listed as the most significant is that of openings to the cargo area, door and scuttles, which is why the inspector must place more emphasis on the inspection of the vessel in this area. Another deficiency that we consider important is that of Oil tanker/Chemical tanker with respect to structural condition, more specifically with that of permanent means of access.

The category *"03 Water/Weathertight condition"* represents 5% of the inspections and is represented in Figure 8. We highlight the relationship between the Ro-Ro Passenger ship and deficiency 3112; livestock carrier with 3106; general cargo with 3102; other special activities with 3107; other more polluting activities such as oil or chemical tanker, container and refrigerated cargo more related to 3103, 3105 and 3108; bulk carrier with 3110, 3104, 3109 and 3199.

It is worth noting from these results that the most polluting ships, such as oil or chemical tankers, container ships and refrigerated cargo ships, are closely related to these types of deficiencies. They presented a greater number of deficiencies related to the watertight integrity of holds and compartments that can compromise the safety of the cargo and increase the risk of marine pollution. Also, the Ro-Ro Passenger ship can be deduced from this representation that it is closely related to this type of deficiencies, more specifically to that of scuppers, inlets and discharges, being another target for supervision by inspectors.

The category *"04 Emergency Systems"* represents 6% of the inspections, and it is represented in Figure 9. Once again, the most polluting ships are located close to each other and are related to deficiencies 4102, 4109 and 4114, more specifically the oil tanker/chemical tanker with 4116, 4117; refrigerated cargo with 4103; Ro-Ro Passenger ship with 4101; general cargo and livestock carrier with 4107 and 4108.

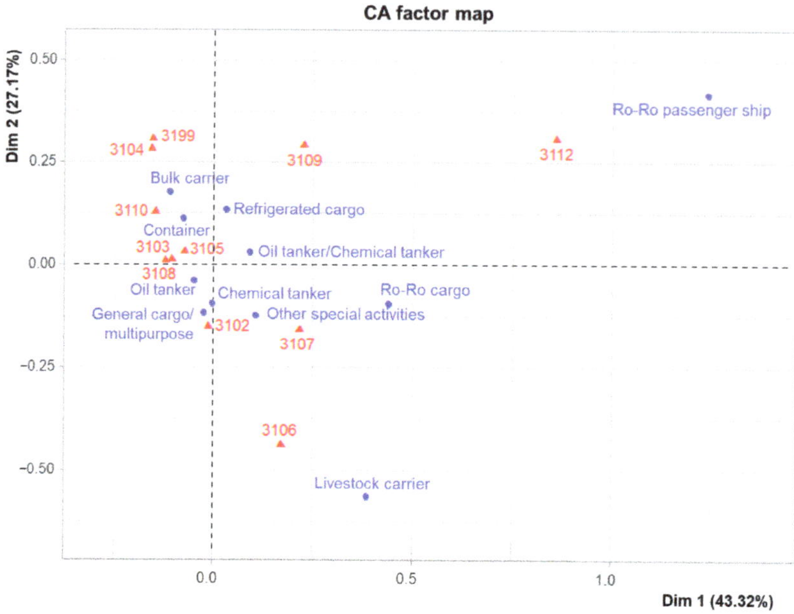

Figure 8. Correspondence analysis of the type of vessel and deficiencies in category 03 Water/Weathertight condition.

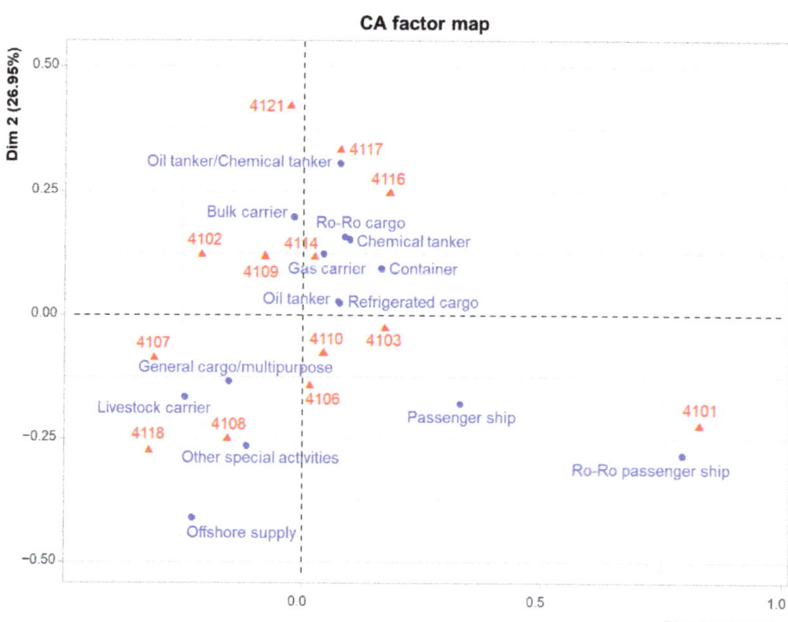

Figure 9. Correspondence analysis of the type of vessel and deficiencies in category 04 Emergency Systems.

One of the most noteworthy results that we can deduce from this figure is that the most polluting ships, such as oil tankers and chemical tankers, had a significantly higher number of deficiencies in essential emergency systems, such as firefighting and life-saving equipment [18]. These deficiencies imply serious risks in the event of incidents on board and, therefore, require exhaustive supervision due to the high risk of explosions and environmental disasters that a fire could trigger. In this representation, we also observe an aspect to consider for Ro-Ro Passenger ships, where the public address system is closely related to this type of ship and constitutes a focal point of attention for inspectors due to its importance in the safety and operation of the ship.

The category *"05 Radio communication"* represents only 2% of the inspections, and we have its representation in Figure 10. In this analysis, we do not highlight any relationship between the type of ship and the deficiencies found after the inspection; this reveals recurring deficiencies in PSC inspections, common to many ships. We name some relatively intense ones such as Ro-Ro Passenger ship with 5109 and 5113; container and other special activities with 5116; chemical tanker with 5118; bulk carrier with 5105; Ro-Ro cargo, livestock carrier and offshore supply with 5115 and general cargo with 5106, 5110, 5111 and 5199.

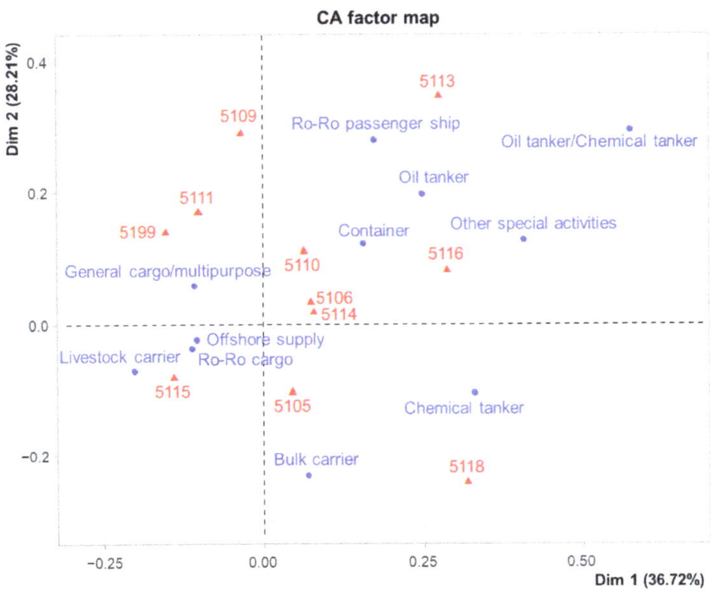

Figure 10. Correspondence analysis of the type of vessel and deficiencies in category 05 Radio communication.

The category *"07 Fire safety"* represents 13% of the inspections and is represented in Figure 11. In this case, we can see that there are no clear relationships; the majority of ships are located in the center of the figure together with the majority of deficiencies, with the most polluting ones standing out. It is the third most frequent category in the study.

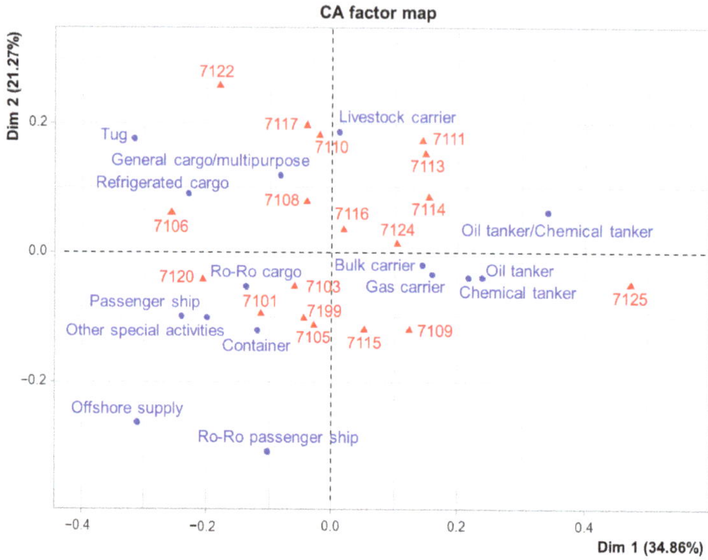

Figure 11. Correspondence analysis of the type of vessel and deficiencies in category 07 Fire safety.

To highlight some of the relationships, we have livestock carrier with 7117, 7110 and 7111; refrigerated cargo with 7106; general cargo with 7108; bulk carrier and gas carrier with 7124; oil and chemical tanker with 7125; Ro-Ro cargo, Passenger ship, other special activities and container with 7101 and 7120.

The category *"10 Safety of Navigation"* represents 11% of the inspections and is represented in Figure 12. As can be seen, this category is among the main deficiencies detected and is made up of a large number of subcategories, as can be seen in the correspondence map. This translates into common deficiencies in PSC inspections and deficiencies common to a large number of ships, so there is no strong relationship between a type of ship and a specific deficiency.

The category *"11 Life saving appliances"* represents 8% of the inspections and is shown in Figure 13. In this category, there are few differences, as we can see all the subcategories represented are located in the center of the figure, as are the majority of ships, indicating that they tend to be deficiencies commonly found in PSC inspections.

The category *"13 Propulsion and auxiliary machinery"* represents 4% of the inspections. Its representation in Figure 14 shows few differences. It is worth noting that deficiency 13108 is not associated with any ship, and the others are located in the center of the figure.

The category *"14 Pollution Prevention"* constitutes 7% of the inspections. Its representation in Figure 15 has been made using only the main categories, so they are not contained in Table S3 and are equivalent to deficiencies found on ships determined through the annexes of the MARPOL Convention. This point is considered very important for preserving the environment, and it should be noted in this analysis that the General Cargo/Multipurpose and Bulk Carrier ships have deficiencies related to MARPOL Annexes I, IV and V. It is also worth mentioning that the deficiencies related to MARPOL Annex VI correspond to the Oil Tanker. [21] With regard to the Ballast Water deficiency, it is closely related to Container, Oil/Chemical Tanker and Ro-Ro Cargo ships.

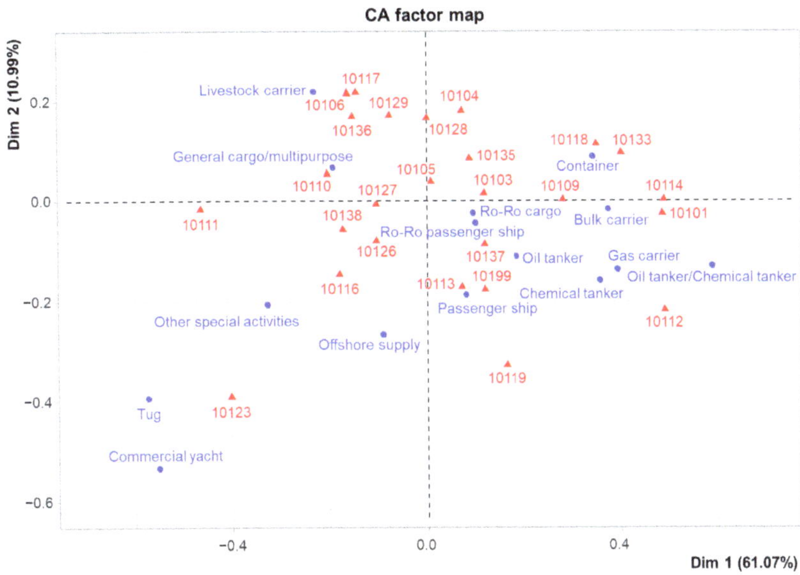

Figure 12. Correspondence analysis of the type of vessel and deficiencies in category 10 Safety of Navigation.

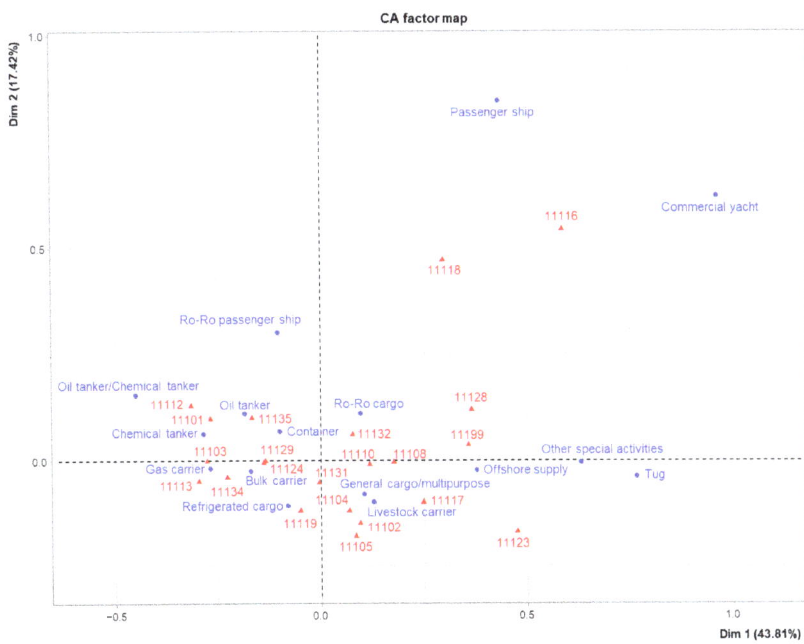

Figure 13. Correspondence analysis of the type of vessel and deficiencies in category 11 Life saving appliances.

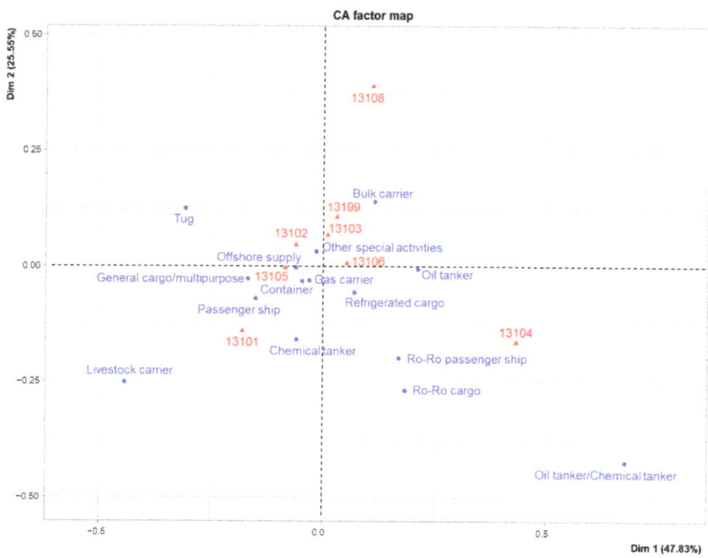

Figure 14. Correspondence analysis of the type of vessel and deficiencies in category 13 Propulsion and auxiliary machine.

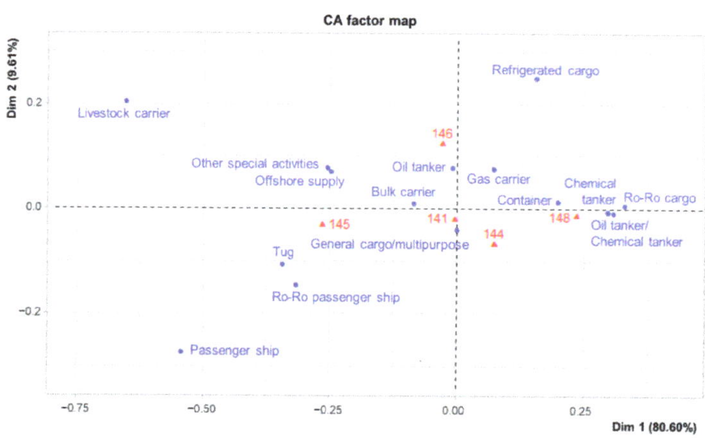

Figure 15. Correspondence analysis of the type of vessel and deficiencies in category 14 Pollution Prevention.

The category *"15—ISM"* represents 5% of the inspections, and since it has no subcategories, we cannot perform the correspondence analysis. Instead, we present the information from Table 2. With this information, it is not possible to draw relevant conclusions, since the highest frequencies correspond to general cargo and bulk carriers, but we remember that these are the ships that appear most in the global sample, Table S1.

Table 2. Frequency distribution of the type of vessel and deficiencies in category 15 ISM.

15150 ISM	Frequency
Bulk carrier	2364
Chemical tanker	427
Commercial yacht	21
Container	783
Gas carrier	94
General cargo/multipurpose	3586
Livestock carrier	121
Offshore supply	114
Oil tanker	364
Oil tanker/Chemical tanker	210
Other special activities	98
Passenger ship	67
Refrigerated cargo	85
Ro-Ro cargo	291
Ro-Ro Passenger ship	191
Tug	12
Total	8828

The category *"18 MLC, 2006"* is the most common category in the study, representing 16% of the sample. Its representation in Figure 16 shows the subcategories grouped into three digits. Subcategory 181 relates to general cargo and Ro-Ro cargo ships. Subcategory 184 includes the above ships as well as container ships and Ro-Ro Passenger ships. Subcategory 183 includes bulk carriers and oil tankers. And finally, subcategory 182 includes offshore supply vessels, chemical tankers and, to a lesser extent, gas carriers and oil/chemical tankers.

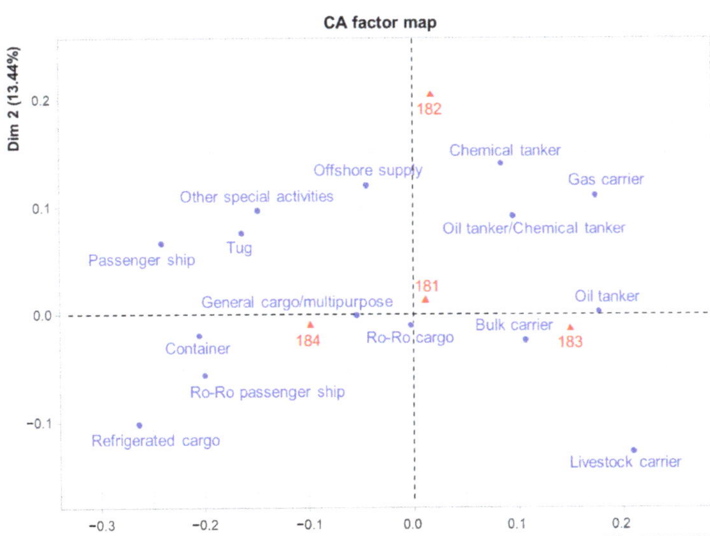

Figure 16. Correspondence analysis of the type of vessel and deficiencies in category 18 MLC.

In this analysis, the most noteworthy aspect is that tankers, Ro-Ro ships and passenger ships presented a high number of deficiencies related to the living conditions on board, according to the inspections carried out. The lack of adequate maintenance of cabins and common areas constitutes a risk to the health and safety of the crew, in addition to affecting their well-being and productivity.

4. Conclusions

The present study has conducted an exhaustive analysis of the deficiencies detected in PSC inspections within the Paris MoU, aiming to identify patterns and trends across different types of vessels. To achieve this, correspondence analysis—a statistical technique used to identify relationships between categorical variables—was applied to 186,255 records from the THETIS platform. The data were normalized according to the Paris MoU classification and processed using statistical software (R) to generate visual representations of the relationships between vessel types and the most frequent deficiencies. However, its implementation posed challenges, such as data heterogeneity, harmonization and the interpretation of statistical results, requiring a meticulous analysis to ensure the coherence and relevance of the findings. Despite these challenges, the applied methodology optimizes resource allocation in PSC inspections and could be integrated with artificial intelligence in the future to enhance deficiency prediction and efficiency in maritime control.

One of the main findings is the heterogeneity in the deficiency profiles between the different types of ships. This variability can be attributed to factors such as design, age, type of cargo carried and the operations carried out in each of them, depending on their use. To improve the efficiency of inspections, a more in-depth study of each subcategory has been carried out. In this way, we will be able to identify the specific deficiencies of each type of vessel and develop personalized inspection strategies that will optimize time and resources.

Here, we present the relationships that have been found between the different subcategories of deficiencies and the types of vessel, presenting the most important conclusions based on the above results. Table 3 is drawn up for this purpose, summarizing the type of vessel and its most common deficiencies detected after inspection.

Ships involved in specific operations, such as the transport of dangerous cargo or the transport of passengers, presented particular deficiency profiles [12], which indicates the importance of adapting inspections to the characteristics of each type of operation.

The results of this study have important implications for maritime safety. By identifying the most common types of deficiencies, it is possible to be proactive and design more effective strategies to prevent accidents and protect the marine environment.

We can propose certain recommendations based on the above:

- Strengthen inspection programs by increasing the frequency of inspections on ships with a history of significant deficiencies [26].
- Promote crew training by implementing more comprehensive and up-to-date training programs that address the specific needs of each type of vessel and operation.
- Encourage preventive maintenance by shipowners, who should prioritize the upkeep of their vessels to avoid major breakdowns and ensure optimal performance.

This study highlights the need for an in-depth analysis of the deficiencies detected in PSC inspections in order to identify specific vulnerabilities by ship type and thus optimize maritime safety control and improvement strategies. The results obtained offer a valuable tool for maritime authorities, shipowners and PSCO inspectors, allowing them to focus their supervision efforts more effectively [20]. As next steps, it is proposed to deepen the practical application of these findings, developing specific guides and protocols for each type of vessel [18]. This will facilitate the implementation of more precise preventive and corrective measures, contributing significantly to risk reduction and the protection of the marine environment. In addition, it is suggested that the research be expanded to include a comparative analysis between different regions and fleets, with the aim of identifying global patterns and promoting the adoption of best practices at an international level.

Table 3. Types of vessel and the most common deficiencies detected.

Vessel Type	Common Deficiencies
General Cargo & Multipurpose Vessels	Certificates and documentation (lack of updated certificates, incomplete documentation) Pollution prevention (MARPOL Annexes I, IV and V)
Oil Tankers	Structural condition Watertight integrity Emergency systems Pollution prevention (MARPOL Annexes I, IV, V and VI) Living conditions on board
Ro-Ro Vessels	Structural condition Pollution prevention (ballast water management) Living conditions on board
Ro-Ro Passenger Vessels	Structural condition (cargo holds and doors)
Chemical Tankers	Watertight integrity Emergency systems Pollution prevention (ballast water management)
Container Ships	Watertight integrity Pollution prevention (ballast water management)
Bulk Carriers	Pollution prevention (MARPOL Annexes I, IV and V)
Passenger Ships	Living conditions on board

Supplementary Materials: The following supporting information can be downloaded at https://www.mdpi.com/article/10.3390/oceans6010015/s1; Table S1: Vessel types by year; Table S2: Type of general categories of deficiencies (2–3 digits) grouped and separated by year; Table S3: Specific categories of deficiencies (4–5 digits). Figure S1: Annual Deficiencies Trend in PSC Inspections (2018–2022).

Author Contributions: Conceptualization, J.M.P. and D.A.; methodology, D.A. and V.A.-E.; software, V.A.-E.; validation, B.J.-A. and J.J.M.-P.; formal analysis, D.A. and V.A.-E.; investigation, J.M.P.; resources, J.M.P., B.J.-A. and J.J.M.-P.; data curation, V.A.-E. and J.M.P.; writing—original draft preparation, J.M.P., V.A.-E. and D.A.; writing—review and editing, J.M.P., B.J.-A. and J.J.M.-P.; visualization, J.M.P., B.J.-A. and J.J.M.-P.; supervision, B.J.-A., J.J.M.-P. and D.A.; project administration, J.M.P., B.J.-A. and J.J.M.-P. All authors have read and agreed to the published version of the manuscript.

Funding: This work was partly support by University of Cádiz.

Institutional Review Board Statement: Not applicable.

Informed Consent Statement: Not applicable.

Data Availability Statement: European Maritime Safety Agency (EMSA), https://portal.emsa.europa.eu/web/thetis (accessed on 18 November 2024).

Conflicts of Interest: The authors declare no conflicts of interest.

Abbreviations

The following abbreviations are used in this manuscript:

PSC	Port State Control
MoU	Memoranda of Understanding
IMO	International Maritime Organization
SOLAS	The International Convention for the Safety of Life at Sea
MARPOL	The International Convention for the Prevention of Pollution from Ships
STCW	The International Convention on Standards of Training, Certification and Watchkeeping for Seafarers
PSCOs	Port State Control Officers
THETIS	The Hybrid European Targeting and Inspection System
EMSA	European Maritime Safety Agency
NIR	New System of Inspections

References

1. Alderton, T.; Winchester, N. Globalisation and De-Regulation in the Maritime Industry. *Mar. Policy* **2002**, *26*, 35–43. [CrossRef]
2. Carmona, A.; Muñoz Pérez, J.J.; Gomez-Pina, G.; Román-Sierra, J. Gestión de La Limpieza de Playas Por Derrame de Hidrocarburos En La Bahía de Algeciras. In *Xi Jornadas Españolas Costas Y Puertos*; University of Cádiz: Cádiz, Spain, 2012; pp. 162–168.
3. *Resolution A. 682(17) Adopted on 6 November 1991 Regional Co-Operation in the Control of Ships and Discharges*; IMO: London, UK, 1991.
4. Paris MoU Paris MoU. Available online: https://parismou.org/ (accessed on 8 October 2024).
5. ES-Español-ES-EMSA-European Maritime Safety Agency. Available online: https://www.emsa.europa.eu/es/ (accessed on 8 October 2024).
6. Ravira, F.J.; Piniella, F. Evaluating the Impact of PSC Inspectors' Professional Profile: A Case Study of the Spanish Maritime Administration. *WMU J. Marit. Aff.* **2016**, *15*, 221–236. [CrossRef]
7. E.U.R.-Lex Directive 2009/16/EU, of the European Parliament and of the Council, of the 23rd April, on the Control of Ships by the Port State Control Amended by Directive 2013/38/EU. 2009. Available online: https://eur-lex.europa.eu/legal-content/EN/TXT/?qid=1377073917502&uri=CELEX:32013L0038 (accessed on 1 February 2025).
8. European Maritime Safety Agency (EMSA), THETIS. Available online: https://portal.emsa.europa.eu/web/thetis (accessed on 6 May 2022).
9. List of Paris MoU Deficiency Codes (1 July 2024) | Paris MoU. Available online: https://parismou.org/2023/07/list-paris-mou-deficiency-codes (accessed on 9 October 2024).
10. Brooks, M.R. The Privatization of Ship Safety. *Marit. Policy Manag.* **1996**, *23*, 271–288. [CrossRef]
11. Havold, J.I. Culture in Maritime Safety. *Marit. Policy Manag.* **2000**, *27*, 79–88. [CrossRef]
12. Cariou, P.; Mejia Jr, M.Q.; Wolff, F.-C. An Econometric Analysis of Deficiencies Noted in Port State Control Inspections. *Marit. Policy Manag.* **2007**, *34*, 243–258. [CrossRef]
13. Li, K.X.; Zheng, H. Enforcement of Law by the Port State Control (PSC). *Marit. Policy Manag.* **2008**, *35*, 61–71. [CrossRef]
14. Hänninen, M.; Kujala, P. Bayesian Network Modeling of Port State Control Inspection Findings and Ship Accident Involvement. *Expert Syst. Appl.* **2014**, *41*, 1632–1646. [CrossRef]
15. Graziano, A.; Cariou, P.; Wolff, F.-C.; Mejia, M.Q.; Schröder-Hinrichs, J.-U. Port State Control Inspections in the European Union: Do Inspector's Number and Background Matter? *Mar. Policy* **2018**, *88*, 230–241. [CrossRef]
16. Wang, S.; Yan, R.; Qu, X. Development of a Non-Parametric Classifier: Effective Identification, Algorithm, and Applications in Port State Control for Maritime Transportation. *Transp. Res. Part B Methodol.* **2019**, *128*, 129–157. [CrossRef]
17. Chen, J.; Zhang, S.; Xu, L.; Wan, Z.; Fei, Y.; Zheng, T. Identification of Key Factors of Ship Detention under Port State Control. *Mar. Policy* **2019**, *102*, 21–27. [CrossRef]
18. Chung, W.H.; Kao, S.L.; Chang, C.M.; Yuan, C.C. Association Rule Learning to Improve Deficiency Inspection in Port State Control. *Marit. Policy Manag.* **2020**, *47*, 332–351. [CrossRef]
19. Knapp, S.; Heij, C. Improved Strategies for the Maritime Industry to Target Vessels for Inspection and to Select Inspection Priority Areas. *Safety* **2020**, *6*, 18. [CrossRef]
20. Fu, J.; Chen, X.; Wu, S.; Shi, C.; Wu, H.; Zhao, J.; Xiong, P. Mining Ship Deficiency Correlations from Historical Port State Control (PSC) Inspection Data. *PLoS ONE* **2020**, *15*, e0229211. [CrossRef] [PubMed]

21. Shen, J.H.; Liu, C.P.; Chang, K.Y.; Chen, Y.W. Ship Deficiency Data of Port State Control to Identify Hidden Risk of Target Ship. *J. Mar. Sci. Eng.* **2021**, *9*, 1120. [CrossRef]
22. Wang, Y.; Zhang, F.; Yang, Z.; Yang, Z. Incorporation of Deficiency Data into the Analysis of the Dependency and Interdependency among the Risk Factors Influencing Port State Control Inspection. *Reliab. Eng. Syst. Saf.* **2021**, *206*, 107277. [CrossRef]
23. Prieto, J.M.; Amor, V.; Turias, I.; Almorza, D.; Piniella, F. Evaluation of Paris Mou Maritime Inspections Using a Statis Approach. *Mathematics* **2021**, *9*, 2092. [CrossRef]
24. Demirci, S.M.E.; Cicek, K. Intelligent Ship Inspection Analytics: Ship Deficiency Data Mining for Port State Control. *Ocean Eng.* **2023**, *278*, 114232. [CrossRef]
25. Lai, C.Y.; Liu, C.P.; Huang, K.M. Optimization of the Concentrated Inspection Campaign Model to Strengthen Port State Control. *J. Mar. Sci. Eng.* **2023**, *11*, 1166. [CrossRef]
26. Yan, R.; Liu, Y.; Wang, S. A Data-Driven Optimization Approach to Improving Maritime Transport Efficiency. *Transp. Res. Part B Methodol.* **2024**, *180*, 102887. [CrossRef]
27. Almorza, D.; Prieto, J.M.; Amor-Esteban, V.; Piniella, F. Port State Control Inspections under the Paris Memorandum of Understanding and Their Contribution to Maritime Safety: Additional Risk Classifications and Indicators Using Multivariate Techniques. *J. Mar. Sci. Eng.* **2024**, *12*, 533. [CrossRef]
28. Prieto, J.M.; Amor-Esteban, V.; Almorza-Gomar, D.; Turias, I.; Piniella, F. Application of Multivariate Statistical Techniques as an Indicator of Variability of the Effects of COVID-19 on the Paris Memorandum of Understanding on Port State Control. *Mathematics* **2023**, *11*, 3188. [CrossRef]
29. Benzécri, J.P. *L'Analyse Des Données*; Presses Universitaires de France: Paris, France, 1973.
30. Greenacre, M. *La Práctica Del Análisis de Correspondencias*; Fundación BBVA: Bilbao, Spain, 2008.
31. Raimo, N.; Vitolla, F.; Esposito, B.; Malandrino, O.; García-Sánchez, I.-M.; Somohano-Rodríguez, F.-M.; Amor-Esteban, V.; Gonzalez-Valdueza, B. Circular Economy Projects and Firm Disclosures in an Encouraging Institutional Environment. *Sustainability* **2022**, *14*, 1149. [CrossRef]
32. Palazzo, M.; Vollero, A.; Siano, A. Intelligent Packaging in the Transition from Linear to Circular Economy: Driving Research in Practice. *J. Clean. Prod.* **2023**, *388*, 135984. [CrossRef]
33. Romero Ferreiro, C.; Cancelas Navia, P.; Lora Pablos, D.; Gómez de la Cámara, A. Geographical and Temporal Variability of Ultra-Processed Food Consumption in the Spanish Population: Findings from the DRECE Study. *Nutrients* **2022**, *14*, 3223. [CrossRef]
34. Parás-Bravo, P.; Salvadores-Fuentes, P.; Alonso-Blanco, C.; Paz-Zulueta, M.; Santibañez-Margüello, M.; Palacios-Ceña, D.; Boixadera-Planas, E.; Fernández-De-Las-Peñas, C. The Impact of Muscle Relaxation Techniques on the Quality of Life of Cancer Patients, as Measured by the FACT-G Questionnaire. *PLoS ONE* **2017**, *12*, e0184147. [CrossRef]
35. Zambrano Bohorquez, B.; Villarín Castro, A.; Muñoz Bodas, F.; Gómez Rey, I. Síndrome de Burnout, Resiliencia y Miedo a La COVID-19 En Personal Sanitario Durante La Pandemia Por SARS-CoV-2. *Rev. Clín. Med. Fam.* **2022**, *15*, 134–142. [CrossRef]
36. Change for: Brito-Carrillo, C.; Pitre-Redondo, R.; Cardona-Arbeláez, D. Organizational Climate and Its Influence on the Performance of Employees in a Service Company. *Inf. Tecnol.* **2020**, *31*, 141–148. [CrossRef]
37. Ledesma, R. Software De Análisis De Correspondencias Múltiples: Una Revisión Comparativa. *Metodol. Encuestas* **2008**, *10*, 59–75.

oceans

MDPI

Article

The Cultural Heritage Framework Programme: Highlighting the Contribution of Marine Cultural Heritage to the UN Decade of Ocean Science for Sustainable Development (2021–2030)

Jon Henderson [1,*], Georgia Holly [1], Arturo Rey da Silva [1] and Athena Trakadas [2]

[1] School of History, Classics and Archaeology, University of Edinburgh, Edinburgh EH8 9JX, UK; georgia.holly@ed.ac.uk (G.H.); arturo.rey@ed.ac.uk (A.R.d.S.)

[2] Ocean Decade Heritage Network, DK 2100 Copenhagen, Denmark; chair@oceandecadeheritage.org

* Correspondence: jon.henderson@ed.ac.uk

Abstract: This paper considers the importance of including Marine Cultural Heritage (MCH) in the UN Decade of Ocean Science for Sustainable Development (2021–2030) in order to fully address the challenges facing the health of our oceans. It outlines the aims and objectives of the Cultural Heritage Framework Programme (CHFP), an official action of the UN Decade of Ocean Science and the only one that specifically deals with cultural heritage. The role of MCH in the Decade is discussed, and its under-appreciation in the UNESCO Cultural Conventions is outlined. The paper then presents the activities of the CHFP, discussing the main themes identified to date, and finishes by outlining the initiative's next steps.

Keywords: tangible and intangible cultural heritage; ocean literacy; sustainable development goals; ocean science; marine archaeology; decolonising; capacity building; traditional ecological knowledge

1. Introduction

The Cultural Heritage Framework Programme (CHFP) was designated in June 2021 as an official, endorsed action of the UN Decade of Ocean Science for Sustainable Development 2021–2030 (hereafter, 'Ocean Decade'). It aims to highlight the role of 'Marine Cultural Heritage' in helping to achieve the key aim of the Ocean Decade–the realisation of a sustainable ocean. CHFP is part of the Ocean Decade Heritage Network (ODHN) [1] and is the first UN-level, global platform that brings together marine heritage practitioners with ocean science and policy stakeholders. At the core of the CHFP is the view that cultural heritage, both past and present, is an essential component of a sustainable and inclusive ocean system, and has so far been under-represented in ocean management, policy, and research. By providing a platform for the consideration of heritage within ocean science, the CHFP contextualises data of past human interactions with the ocean, and fills knowledge gaps through elevating previously unrecorded voices. By inserting heritage data from the past and present into the sustainable development goals for the ocean, we provide vital insights—currently missing from the present literature—that can inform local solutions to global challenges. This has the particular aim of providing inclusive, informed, and adaptable coastal community resilience to the intersectional impacts of climate change on culture, heritage, and ocean sustainability. This paper presents the background and activities of the CHFP to date and outlines the initiative's next steps.

2. Background

The primary aim of the Ocean Decade is to apply scientific approaches to foster the well-being, health, and sustainable development of the ocean [2]. Following the 2030 Agenda, this entails meeting the economic, social, and environmental needs of current and future generations whilst protecting and conserving the health and diversity of ocean ecosystems [3]. Although tangible and intangible cultural heritage, both under water and along the coasts, are essential elements of ocean ecosystems and knowledge transfer, the role of culture and heritage in sustainable management approaches has been largely underestimated or, worse, completely overlooked [1,4–6]. This is surprising, given the Ocean Decade, when it was first proposed in 2017, set out to foster cross-disciplinary collaboration and knowledge sharing to achieve 'the science we need for the ocean we want' [7].

With this in mind, ODHN was established in 2019 by an international group of cultural heritage practitioners with a two-fold agenda: to promote understanding in the cultural heritage sector about the Ocean Decade and to integrate cultural heritage resources and their management into ocean sciences during both the Preparatory Phase (2018–2020) and the Implementation Phase (2021–2030) of the Ocean Decade [1]. As a recognised partner of the Ocean Decade, ODHN developed, in 2021, and chairs the CHFP [8], which provides actionable and accountable steps towards the integration of 'Marine Cultural Heritage' into the ocean sciences. To do so, CHFP has developed a platform to coordinate a diverse network of local and regional projects which align with CHFP's goals, from engaging with traditional knowledge holders to understanding the complex dynamics between reefs and shipwrecks. This is done to drive a coordinated, informed, and inclusive interdisciplinary response to integrating cultural heritage into the sustainable ocean development narrative. In line with the aims of the Ocean Decade, it promotes collaboration and knowledge sharing, improving ocean data and observations, and enhancing the links between ocean science, policy, and society. The CHFP is currently the only recognised Decade Action that addresses cultural heritage.

The CHFP is funded by the Lloyd's Register Foundation from July 2022 to December 2026 and is co-ordinated at the University of Edinburgh. This funding period represents the development phase of CHFP, whereby the network of associated projects will grow and together develop action points for coordinating and inserting the network's knowledge base into the final phases of the Ocean Decade and beyond. ODHN is focused on enabling the Ocean Decade's aim of a 'paradigm shift' in how ocean sciences and ocean resources are studied and preserved in legislation and policy at the national and international levels. In contrast, the CHFP is focused on regional and local actions, engagement practices, and resourcing. These actions can also be conceived as ODHN engaging top-down, whilst CHFP engages bottom-up, with a variety of Ocean Decade stakeholders.

3. Cultural Heritage and the Decade of Ocean Science

A deep and holistic appreciation of what 'Marine Cultural Heritage' (MCH) entails is key to the CHFP approach. We understand MCH to encompass all human activities on, and interaction with, the ocean, including activities along the coasts. MCH therefore refers to both tangible remains like shipwrecks, submerged settlements, coastal villages, harbours, and ocean biocultural heritage, as well as intangible elements such as cultural customs, artistic expressions, local skills, and historical and traditional knowledge [4,5]. The scope of MCH is far-reaching, encompassing links to economic growth and poverty reduction, environmental conservation, social equity, education, and cultural identity. MCH is an intentionally broader and more holistic concept than 'Underwater Cultural Heritage' (UCH), the definition typically used by UNESCO following the UNESCO 2001

Convention on the Protection of the Underwater Cultural Heritage. MCH includes the range of underwater heritage covered in UCH but extends the definition to include all forms of heritage (tangible and intangible) associated with the sea, including the heritage of coastal communities. Furthermore, replacing 'underwater' with 'marine' brings the discipline in parallel with the nomenclature used by the other marine sciences with which the CHFP actively engages, such as marine biology, marine ecology, marine engineering, etc., as well as with marine stakeholders more widely, including developers and policy makers.

The CHFP responds to capacity and knowledge gaps within the Ocean Decade, including a lack of awareness, utilisation, and protection of MCH within most significant ocean policies and frameworks, and a lack of integration and collaboration between the natural and cultural heritage sciences both within and outside of academia [8]. Explored during the UNESCO Conference *Identifying Critical Science Gaps at Marine World Heritage Sites* in 2021, cultural heritage was highlighted as a key knowledge gap impeding the sustainable conservation of marine World Heritage Sites [9,10] see also below. The lack of cultural heritage within integrated and place-based marine management such as Marine Protected Areas (MPAs), Integrated Coastal Zone Management, and Marine Spatial Planning (MSP) was deemed to be actively impeding the management of the marine zone. Despite this, cultural heritage was not then included in the resulting roadmap for action for marine World Heritage Sites throughout the Ocean Decade [11], with the result that the gap remains.

The insufficient integration of culture (as human society) and heritage (encompassing both tangible and intangible elements) into marine resource management has hindered the effectiveness and inclusivity of frameworks like Environmental Impact Assessments (EIAs) and climate mitigation strategies. While many 'top-down' management approaches have had some success, their impact could have been significantly enhanced from deeper engagement with coastal communities and their culture. Expanding the integration of cultural heritage into research and policy development offers a direct path to achieving meaningful engagement and ensuring that these strategies are more sustainable and widely embraced. Notably, many coastal management frameworks worldwide already emphasise the importance of broadening stakeholder engagement, underscoring the critical opportunity for researchers and managers to align with these expectations and further enhance the efficacy of marine resource management.

Aligning cultural heritage with coastal and marine environmental management remains complex and under-researched. This is particularly the case when it comes to the integration of cultural heritage within large-scale resource management policies that deal with marine industry and development, such as in Blue Economy initiatives and MSP designations, which primarily monitor and measure development using economic input. The complexity associated with accurate and representative Culture and Heritage economies within these systems has resulted in a slower uptake of these methodologies within environmental management frameworks [12,13]. This is particularly the case for the intangible values of cultural heritage, which are rarely extrapolated proportionately for the benefit of coastal communities, and as such are often represented solely by tourism—which, although easily measured, does not reflect the vast importance of cultural heritage to local people or to the global ocean challenges [14–16].

As we begin to see an increasing recognition of the centrality of culture (including heritage) to inclusive and equitable ocean resource management, it is essential that inclusion goes beyond mere nominal acknowledgment. To this end, the CHFP aims to guide collaboration between cultural and natural heritage bodies to effectively protect, conserve, and manage marine cultural heritage (MCH) as an integral part of the broader marine environment. To do so, the CHFP is building a network of projects that address long-term human behaviour; the relationship between people and the marine environment over time; cultural and heritage economies and management; coastal communities; evolving social views; sea blindness; and ocean literacy. This effort translates into activities such as showcasing the integration of heritage and ocean science; facilitating co-design processes; managing data and knowledge; building capacities within the ocean sciences; enabling greater diversity and representation; supporting public engagement and outreach; and evaluating impact. A core principle of CHFP is bringing together a range of projects to highlight their impact, identify mutual support opportunities, and ensure meaningful collaboration.

A critical aspect of this approach is the engagement of Indigenous voices and perspectives—not as an afterthought but as a foundational element of the discussion from the inception of ideas through the co-production of knowledge and the development of management strategies. Indigenous and coastal community voices are integrated into CHFP's development through ongoing consultation with existing projects and participation in wider network meetings. Through the process of assisting projects toward endorsement by the Ocean Decade Secretariat, and in turn, affiliation with the CHFP, the network brings together and evaluates the overlaps, gaps, and insights of key projects that are actively engaging with the CHFP's goals from the ground up. The CHFP provides a platform for guiding the collection's shared goals towards actionable change, through a coordinated approach to informing policy, enhancing ocean literacy through shared approaches to public engagement, and by creating links between researchers and communities with common aims, approaches, and methodologies. Decolonising the processes of ocean research and management is a key goal, and CHFP strives to create a safe and accountable space that upholds shared principles of cooperative and long-term engagement with Indigenous and coastal populations, ensuring their active leadership and representation in shaping the future of ocean resource management.

4. Awareness of Marine Cultural Heritage in the UNESCO Culture Conventions

The lack of consideration of Marine Cultural Heritage (MCH) within existing UNESCO Culture Conventions has acted as a barrier to the awareness and inclusion of it in the Ocean Decade. Instead, UNESCO has focussed on Underwater Cultural Heritage (UCH), a much more restricted term that covers only submerged archaeological sites that have been '*been partially or totally underwater, periodically or continuously, for at least 100 years*' [17]. This is the approach taken in the UNESCO 2001 Convention on the Protection of the Underwater Cultural Heritage (2001 Convention), which does not account for the value of the intangible heritage of coastal societies and environments and perpetuates a divide in policy and management between terrestrial and underwater heritage. As a result, despite its centrality and importance for the development of protective policies and the establishment of research standards for underwater archaeological work worldwide, the 2001 Convention does not consider, nor protect, the full range of heritage identified in the marine zone currently under threat from infrastructural development and climate change [18,19].

An internal UNESCO evaluation in 2019 highlighted the potential areas for improvement of the 2001 Convention and underlined the importance of the 2001 Convention for other processes at the United Nations level regarding oceans, including the United Nations Law of the Sea (UNCLOS), UN Oceans, and the Ocean Decade [20]. The evaluation recommended the Secretariat and States Parties to 'Advocate for strengthening the integration of the protection of Underwater Cultural Heritage (UCH) into the Roadmap of the UN Decade of Ocean Science. In particular, facilitating the cooperation between the UCH Unit (*the Secretariat of the Convention in Paris*) and the Intergovernmental Oceanographic Commission

(IOC) in the implementation of the Roadmap' [20]. The audit also identified that one of the main challenges is the need to have specific tools to measure the level of implementation of the 2001 Convention, and its contribution to the Sustainable Development Goals, to the Ocean Decade and other Internationally Agreed Development Goals (IADGs). The 2001 Convention is the only UNESCO Culture Convention without a reporting mechanism, or a defined strategy outlining its expected results, measuring indicators, and baselines. As a result, there is not an existing tool that can analyse the contribution of the 2001 Convention to the Ocean Decade Outcomes (UNESCO is currently working to develop a Results-based framework for the 2001 Convention that, once approved by the Meeting of States Parties, will be the basis to establish a reporting mechanism).

Within the UNESCO World Heritage Centre, the World Heritage Marine Programme was launched in 2005 to guarantee the protection of marine areas of Outstanding Universal Value and to follow up on World Heritage Committee decisions. The Marine Programme focuses mainly on monitoring the state of conservation of sites, boosting a network of site managers, and building resilience and climate adaptation capacity [21,22]. Despite focusing on all ocean places inscribed in the World Heritage list, many combining cultural and natural heritage in the marine context (i.e., St. Kilda in Scotland, or Ibiza, Biodiversity and Culture in Spain), the major work of this programme excludes cultural heritage within inscribed properties [21].

The implementation of the UNESCO Culture Conventions alone does not guarantee the preservation of MCH in all its manifestations, nor does it enhance its contribution to achieve the goals of the Ocean Decade. Important scientific and policy gaps have been identified, and the need to provide a cultural perspective to sustainable development has been recognised as one of the main priorities of the Cultural Sector. Addressing this requires collaborative efforts among governments, policy makers, scientists, local communities, and other stakeholders. Integrated and interdisciplinary approaches, informed by peer-reviewed science, local knowledge, and inclusive decision-making processes, are essential to overcome these challenges and achieve sustainable ocean management and conservation. The recent adopted strategies call to establish synergies between the Conventions, advance innovative scientific approaches, and enable infrastructures according to the way we and others understand, study, preserve, and use heritage in the context of sustainable development.

The adoption of the CHFP as one of the Ocean Decade-endorsed Actions builds on the increasing acknowledgement of the international community for more social and humanities perspectives within the wider marine sciences and policies preserving our oceans. The CHFP has been created to fill the gap identified in the implementation strategies of UNESCO programmes, and to create a space for different ways of knowing in the marine sciences. Essentially, the CHFP is the *enabling environment* to (re)connect cultural heritage to wider ocean science initiatives and marine policies, with the belief that without engaging with cultural heritage, science-based solutions are unlikely to be sustainable.

5. The Cultural Heritage Framework Programme: Aims and Objectives

The aim of the CHFP is to transform engagement with cultural heritage relating to the ocean, the marine environment, and the people who depend on it. Ultimately, by the end of the Ocean Decade, the historical and cultural dimensions of people's relationships with the sea should be integrated within ocean science and policy. The 'Ocean We Want' will be inspired and informed by the long and diverse histories and living heritage of people and the sea [23]. Considering this, the programme will actively highlight cultural heritage contributions to the sustainable development agenda over the course of the Ocean Decade, for the benefit of both heritage, environment, and coastal communities. Through the integration of cultural heritage, the impacts of Decade Actions are linked by human stories and heritage, thus being more significant together than they would be individually and will fill social and cultural research gaps identified for the functioning of the Ocean Decade.

Operationally, the CHFP has a number of objectives throughout the course of the Ocean Decade, which are categorised as *Primary Objectives* (Figure 1), which describe the actions that will be undertaken by the CHFP for the purposes of integrating cultural heritage within the ocean sciences, and *Supporting Objectives*, which describe actions which aim to aid in supporting individual Decade Actions, and other complementary projects (Figure 2).

1. Showcase integration of heritage and ocean science.
2. Encourage ocean literacy, public engagement, and outreach.
3. Develop capacity/early career support.
4. Enable greater diversity and representation.
5. Facilitate co-design.
6. Encourage best practice in managing data and knowledge.
7. Encourage effective and efficient communication between:
    a. Cultural heritage stakeholders.
    b. Other marine science stakeholders, policy and decision makers.
    c. Traditional communities and the general public.
    d. Decade institutions.
8. Encourage effective evaluation of impact.

Figure 1. The primary Cultural Heritage Framework Programme objectives (authors).

The scope of the CHFP's activities and services are likely to evolve over the course of the Decade in response to need. However, these are expected to include:

• Provision of a dedicated digital platform for resources, news, dissemination, etc.

• Organising and promoting webinars on specific research topics (e.g., potentially polluting wrecks, Indigenous and traditional knowledge) and themes such as capacity building, data management, citizen science, managing Decade Actions, ocean literacy, etc.

• Hosting matchmaking meetings to stimulate future Decade Actions.

• Hosting co-design workshops to encourage the development of interdisciplinary concepts into viable, effective Decade Actions.

• Supporting the development of training workshops and other capacity-building measures, including addressing imbalances in gender, generational, and geographic diversity.

• Providing and sourcing technical advice for stakeholders (e.g., specialist knowledge on archaeological periods, regions, themes, methodologies, etc.).

• Supporting outputs from Decade Actions, including publication and through digital media, while using the engaging character of cultural heritage to disseminate the vision, mission, and outcomes of the Decade among the public. Inter-related activities are designed to achieve the outcomes listed below:

- Inter-related activities will achieve the CHFP's outcomes.
- Digital platforms.
- Webinar series.
- Matchmaking meetings.
- Co-design workshops.
- Training workshops.
- Technical advisory services.
- Supporting outputs including publication and digital media.

Figure 2. The outputs of the Cultural Heritage Framework Programme (authors).

These aims and objectives are designed broadly, so as to evolve over time in reaction to the needs of target areas, industries, partners, and projects. The objectives will be delivered by the CHFP Secretariat, and associated projects and will contribute to the greater aims and objectives of the Ocean Decade.

6. The Cultural Heritage Framework Programme: Methodologies

As part of the first stage of the programme, a series of key themes have been identified in order to achieve the objectives outlined above (see Figure 1). The second stage detailed in the following section was co-designed with CHFP's associated projects through an online survey and selected follow-up interviews. This process of co-design will recur annually, with an end-of-year workshop to assess impact and re-design themes for the following year. As such, it is important to note that the highlighted themes and sub-themes in this paper are constantly evolving through a reiterative process of re-evaluation and co-design.

The following themes were identified by the CHFP Secretariat in the first year of its implementation. Regarding the objective outlining the 'showcasing of the integration of heritage and ocean science' (Objective 1), the primary limitation identified was 'interdisciplinary engagement', with the need for co-ordinated and multi-disciplinary representation of projects that engage with cultural heritage in the ocean sciences. To 'encourage ocean literacy, public engagement, and outreach' and 'develop capacity/early career support' (Objectives 2, 3), the main limiting factor identified was a lack of an engaged, dedicated platform by which inspirational culture–nature ocean stories could be shared and disseminated by the scientists, community members, and students who run them. Regarding 'enabling greater diversity and representation', 'facilitating co-design', and 'encouraging best practice in managing data and knowledge' (Objectives 4, 5, 6), the theme of 'advocating and integrating for the voices of coastal communities and their livelihoods' was identified as the first necessary step for more inclusive co-design, representation, and data management. Finally, to 'encourage effective and efficient communication between cultural heritage stakeholders, other marine science stakeholders, policy and decision makers, traditional communities and the general public, and Decade institutions' (Objective 7), the theme of decolonising ocean resource management was deemed the most urgent next step in diversifying multi-stakeholder ocean resource engagement. The final objective, 'encourage effective evaluation of impact' is applied throughout the process of co-operative re-design of CHFP's annual themes. The following section provides further details on the context driving the CHFP Secretariat's first identified themes.

6.1. Theme 1: Interdisciplinary Engagement: Including Showcasing Integration of Heritage and Ocean Science

It is the main contention of the CHFP that nature and culture should be considered together, acknowledging the mutual impact of cultural traditions on biological diversity and emphasizing the interconnectedness between human communities and their environments. Considering the past and ongoing human impact and interaction with the natural environment is an essential aspect of achieving sustainable ocean practices [24]. This involves considering both tangible elements, like shipwrecks fostering reef ecosystems, and intangible ones, such as traditional ecological knowledge (Figure 3) [25]. Biocultural heritage emphasises the interconnectedness between human communities and their environments, acknowledging the mutual impact of cultural traditions and biological diversity, and as such, is considered a key ocean resource and CHFP theme.

Figure 3. The cargo wreck (*Pecheur Breton*) in Sri Lanka, surveyed as part of the Shipwrecks as Artificial Reef Structures Project. Shipwrecks are not only heritage assets; they can also improve fish biodiversity and enhance marine ecosystem health (photo credit: Ales Reich).

Considering that the primary role of the CHFP is to advocate for the inclusion of heritage within the sustainable development frameworks, the marine sciences, and marine policy and management, a key aspect of the CHFP's work is sharing and evolving the understanding of how the past may inform the future of the ocean. As such, an interdisciplinary Secretariat with expertise in natural and cultural heritage, policy, and management oversees the programme and works as the primary point of contact to implement the key outcomes and objectives described above.

To do so, CHFP provides infrastructure to encourage, share, and support Decade Actions relating to cultural heritage, as well as partners with international research projects which lead in developing heritage solutions for global challenges, so as to share and expand the CHFP knowledge base practically. Further to providing the infrastructure and guidelines for marine cultural heritage projects, the programme provides advice and assistance to other Decade Actions (programmes, projects, and activities) that engage with, or have the potential to engage with, ocean heritage. Through collaborating with other networks and initiatives relating to the Ocean Decade that share common strands with ODHN and CHFP, the programme aims to extend the reach of MCH outside of the heritage and archaeology discipline.

The precedent for this type of interdisciplinary engagement has been set through early involvement of ODHN with international forums such as the COP27, COP28, UN Oceans, and the CHFP with Ocean Science Networks such as Edinburgh Oceans. Over the course of the CHFP, this work will be extended through the following:

(i) Stakeholder mapping, audience identification, and engagement with environment and sustainable development conferences and events internationally.
(ii) Working to expand cultural heritage capacities and understandings within this network, through facilitating interdisciplinary project design, training, and management plans.

(iii) Directly applying MCH solutions to global ocean challenges by connecting the large-scale, interdisciplinary results identified through the CHFP network of partners and associated projects to the Ocean Decade platform.

Furthermore, by building an interdisciplinary network of endorsed projects that engage with the shared goals of ODHN, CHFP, and the Ocean Decade, CHFP aims to showcase and develop the role of heritage in the ocean sciences. Projects are endorsed through a rigorously assessed application which passes through both the Ocean Decade Secretariat and the CHFP Secretariat before endorsement can proceed. In doing so, each project must answer specific questions regarding the relevance of their project to both the goals of the Ocean Decade, and the goals of CHFP. Further to this, interviews are typically held by CHFP to discuss both the relevance and ethical rigidity of each project, so as to guarantee the highest research and ethics standards within the CHFP network.

6.2. Theme 2: Developing an Engaged, Dedicated Platform by Which Inspirational Culture-Nature Ocean Stories Can Be Disseminated

The CHFP has a dedicated online platform (https://oceandecadechfp.org/, accessed on 18 December 2024), a key part of which is a global online magazine, *SeaVoice* (https://www.seavoice.online/, accessed on 18 December 2024), representing a public-facing element of the CHFP aimed at sharing engaging stories about culture and the ocean (Figure 4). The magazine showcases the work of a diverse range of voices working in the intersection between culture, climate, and the ocean. It has been designed to be accessible and inspirational, and presents the work of a range of ocean practitioners including researchers, activists, advocates, community members, and artists. To date, five online volumes and one annual printed edition have been released, containing 45 papers by authors from 24 different countries and multiple demographics, from policy makers to Indigenous community members [26].

Figure 4. *SeaVoice* magazine poster, presented at the UNESCO Ocean Decade Conference organised in Barcelona, Spain, 10–12 April 2024 (photo credit: SeaVoice, CHFP/University of Edinburgh).

6.3. Theme 3: Advocating for and Integrating the Voices of Coastal Communities and Their Livelihoods

CHFP's early partnership with the Rising from the Depths Network (RftD), a project that aimed to enhance the utilisation of MCH for the sustainable development of local communities in East Africa, provided a key knowledge base for advocating for the voices of coastal communities and their livelihoods (Figure 5). From 2018 to 2022, RftD funded 27 projects throughout Kenya, Tanzania, Mozambique, and Madagascar, which co-produced community-led projects focussing on MCH livelihoods [27]. Various themes within the CHFP were exemplified in this work, including community perception of ocean hazards; integration of scientific and traditional MCH knowledge; climate change mitigation through cultural practices; cultural livelihoods; gender equity; and engagement, capacity building, and education [27]. In collaboration with CHFP, RftD is working towards the production of various best-practice models within these themes. Furthermore, through connecting this project with others within the ocean sciences, CHFP can continue to facilitate the platform by which this research can be shared and expanded upon on a global stage.

Figure 5. Bidii na Kazi women fencing plots for mangrove nurseries as part of the Rising from the Depths project 'MUCH to Discover in Mida Creek: creating pathways to community resilience and sustainable development through the maritime cultural landscape in Kenya' (photo credit: Rising from the Depths network).

6.4. Theme 4: Developing Capacities and Interactions Between Ocean Leaders, Policy Makers, Scientists, Academics, and the Public, and Enabling Greater Diversity and Representation

Beyond establishing a MCH-based framework for large-scale ocean challenges, it is also imperative that results are communicated and applicable to a broad range of audiences (Figure 6). To do so, CHFP shares access to a global network of maritime heritage initiatives, each with its own potential to inspire public engagement through supporting plans for promoting different ocean 'voices' and ocean 'citizenship', as a way to improve ocean safety and sustainability. ODHN's membership has already established an extensive list of members and partners which represent a global network of maritime archaeologists and heritage experts. The work of CHFP is actively expanding this list, with particular focus on increasing diversity and interdisciplinarity within ocean disciplines; higher education and non-higher education institutions; and governments and NGOs. Of particular importance is the establishment through ODHN of regional representatives, through which local gatekeepers may be approached in areas which have been typically excluded, or have proved inaccessible previously. Furthermore, particular focus is placed on organisations which engage and represent local and Indigenous communities, that engage with traditional knowledge, and provide a multivocal platform to have access to ocean stakeholders.

Figure 6. School children in Malta experience an immersive 360 virtual dive on a historic wreck surveyed as part of the WreckLife project (photo credit: Heritage Malta).

6.5. Theme 5: Decolonising Ocean Resource Management

Scientific research, aid and capacity-building practices between the Global North and South have a significant history in colonial methodologies and practice [27]. As such, CHFP methodologies focus a spotlight on decolonising the methodologies of top-down, Western-developed marine management frameworks, policies, and practices. Through engagement with local stakeholders, representatives, and projects, CHFP prioritises research practices developed with cultural sensitivity, inclusivity, and equity; participatory, bottom-up methodologies; and community-centred objectives (Figure 7). Research projects are prioritised that aim to decolonise living narratives and histories; ocean management practices and implementation; and knowledge representation and diversity. Themes that engage with living heritage in the form of traditional knowledge, culture, stories, and traditions alongside traditional customary practices, taboos, and frameworks are key to this work, and will be a core theme of future activities and endorsed projects.

Figure 7. Indigenous leaders from Malawig, north of Coron, Palawan in the Philippines, identify illegal fishing hotspots within their ancestral territories during a workshop facilitated by Community Centred Conservation (C3) Philippines, Inc. on 9 September 2024. The threat maps from the workshop guided the Indigenous community in designating Indigenous Marine Managed Areas within their ancestral territory (photo credit: C3 Philippines, Replicating a Successful Model to Develop a Network of Indigenous Marine Managed Areas in the Philippines).

Once network-building activities have established a strong base, CHFP aims to produce active results in collaboration with associated projects and actions. These actionable results will be focussed on addressing global ocean challenges and will be brought together with partners and projects to enhance the outcomes of the Ocean Decade. Outcomes will be delivered both on a global scale, in relation to the Ocean Decade itself, and a local scale, in relation to the challenges faced by the coastal communities on the frontline of the climate and development crisis.

7. Expected Outcomes

The primary outcome of CHFP is to increase the impact of Decade Actions through engagement with MCH, so Decade Actions relating to cultural heritage will have greater overall coherence over the duration of the Ocean Decade than if this framework were absent. To do so, a number of outcomes must be achieved (Figure 8).

1. Increased impact of Decade Actions due to the active support of the CHFP in terms of:
    a. Integration between cultural heritage and marine science;
    b. The quality of co-design and the diversity and representation of participants (scientists; stakeholders; communities);
    c. Capacity building, ocean literacy, and public engagement and outreach.
2. Decade Actions relating to cultural heritage will have greater overall coherence over the duration of the Decade than if this framework were absent.
3. Communication will be more effective and efficient between cultural heritage stakeholders, other marine science stakeholders, policy and decision makers, traditional communities, and the general public, and with the Decade's institutions (Decade Co-ordination Unit; Co-ordination Offices and Collaborative Centres; National Decade Committees; Stakeholder Platforms; Ocean Decade Alliance).
4. Duplication will be avoided among networks, programmes, projects, and activities relating to cultural heritage under the umbrella of the Ocean Decade and beyond. Expertise will be promptly available for responding to enquiries from stakeholders regarding the Ocean Decade, Decade Actions, Calls, institutions, and contributions.

Figure 8. The primary outcomes of the Cultural Heritage Framework Programme (authors).

The above outcomes are aimed at a number of stakeholders: marine science researchers; policy and decision makers; traditional communities and the public; and Ocean Decade co-ordination structures (Decade Co-ordination Unit, Co-ordination Offices and Collaborative Centres; National Decade Committees; Stakeholder Platforms; Ocean Decade Alliance). Part of this work will be ensuring that duplication is avoided among networks, programmes, projects, and activities relating to cultural heritage under the umbrella of the Ocean Decade and beyond, and so, expertise will be promptly available for responding to enquiries from stakeholders regarding the Ocean Decade, Decade Actions, Calls for Actions, institutions, and contributions.

Ultimately, the CHFP hopes to begin decreasing the gap between, and increase representation of MCH within, the rest of the ocean sciences in relation to research, policy, legislation, and community. This outcome will develop over the course of the Ocean Decade, but will remain the primary, measurable indicator for CHFP.

8. Endorsement Process

Part of the aforementioned outcomes can already be measured through Ocean Decade endorsed programmes associated with the CHFP (long-term initiatives focussed on a number of global ocean challenges); projects (focussed regimes at regional, national, or local levels, which may appear within a greater programme); and activities (individual, smaller initiatives which address or contribute to a specific challenge). The CHFP, at the time of writing, has 19 endorsed actions, as can be seen in Table 1.

Table 1. The current Cultural Heritage Framework Programme-affiliated Decade Actions (authors).

Name of the Decade Action	Lead Institution	Project's Website	Country/Region Action Is Implemented
Unpath'd Waters	Historic England	https://unpathdwaters.org.uk/ (accessed on 18 December 2024)	UK
Indigenous People, Traditional Ecological Knowledge, and Climate Change: the iconic underwater cultural heritage of stone tidal weirs	Tokyo University of Marine Science and Technology [UNESCO UNITWIN Network for Underwater Archaeology]	https://oceandecade.org/actions/indigenous-people-traditional-ecological-knowledge-and-climate-change-the-iconic-underwater-cultural-heritage-of-stone-tidal-weirs/ (accessed on 18 December 2024)	East Asia
Community Engaged Ocean Science in Canada	Canadian Commission for UNESCO–Canada	https://oceandecade.org/actions/community-engaged-ocean-science-in-canada/ (accessed on 18 December 2024)	Canada
Rising from the Depths	University of Edinburgh	https://risingfromthedepths.com/ (accessed on 18 December 2024)	East Africa
Linking Nature and Culture to Support Sustainable Coastal Livelihoods: Establishing a Marine Protected Area at the Island of Mozambique, East Africa (LiNaCuMoz)	University of Edinburgh	https://edmarinearch.com/ (accessed on 18 December 2024)	Mozambique
Replicating a Successful Model to Develop a Network of Indigenous Marine Managed Areas in the Philippines	C3 Philippines	https://c-3.org.uk/darwin-initiative/ (accessed on 18 December 2024)	Philippines
Threats to our Ocean Heritage: A book series	The Ocean Foundation	https://oceandecadechfp.org/activities/threats-to-our-ocean-heritage-a-book-series/ (accessed on 18 December 2024)	Global
Erasmus Mundus: Marine Spaces Planning/Culture/Climate (MSPCC)	Koç University	https://mspcc.ku.edu.tr/ (accessed on 18 December 2024)	Europe
Shipwrecks as Artificial Reef Structures	University of Bremen	https://shipwrecksasartificialreefstructures.com/ (accessed on 18 December 2024)	Sri Lanka
Enhancing Estuary Resilience in Coastal BC	The Nature Trust of British Columbia	Estuaryresilience.ca/Naturetrust.bc.ca (accessed on 18 December 2024)	Canada
SeaVoice	University of Edinburgh	https://www.seavoice.online/ (accessed on 18 December 2024)	Global
WreckLife	Heritage Malta	https://heritagemalta.mt/departments/underwater-cultural-heritage-unit/ (accessed on 18 December 2024)	Malta
Women, heritage, & navigation in the Pacific	Universitat Pompeu Fabra	https://oceandecade.org/actions/women-heritage-navigation-in-the-pacific/ (accessed on 18 December 2024)	Spain/USA
Yagarrajalajalan nagula buru	Murdoch University	https://www.yagarrajalajalan-nagula-buru.org/ (accessed on 18 December 2024)	Australia
Vaka Oceanstorylab	Okeanos Foundation	https://okeanos-foundation.org/ (accessed on 18 December 2024)	Cook Islands; France; Nauro
Recovering and reusing ghost nets	Marulho	https://oceandecade.org/actions/recovering-and-reusing-ghost-nets-nets-for-the-ocean/ (accessed on 18 December 2024)	Brazil

Table 1. *Cont.*

Name of the Decade Action	Lead Institution	Project's Website	Country/Region Action Is Implemented
Gathering Information via Recreational and Technical (GIRT) Scientific Divers	GIRT Scientific Divers	www.girtsd.org (accessed on 18 December 2024)	Australia
Community Waterscapes	University of Hull	https://storymaps.com/stories/c2a7a2206953451ba59eb2cda428c9f0 (accessed on 18 December 2024)	UK
MARGAIN	University of Stavanger	https://www.uis.no/en/museum-of-archaeology/research/margain-marine-resource-gathering-and-infrastructure-in-the-norse (accessed on 18 December 2024)	Norway

As discussed above, these actions are co-creating and formulating the evolving research themes of the programme. While the themes are intersectional across the actions, certain projects exemplify particular themes. For example, Theme 1 is highlighted both by projects which engage with tangible heritage, such as shipwrecks as artificial reef structures, and intangible heritage, such as traditional ecological and heritage knowledge, such as the *Community Engaged Ocean Science in Canada Project or Enhancing Estuary Resilience in Coastal BC*. As discussed above, the CHFP magazine, *SeaVoice*, provides a platform for disseminating inspirational culture–nature ocean stories to a wider public (Theme 2).

Themes 3 and 5 are represented by projects which engage with coastal community resilience and decolonising marine resource management, such as *Replicating a Successful Model to Develop a Network of Indigenous Marine Managed Areas in the Philippines*, and *Linking Nature and Culture in Mozambique*. Theme 4 is represented by *Unpath'd Waters*, which seeks to allow the general public to access UK maritime collections, the *Indigenous People, Traditional Ecological Knowledge, and Climate Change* project which showcases the interdisciplinarity of stone tidal weir research to the wider scientific community, and *Rising from the Depths*, which is a network of arts and humanities researchers, community groups, heritage professionals, NGOs, ocean scientists, UN officials, government policy practitioners, coastal developers, offshore companies, overseas aid specialists, ICT specialists, and artists drawn from the diverse components of MCH.

To continue growing and evolving CHFP-listed outcomes, a primary goal for the future is to engage with further projects, programmes, and activities, such as those listed in Table 1. These projects exemplify a number of representative characteristics which actively complement the overall outcomes and objectives of CHFP and the Ocean Decade, such as interdisciplinarity; creating communication pathways between community, science, industry, and policy; research that engages with traditional knowledge and methods; community engagement; and integrated management techniques.

Through the process of endorsement by the Ocean Decade Secretariat, CHFP can provide multiple benefits to associated projects. The CHFP platform can facilitate projects and research on an interdisciplinary scale, the Secretariat can deliver expert advice, and the Ocean Decade can provide accountability for objectives. On receiving endorsement, a project receives official recognition that they are within the global network of the Ocean Decade Vision, and as such, will have access to increased opportunities, partnerships, opportunities, visibility, and support.

The process of endorsement is as follows. The first stage of endorsement involves *communication* between CHFP and the potential programme, project, or activity (henceforth discussed simply as 'action'). At this stage, the appropriateness of an action may be discussed, and advice and next steps will be given and adjusted depending on the action itself. Further to this, the action will be asked to submit a proposal (Figure 9), which will go through technical review by the Ocean Decade Unit, in consultation with CHFP, and if successful, will be endorsed and announced. Decade actions can be submitted for endorsement at any time, and can take a period of six weeks to be accepted. For programmes and projects, official calls with various themes are announced every six months, and can take a period of 3–6 months to be accepted. Further information on this process can be found on CHFP website and alongside each new call for Decade Actions.

1. Contribute to fulfilling the Ocean Decade Challenges and to achieving the Decade objectives and the associated sub-objectives.
2. Accelerate the generation or use of knowledge and understanding of the ocean, with a specific focus on knowledge that will contribute to SDGs.
3. Are co-designed or co-delivered by knowledge generators and users, and facilitate the uptake of knowledge for policy, decision making, management and/or innovation.
4. Ensure that all data and resulting knowledge are provided in an open access, shared, discoverable manner, and are deposited in recognized data repositories.
5. Strengthen existing or create new partnerships.
6. Contribute toward capacity development.
7. Overcome barriers to diversity and equity.
8. Collaborate with and engage local and Indigenous knowledge holders.

Figure 9. Endorsement Proposal Criteria (Ocean Decade Guidance Notes for Applicants, Accessed 2023).

9. Cultural Heritage Framework Programme: Next Steps

The scope of CHFP will continue to evolve over the course of the Ocean Decade in response to need, and through the re-evaluation of annual themes and co-design with marine stakeholders and communities. The actionable next steps currently in progress are particularly focussed on network generating developments. These include the continued development of the *SeaVoice* magazine and the launch of a new dedicated digital platform for resources, news, and dissemination; various informative webinars on specific research topics (e.g., potentially polluting wrecks, Indigenous and traditional knowledge) and themes (e.g., capacity building, data management, citizen science, managing Decade Actions, ocean literacy); matchmaking meetings to stimulate future Decade Actions; the provision and sourcing technical advice for stakeholders (e.g., specialist knowledge on archaeological periods, regions, themes, methodologies, etc.); continuous support for associated and future Decade Actions including publication and through digital media; and dissemination of the vision, mission, and outcomes of the Ocean Decade throughout the public. With specific regard to supporting the growing network of ocean scientists associated with the CHFP, and in consultation with current associated projects, the CHFP Secretariat is developing a reporting framework for associated actions, and a framework for a global network of Early Career Ocean Professionals (ECOPs), representatives, and stakeholders.

10. Conclusions

The future of the CHFP lies in the pathways built between MCH and the ocean sciences. Significant work has already been done to engage cultural heritage with the global ocean challenges, particularly regarding work with coastal communities, climate change, sustainable development, MSP, and MPAs [1,4,15,28–31]. As has been evidenced in the UNESCO Ocean Decade Conference, held in Barcelona, in April 2024, although 'cultural engagement' was duly recognised in the final conference statement, much remains to be done to achieve a global recognition for these contributions, and to shape the progress of MCH research towards challenge-led, inclusive, equitable, and solution-driven goals. The CHFP will build upon this work by facilitating the platform by which this type of research can be shared, communicated, applied, and built upon, to provide space for culture, heritage, and coastal communities within the ongoing Ocean Decade.

The overall objective of the CHFP is to ensure that MCH, and most importantly, the communities related to that heritage, have a voice in the Ocean Decade. The growing range of endorsed CHFP projects showcases how MCH contributes to achieving the 'Ocean We Want'. These projects support conservation, involve communities, promote sustainable tourism, enhance ocean literacy, and help to build resilience (through the use of historic data in mitigation and adaption strategies). They illustrate the value of a truly interdisciplinary approach that creates communication pathways between local communities, science, industry and policy, to co-create viable initiatives that are supported on the ground and respect traditional knowledge and practices.

Engaging with CHFP enables us to highlight projects and research onto an interdisciplinary platform, offer expert guidance, and ensure accountability for Ocean Decade objectives. It is expected that the achievement of CHFP outcomes can help to transform current policies and implementation strategies concerning cultural heritage and the marine environment. The aim is to provide science-based recommendations aligning international frameworks (such as the UNESCO 2001 Convention) with ocean sustainability, fostering self-sustained ocean heritage practice well beyond the Ocean Decade.

It is widely recognised that science alone cannot bring about change in public policy or transformations in the behaviour of communities and individuals. Humans are the driving force in sustainable development—they sit at the centre of any mitigation strategies or management programmes put in place, and it is clear that if these initiatives do not respect their cultural norms or beliefs, they will not be successful nor sustainable. By recognising and valuing the role of MCH in the ocean, the CHFP is helping to ensure the human dimension of the Ocean Decade and that the benefits of the ocean are shared fairly and sustainably.

Author Contributions: Conceptualisation, J.H., G.H., A.R.d.S. and A.T.; writing—original draft preparation, J.H., G.H., A.R.d.S. and A.T.; writing—review and editing, J.H., G.H., A.R.d.S. and A.T.; visualisation, J.H. and G.H.; supervision, J.H. and A.T.; project administration, G.H.; funding acquisition, A.T. and J.H. All authors have read and agreed to the published version of the manuscript.

Funding: This research is funded by the Lloyd's Register Foundation, grant number GA\100473.

Institutional Review Board Statement: Not applicable.

Informed Consent Statement: Not applicable.

Data Availability Statement: The data is contained within the article.

Acknowledgments: ODHN is funded by the Honor Frost Foundation, and the CHFP is funded by the Lloyds Register Foundation. The authors would like to thank all the investigators, communities, and participants involved in CHFP-affiliated Ocean Decade-endorsed activities.

Conflicts of Interest: The authors declare no conflicts of interest.

References

1. Trakadas, A.; Firth, A.; Gregory, D.; Elkin, D.; Guerin, U.; Henderson, J.; Kimura, J.; Scott-Ireton, D.; Shashoua, Y.; Underwood, C.; et al. The Ocean Decade Heritage Network: Integrating Cultural Heritage Within the UN Decade of Ocean Science 2021–2030. *J. Marit. Archaeol.* **2019**, *14*, 153–165. [CrossRef]
2. UNESCO-IOC. *The United Nations Decade of Ocean Science for Sustainable Development (2021–2030) Implementation Plan*; IOC Ocean Decade Series; UNESCO: Paris, France, 2021.
3. Claudet, J.; Bopp, L.; Cheung, W.W.; Devillers, R.; Escobar-Briones, E.; Haugan, P.; Heymans, J.J.; Masson-Delmotte, V.; Matz-Lück, N.; Miloslavich, P.; et al. A Roadmap for Using the UN Decade of Ocean Science for Sustainable Development in Support of Science, Policy, and Action. *One Earth* **2019**, *2*, 34–42. [CrossRef]
4. Henderson, J. Oceans without history? Marine cultural heritage and the sustainable development agenda. *Sustainability* **2019**, *11*, 5080. [CrossRef]
5. Trakadas, A. The importance of integrating cultural heritage into the UN Ocean Decade. In *Threats to Our Ocean Heritage, Volume I: Bottom Trawling*; Jarvis, C., Ed.; Springer: Berlin/Heidelberg, Germany, 2024.
6. Mark, L. Marine Cultural Heritage and Sustainable Development. In *Critical Approaches to Heritage for Development*; Charlotte, C., Giblin, J.D., Eds.; Routledge: London, UK, 2023; pp. 144–158. [CrossRef]
7. UNESCO-IOC 2020. *The Science We Need for the Ocean We Want. United Nations Decade of Ocean Science for Sustainable Development 2021–2030*; IOC Brochure 2020-4 (IOC/BRO/2020/4); UNESCO: Paris, France, 2020. Available online: https://unesdoc.unesco.org/ark:/48223/pf0000265198 (accessed on 20 March 2024).
8. Trakadas, A. The Cultural Heritage Framework Programme: Ensuring a Place for Cultural Heritage's Contribution to the UN Decade of Ocean Science. *Mar. Technol. Soc. J.* **2022**, *56*, 110–111. [CrossRef]
9. UNESCO 2021. UNESCO Conference: Identifying Critical Science Gaps at Marine World Heritage Sites. Available online: https://whc.unesco.org/en/events/1613/ (accessed on 20 March 2024).
10. Trakadas, A. Cultural Heritage: A Critical Science Gap of UNESCO Marine World Heritage Sites? Ocean Decade Heritage Network. 5 October 2021. Available online: https://www.oceandecadeheritage.org/cultural-heritage-a-critical-science-gap-of-unesco-marine-world-heritage-sites/ (accessed on 20 May 2024).
11. UNESCO. *Ocean Science Roadmap for UNESCO Marine World Heritage in the Context of the United Nations Decade of Ocean Science for Sustainable Development (2021–2030)*; UNESCO: Paris, France, 2021.
12. Childs, J.R.; Hicks, C. Securing the blue: Political ecologies of the blue economy in Africa. *J. Political Ecol.* **2019**, *26*, 323–340.
13. Historic England. Heritage and the Economy: Heritage Counts. 2019. Available online: https://historicengland.org.uk/content/heritage-counts/pub/2019/heritage-and-the-economy-2019/ (accessed on 18 December 2024).
14. Altvater, S. Integrating Cultural Heritage into Maritime Spatial Planning in the BSR. Final Publication of the Baltic Sea Region Integrated Maritime Cultural Heritage Management Project (BalticRIM). 2020. Available online: https://stmuseovirastoprod.blob.core.windows.net/museovirasto/Palvelut_ja_ohjeet/BalticRIM_Final_Publication.pdf (accessed on 18 December 2024).
15. Holly, G.; Rey da Silva, A.; Henderson, J.; Bita, C.; Forsythe, W.; Ombe, Z.; Poonian, C.; Roberts, H. Utilising Marine Cultural Heritage for the Preservation of Coastal Systems in East Africa. *J. Mar. Sci. Eng.* **2022**, *10*, 693. [CrossRef]
16. Strand, M.; Rivers, N.; Snow, B. The complexity of evaluating, categorising and quantifying marine cultural heritage. In *Marine Policy*; Elsevier: Amsterdam, The Netherlands, 2023; Volume 148, p. 105449.
17. UNESCO. Convention on the Protection of the Underwater Cultural Heritage. In Proceedings of the Records of the General Conference: 31st Session, Paris, France, 15 October–3 November 2001.
18. Holly, G. Integrating the Protection of Marine Cultural Heritage into Marine Policy, for the Sustainable Development of the Ocean. Ph.D. Thesis, University of Southampton, Southampton, UK, 2022.
19. Sarah, D. 2001 UNESCO Convention on the Protection of the Underwater Cultural Heritage. *Int. J. Mar. Coast. Law* **2003**, *18*, 59–108. [CrossRef]
20. Sediakina Rivière, E.; Dlamini, T.; Varmer, O.; Kourkoumelis, D. *Evaluation of UNESCO's Standard-Setting Work of the Culture Sector Part VII—2001 Convention on the Protection of Underwater Cultural Heritage*; UNESCO Internal Oversight Office Evaluation Office: Paris, France, 2019.
21. Martin, J.B. Protecting Outstanding Underwater Cultural Heritage through the World Heritage Convention: The Titanic and Lusitania as World Heritage Sites. *Int. J. Mar. Coast. Law* **2018**, *33*, 116–165. [CrossRef]
22. UNESCO. *UNESCO World Heritage Marine Sites. 50 Beacons of Hope in a Changing Ocean*; UNESCO: Paris, France, 2020.
23. Trakadas, A. Maritime Living Heritage and the Decade of Ocean Science for Sustainable Development 2021–2030. In *Maritime Living Heritage: Building Sustainable Livelihood and Ecosystems in the Asia-Pacific Region*; ICH Webinar Series on Maritime ICH 2020;

해양 무형유산: 아태지역의 지속가능한 생활과 생태계 구축. 해양 무형유산 웨비나 시리즈; Park, W., Ed.; International Information and Networking Centre for Intangible Cultural Heritage in the Asia-Pacific Region under the auspices of UNESCO (ICHCAP): Jeollabuk-do, Republic of Korea, 2020; pp. 13–19. Available online: https://webinar.unesco-ichcap.org/wp-content/uploads/2021/03/1-0-Athena-Lynn-Trakadas_EN.pdf (accessed on 18 December 2024).

24. Poole, A.K. Where is goal 18? The need for biocultural heritage in the sustainable developmentgoals. *Environ. Values* **2018**, *27*, 55–80. [CrossRef]
25. Paxton, A.B.; McGonigle, C.; Damour, M.; Holly, G.; Caporaso, A.; Campbell, P.B.; Meyer-Kaiser, K.S.; Hamdan, L.J.; Mires, C.H.; Taylor, J.C. Shipwreck ecology: Understanding the function and processes from microbes to megafauna. *BioScience* **2024**, *74*, 12–24. [CrossRef]
26. Holly, G.; Cocks, H.; Edwards, A.; Carter, A.; Leone, C. (Eds.) *SeaVoice Annual: Issue 01*; University of Edinburgh: Edinburgh, UK, 2024. [CrossRef]
27. Henderson, J.; Breen, C.; Esteves, L.; La Chimia, A.; Lane, P.; Macamo, S.; Marvin, G.; Wynne-Jones, S. Rising from the Depths Network: A Challenge-Led Research Agenda for Marine Heritage and Sustainable Development in Eastern Africa. *Heritage* **2021**, *4*, 1026–1048. [CrossRef]
28. Fawcett, L.; Havice, E.; Zalik, A. Frontiers: Ocean epistemologies–privatise, democratise, decolonise. In *The Routledge Handbook of Ocean Space*; Routledge: London, UK, 2022; pp. 70–84.
29. Cornelius, H. The Climate Heritage Paradox—How rethinking archaeological heritage can address global challenges of climate change. *World Archaeol.* **2024**, *55*, 268–281. [CrossRef]
30. Bita, C. Integration of Underwater Cultural Heritage in Marine Spatial Planning. In *Capacity Needs Assessment of Institutions Involved in the Kenyan MSP Process*; Kenya Bureau of Statistics: Nairobi, Kenya, 2020.
31. Breen, C.; El Safadi, C.; Huigens, H.; Tews, S.; Westley, K.; Andreou, G.; Vazquez, R.O.; Nikolaus, J.; Blue, L. Integrating cultural and natural heritage approaches to Marine Protected Areas in the MENA region. *Mar. Policy* **2021**, *132*, 104676. [CrossRef]

MDPI AG
Grosspeteranlage 5
4052 Basel
Switzerland
Tel.: +41 61 683 77 34

Oceans Editorial Office
E-mail: oceans@mdpi.com
www.mdpi.com/journal/oceans

MDPI

www.ingramcontent.com/pod-product-compliance
Lightning Source LLC
Chambersburg PA
CBHW061550210926
51850CB00009B/2065
* 9 7 8 3 7 2 5 8 4 8 9 7 3 *